The Influence of Polymer Additives on Velocity and Temperature Fields

International Union of Theoretical
and Applied Mechanics

Deutsche Rheologische Gesellschaft

The Influence of Polymer Additives on Velocity and Temperature Fields

Symposium Universität - GH - Essen, Germany
June 26 - 28, 1984

Editor
B. Gampert

Springer-Verlag
Berlin Heidelberg New York Tokyo

Prof. Dr.-Ing. Bernhard Gampert

Strömungsmechanik. Universität - GH - Essen
Schützenbahn 70
D - 4300 Essen 1 - FRG

ISBN-13: 978-3-642-82634-4 e-ISBN-13: 978-3-642-82632-0
DOI: 10.1007/978-3-642-82632-0

Library of Congress Cataloging-in-Publication Data

Main entry under title:
The Influence of polymer additives on velocity and temperature fields.

At head of title: International Union of Theoretical and Applied Mechanics [and] Deutsche Rheologische Gesellschaft.
Symposium sponsored by the International Union of Theoretical and Applied Mechanics, Universität-GH-Essen, and Gesellschaft von Freunden und Förderern der Universität-GH-Essen.
1. Frictional resistance (Hydrodynamics)--Congresses.
2. Turbulence--Congresses.
3. Polymers and polymerization--Additives--Congresses.
I. Gampert, B. (Bernhard)
II. International Union of Theoretical and Applied Mechanics.
III. Deutsche Rheologische Gesellschaft.
IV. Universität Essen.
V. Gesellschaft von Freunden und Förderern der Universität-GH-Essen.
TP156.F6I54 1985 620.1'06 85-26217
ISBN-13: 978-3-642-82634-4 (U.S.)

2161/3020-5 4 3 2 1 0

Scientific Committee

B. Gampert, FRG (Chairman)	V.A. Ioselevich, USSR
F. Durst, FRG	D.F. James, Canada
H. Giesekus, FRG	J.L. Lumley, USA
A. Gyr, Switzerland	O. Scrivener, France
E.J. Hinch, England	I. Tani, Japan

Symposium Sponsors

International Union of Theoretical and Applied Mechanics (IUTAM)

Universität-GH-Essen

Gesellschaft von Freunden und Förderern der Universität-GH-Essen e.V.

Preface

The Symposium on "The Influence of Polymer Additives on Velocity and Temperature Fields" was proposed to the General Assembly of the International Union of Theoretical and Applied Mechanics (IUTAM) by the "Gesellschaft für Angewandte Mathematik und Mechanik" (GAMM). The Symposium was held under the auspices of IUTAM in association with the "Deutsche Rheologische Gesellschaft" (DRG) with responsibility for the organization lying with B. Gampert (Universität-GH-Essen).

The main aim of this IUTAM Symposium was to consider the fundamental aspects of the phenomena that occur when small amounts of polymers are added to turbulent flows (turbulent drag reduction) and laminar porous media flows. In particular attention was devoted to

- the influence of molecular parameters of the polymers and solution properties, especially the elongational viscosity, on turbulent flow and laminar porous media flow;

- the influence of polymers on the turbulence structure in polymer drag reduction.

A thematical structure was provided to the Symposium by seven invited lectures which reviewed various aspects of the flow of polymer additives. These reviews were supplemented by twenty-nine contributed papers. In accordance with the IUTAM regulations, there was no general call for papers. The contributed papers were selected by the Scientific Committee from extended abstracts whereby each paper was reviewed by all members of the Scientific Committee. Unfortunately, it was not possible to include all papers in the Symposium. The papers finally accepted were more relevant and fitted into the planned program of the Symposium. Fifty minutes were allotted for the invited lectures and thirty minutes for the selected papers; this included ten minutes for questions and discussion.

The Symposium brought together many scientists from all over the world who are active in the field of polymer additive flow research. The conference had quite an interdisciplinary character as chemists, physico-chemists and fluid dynamicists specializing in polymer science, rheology and turbulence took part. Such a group is certainly necessary if the flow of polymer solutions is to be dealt with succesfully.

It was intended that the Symposium should present the state of the art and develop concepts for achieving further progress in this subject, since the mechanism by which polymer additives influence velocity and temperature fields is still poorly understood. The latter goal was the topic of a very lively final discussion, which was chaired by D. James. The papers submitted showed that little work has been undertaken on the influence of polymer additives on temperature fields. This subject should be an objective of further research in the near future.

As always many people were involved in the preparation and running of the Symposium. We are particularly grateful to the members of the Scientific Committee who at all stages during the preparation of the conference, maintained a very close rapport, in spite of the large geographical distances between them.

In addition to the members of the Scientific Committee we would like to thank especially Dr. M. Hibberd (Universität Dortmund), Prof. Dr. K. Kirschke (Bundesanstalt für Materialprüfung, Berlin) and Dr. P. Wagner (Universität-GH-Essen) for their help. The support by many offices of the Universität-GH-Essen and by all the members of the "Fachgebiet Strömungsmechanik", especially during the conference, is very much appreciated. Thanks are also due to D. Fruman, A. Gyr, J. Hoyt and R.H.J. Sellin for their contributions which were not part of the official program. We are much indebted for the financial support provided by the sponsors which made this Symposium possible.

The ready and efficient cooperation of Herr von Hagen and the editorial staff of Springer Verlag during the preparation for the printing of the Proceedings is specially acknowledged.

It is our hope that this book will stimulate further research efforts in this exciting field of fluid mechanics.

B. Gampert

Participants

Abernathy, F.H.	Division of Applied Sciences, Pierce Hall, Harvard University, Cambridge, Massachusetts o2138, USA
Bartels, P.V.	Laboratory for Physical Technology, Delft University of Technology, Prins Bernhard Laan 6, 2628 BW Delft, The Netherlands
Berman, N.S.	Department of Chemical and Bio-Engineering, Arizona State University Tempe, Arizona 85287, USA
Bewersdorff, H.W.	Universität Dortmund, Abteilung Chemietechnik, Lehrstuhl für Strömungsmechanik, Postfach 5oo5oo 46oo Dortmund 6o, West Germany
Bismuth, D.	E.N.S.T.A., Centre de l'Yvette, Chemin de la Hunière, 9112o Palaiseau, France
Darby, R.	Chemical Engineering Department, Texas A & M University, College Station, Texas 778 43, USA
Davidson, A.	Dow Chemical Europe, Industriestr. 1, 7587 Rheinmünster 2, West Germany
Delgado, A.	Universität-GH-Essen, FB 12, Strömungsmechanik, Schützenbahn 7o, 43oo Essen 1, West Germany
Desch, H.J.	Fachgebiet Technische Mechanik und Strömungslehre, - Produktionstechnik - Universität Bremen, Langemarkstr. 116, 28oo Bremen, West Germany

Deslouis, C.

Laboratoire Physique des Liquides
et Electrochimie,
Université Pierre et Marie Curie,
4, Place Jussieu, Tour 22 - 5e étage,
7523o Paris Cedex o5,
France

Durst, F.

Lehrstuhl für Strömungsmechanik,
Technische Fakultät,
Friedrich-Alexander-Universität,
Erlangen-Nürnberg,
Egerlandstr. 13,
852o Erlangen,
West Germany

Eckel, A.

BASF Aktiengesellschaft,
VIT/ET J532,
67oo Ludwigshafen/Rhein,
West Germany

Eckelmann, H.

Max-Planck-Institut für
Strömungsforschung,
Göttinger Str. 6/8,
34oo Göttingen,
West Germany

Elata, C.

Department of Mechanical Engineering,
Ben Gurion University of the Negev,
PO Box 2o53,
Beer-Sheba 8412o,
Israel

Falco, R.E.

Turbulence Structure Laboratory,
Michigan State University,
East Lansing, MI 48824,
USA

Fortuin, J.M.H.

University of Amsterdam,
DSM Geleen,
Central Laboratory DSM,
P.O. Box 18,
616o MD Geleen,
The Netherlands

Frings, B.

Universität Dortmund,
Abteilung Chemietechnik,
Lehrstuhl für Strömungsmechanik,
Postfach 5o o5 oo, 46oo Dortmund 6o,
West Germany

Fruman, D.H.

E.N.S.T.A.,
Centre de l'Yvette,
Chemin de la Hunière,
9112o Palaiseau,
France

Gampert, B.

Universität-GH-Essen,
FB 12, Strömungsmechanik,
Schützenbahn 7o
43oo Essen 1,
West Germany

Giesekus, H.

Lehrstuhl für Strömungsmechanik,
Universität Dortmund,
Postfach 5o o5 oo,
August-Schmidt-Straße,
46oo Dortmund,
West Germany

Gyr, A.

Institut für Hydromechanik und
Wasserwirtschaft,
ETH-Hönggerberg,
CH-8o93 Zürich,
Switzerland

Haas, R.

Fa. Hoechst AG,
Abteilung Angewandte Physik,
Postfach 8oo 32o,
623o Frankfurt/Hoechst,
West Germany

Hackler, L.

Universität-GH-Essen,
FB 12, Strömungsmechanik,
Schützenbahn 7o,
43oo Essen 1,
West Germany

Halasz von, S.-P.

Hoechst Aktiengesellschaft,
ATA TH, Gebäude D561,
623o Frankfurt am Main 8o,
West Germany

Hibberd, M.

Universität Dortmund,
Abteilung Chemietechnik,
Lehrstuhl für Strömungsmechanik,
Postfach 5o o5 oo,
46oo Dortmund 5o,
West Germany

Holtum, D.	Universität-GH-Essen, FB 12, Strömungsmechanik, Schützenbahn 7o, 43oo Essen 1, West Germany
Hoyt, J.W.	Department of Mechanical Engineering, College of Engineering, San Diego State University, San Diego, California 92182-o191, USA
Interthal, W.	Hoechst AG, Angewandte Physik G 864, Postfach 8oo 32o, 623o Frankfurt/M-Hoechst, West Germany
James, D.F.	University of Toronto, Department of Mechanical Engineering, 5, King's College Road, Toronto, Ontario, M5S 1A4, Canada
Jischa, M.	Institut für Technische Mechanik der Technischen Universität Clausthal, Graupenstr. 3, 3392 Clausthal-Zellerfeld, West Germany
Korte, J.-G.	Universität-GH-Essen, FB 12, Strömungsmechanik, Schützenbahn 7o. 43oo Essen 1, West Germany
Kulicke, W.-M.	Institut für Chem. Technologie, TU Braunschweig, Hans-Sommer-Str. 1o, 33oo Braunschweig, West Germany
Lagerstedt, T.	The Royal Institute of Technology, Department of Mechanics, S-1oo 44 Stockholm 7o, Sweden

Lasek, A.

Lab. d'Aérothermique,
C.N.R.S.,
4 route des Gardes,
F-9219o - Meudon,
France

Layec-Raphalen, M.N.

Laboratoire d'Hydrodynamique
Moléculaire,
Faculté des Sciences,
6, Avenue Le Gorgeu,
29283 Brest-Cedex,
France

Leal, L.G.

Department of Chemical Engineering,
California Institute of Technology,
Pasadena,
California 91125,
USA

Lee, H.S.

Hyundai Motor Co.,
Seoul, Korea

Lewalle, J.

Syracuse University,
Department of Mechanical and
Aerospace Engineering,
139 E.A. Link Hall,
Syracuse, New York 1321o,
USA

Lumley, J.L.

Sibley School of Mechanical
and Aerospace Engineering,
Cornell University,
Upson and Grumman Halls,
Ithaca, New York 14853,
USA

Maksimović, C.

Institute of Hydraulic Engineering,
Faculty of Civil Engineering,
University of Belgrade,
Bulevar Revolucije 73,
P.O. Box 895,
YU-11ooo Belgrade,
Jugoslavia

Martischius, F.-D.

BASF Aktiengesellschaft,
Grundstofflabor,
D-ZKM, G 2o1,
D-67oo Ludwigshafen,
West Germany

Meyer, J.	Institut für Erdölforschung, Walter-Nernst-Str. 7, D-3392 Clausthal-Zellerfeld, West Germany
Motier, J.	Atlantic Richfield Company, 15oo Market Street, Philadelphia, Pennsylvania, 19 1o 11, USA
Neubauer, H.	Lehrstuhl für Strömungsmechanik, Technische Fakultät, Friedrich-Alexander-Universität, Erlangen-Nürnberg, Egerlandstr. 13, 852o Erlangen, West Germany
Nienass, D.	Institut für Verfahrenstechnik, RWTH Aachen, Turmstr. 46, 51oo Aachen, West Germany
Patterson, G.K.	The University of Arizona, College of Mines, Department of Chemical Engineering, Tuscon, Arizona 85721, USA
Paul, D.D.	Batelle-Columbus Laboratories, 5o5 King Avenue, Columbus, OH 4322o, USA
Payer, W.	Ruhrchemie Aktiengesellschaft, Postfach 13 o1 6o, D-43oo Oberhausen 11, West Germany
Petty, C.A.	Department of Chemical Engineering, Michigan State University, East Lansing, Michigan, USA

Pollert, J.	C.V.U.T., Faculty of Civil Engineering, Department of Hydraulics and Hydrology, 166-29 Praha 6, Thakurova 7, Czechoslovakia
Rath, H.J.	Fachgebiet Technische Mechanik und Strömungslehre, - Produktionstechnik - Universität Bremen, Langemarkstr. 116, 28oo Bremen, West Germany
Reischmann, M.M.	Scientific Officer, Office of Naval Research, Code 432F, Arlington VA 22217, USA
Rieke, H.-B.	Universität-GH-Essen, FB 12, Strömungsmechanik, Schützenbahn 7o, 43oo Essen 1, West Germany
Rose, G.D.	Central Research-Specialty Products Research, The Dow Chemical Company, 1712 Building, Midland, Michigan 4864o, USA
Schmitt, K.	Lehrstuhl für Strömungsmechanik, Technische Fakultät, Friedrich-Alexander-Universität, Erlangen-Nürnberg, Egerlandstraße 13, 852o Erlangen, West Germany

Schwarz, F.-J.	FB Mathematik, Arbeitsgruppe Technomathematik, Erwin-Schrödinger-Straße, Universität Kaiserslautern, Pfaffenbergstr. 95, Postfach 3o 49, D-675o Kaiserslautern, West Germany
Scrivener, O.	Institut de Mécanique des Fluides, Université Louis Pasteur de Strasbourg, ERA CNRS o594, 2, rue Boussingault, 67o83 Strasbourg Cedex, France
Sedov, L.I.	Institute of Mechanics, Moscow University, Michursinskii Pr. 1 117234 Moscow V-234, USSR
Sellin, R.H.J.	University of Bristol, Department of Civil Engineering, Queen's Building, University Walk, Bristol, BS8 1TR United Kingdom
Singh, R.P.	Indian Institute of Technology, Materials Science Centre, Kharagpur - 7213o2, India
Tomita, Y.	Dept. Mechanical Engineering, Kyushu Institute of Technology, Tobata, Kitakyushu-shi, 8o4 Japan
Usui, H.	Dept. Chemical Engineering, Yamaguchi University, Ube, Yamaguchi-ken, 755 Japan
Wagner, P.	Universität-GH-Essen, FB 12, Strömungsmechanik, Schützenbahn 7o, 43oo Essen 1, West Germany

Waid, R.	O/11-13, B/568, Lockheed Advanced Marine Systems, P.O. Box 4ooo, Santa Clara, California 95o54, USA
Walters, K.	University College of Wales, Department of Applied Mathematics, Penglais, Aberystwyth SY23 3BZ, United Kingdom
Wellinghoff, G.	Universität Dortmund, Lehrstuhl für Therm. Verfahrens- technik, Postfach 5oo 5oo, 46oo Dortmund 5o, West Germany
Wishart, A.J.	Dow Chemical Europe, Industriestraße 1, 7587 Rheinmünster 2, West Germany
Wolff, C.	Laboratoire de Mécanique et Rhéologie, École Nationale Supérieure des Industries Textiles de Mulhouse, 11 rue Alfred Werner, 68o93 Mulhouse Cedex, France
Zakin, J.L.	The Ohio State University, Department of Chemical Engineering, 121 Koffolt Laboratories, 14o West 19th Avenue, Columbus, Ohio 4321o, USA
Zimmermann, R.U.	Institut für Mechanik, Universität Hannover, Appelstr. 11, 3ooo Hannover 1, West Germany

Contributors

Abdel-Khalik, S.I. — University of Wisconsin-Madison, USA

Aflalo, S. — Ecole Nationale Supérieure de Techniques Avancées, Paris, France

Ambari, A. — Université Pierre et Marie Curie, Paris, France

Anisimov, I.A. — Institute of Thermophysics, Novosibirsk, USSR

Bartels, P.V. — Laboratory of Physical Technology, Delft University of Technology, Prins Bernhard Laan 6, 2628 BW Delft, The Netherlands

Berman, N.S. — Department of Chemical and Bio-Engineering, Arizona State University Tempe, Arizona 85287, USA

Bewersdorff, H.W. — Universität Dortmund, Abteilung Chemietechnik, Lehrstuhl für Strömungsmechanik, Postfach 500 500, 4600 Dortmund 60, West Germany

Bismuth, D. — E.N.S.T.A., Centre de l' Yvette, Chemin de la Hunière, 91120 Palaiseau, France

Burov, A.N. — USSR Academy of Sciences, Novosibirsk, USSR

Chang, R.C. — Ohio State University, Columbus, USA

Darby, R. — Chemical Engineering Department, Texas A&M University, College Station, Texas 778 43, USA

Desch, H.J.	Fachgebiet Technische Mechanik und Strömungslehre - Produktionstechnik - Universität Bremen, Langemarkstr. 116, 2800 Bremen, West Germany
Deslouis, C.	Laboratoire Physique des Liquides et Electrochimie, Université Pierre et Marie Curie, 4, Place Jussieu, Tour 22 - 5e étage, 75230 Paris Cedex 05, France
Deshmukh, S.R.	Indian Institute of Technology, Kharagpur, India
Durst, F.	Lehrstuhl für Strömungsmechanik, Technische Fakultät, Friedrich-Alexander-Universität, Erlangen-Nürnberg, Egerlandstr. 13, 8520 Erlangen, West Germany
Eckelmann, H.	Max-Planck-Institut für Strömungsforschung, Göttinger Str. 6/8, 3400 Göttingen, West Germany
Falco, R.E.	Turbulence Structure Laboratory, Michigan State University, East Lansing, MI 48824, USA
Frings, B.	Universität Dortmund, Abteilung Chemietechnik, Lehrstuhl für Strömungsmechanik, Postfach 500 500, 4600 Dortmund 60, West Germany
Fruman, H.	E.N.S.T.A., Centre de l' Yvette, Chemin de la Hunière, 91120 Palaiseau, France
Gampert, B.	Universität-GH-Essen, FB 12, Strömungsmechanik, Schützenbahn 70, 4300 Essen 1, West Germany

Gyr, A.	Institut für Hydromechanik und Wasserwirtschaft, ETH-Hönggerberg, CH-8093 Zürich, Switzerland
Haas, R.	Fa. Hoechst AG, Abteilung Angewandte Physik, Postfach 800 320, 6230 Frankfurt/Hoechst, West Germany
Hibberd, M.	Universität Dortmund, Abteilung Chemietechnik, Lehrstuhl für Strömungsmechanik, Postfach 500 500, 4600 Dortmund 50, West Germany
Interthal, W.	Hoechst AG, Angewandte Physik G 864, Postfach 800 320, 6230 Frankfurt/M-Hoechst, West Germany
Ioselevich, V.A.	Institute of Mechanics, Moscow University, Michursinskii Pr. 1, 117234 Moscow V-234, USSR
Irvine. T.F.	SUNY at Stony Brook, Stony Brook, New York, USA
Iwata, N.	Yamaguchi University, Ube, Japan
James, D.F.	University of Toronto, Department of Mechanical Engineering, 5, King's College Road, Toronto, Ontario, M5S 1A4, Canada
Janssen, L.P.B.M.	Delft University of Technology, Delft, Netherlands
Kit, Z.	Israel Oceanographic & Limnological Research Ltd., Haifa, Israel

Klein. J.	Institut für Chemische Technologie TU Braunschweig, Hans-Sommer-Str. 10, 3300 Braunschweig, West Germany
Kolobov, B.P.	USSR Academy of Sciences, Novosibirsk, USSR
Kubo, I.	Sibley School of Mechanical and Aerospace Engineering, Cornell University, Ithaca NY 14853 USA
Kulicke, W.M.	Institut für Chemische Technologie TU Braunschweig, Hans-Sommer-Str. 10, 3300 Braunschweig, West Germany
Kumar, A.	Indian Institute of Technology, Kharagpur, India
Lagerstedt, T.	The Royal Institute of Technology, Department of Mechanics, S-100 44 Stockholm 70, Sweden
Layec, Y.	Laboratoire d' Hydrodynamique Moléculaire, Faculté des Sciences, Brest, France
Layec-Raphalen, M.N.	Laboratoire d' Hydrodynamique Moléculaire, Faculté des Sciences, 6, Avenue Le Gorgeu, 29283 Brest- Cedex, France
Leal, L.G.	Department of Chemical Engineering, California Institute of Technology, Pasadena, California 91125, USA
Lee, H.S.	Hyundai Motor Co., Seoul, Korea
Lookeren van, G.J.	Delft University of Technology, Delft, Netherlands

Lumley, J.L.	Sibley School of Mechanical and Aerospace Engineering, Cornell University, Upson and Grumman Halls, Ithaca, New York 14853, USA
Lyons, S.	Michigan State University, East Lansing, USA
Mahrenholtz, O.H.	Technical University of Hamburg-Harburg, West Germany
Maksimović, C.	Institute of Hydraulic Engineering, Faculty of Civil Engineering, University of Belgrade, Bulevar Revolucije 73, P.O. Box 895, YU-11000 Belgrade, Jugoslavia
Markus, A.	Delft University of Technology, Delft, Netherlands
Matjukhov, A.P.	Institute of Thermophysics, Novosibirsk, USSR
Mironov, B.P.	Institute of Thermophysics, Novosibirsk, USSR
Patterson, G.K.	The University of Arizona, College of Mines, Department of Chemical Engineering, Tuscon, Arizona 85721, USA
Paul, D.D.	Batelle-Columbus Laboratories, 505 King Avenue, Columbus, OH 43220, USA
Petty, C.A.	Department of Chemical Engineering, Michigan State University, East Lansing, Michigan, USA
Pilipenko, V.N.	Institute of Mechanics, Moscow University, Michurinskii Pr. 1 117234 Moscow V-234

Pollert, J.	C.V.U.T., Faculty of Civil Engineering, Department of Hydraulics and Hydrology, 166-29 Praha 6, Thakurova 7, Czechoslovakia
Popkov, A.N.	USSR Academy of Sciences, Novosibirsk, USSR
Rath, H.J.	Fachgebiet Technische Mechanik und Strömungslehre - Produktionstechnik - Universität Bremen, Langemarkstr. 116, 2800 Bremen, West Germany
Rosentraub, E.	Israel Oceanographic & Limnological Research Ltd., Haifa, Israel
Sano, Y.	Yamaguchi University, Ube, Japan
Sedov, L.I.	Institute of Mechanics, Moscow University, Michursinskii Pr. 1, 117234 Moscow V-234, USSR
Singh, R.P.	Indian Institute of Technology, Materials Science Centre, Kharagpur - 721302, India
Singh, J.	Indian Institute of Technology, Kharagpur, India
Smith, J.M.	Delft University of Technology, Delft, Netherlands
Tomita, Y.	Dept. Mechanical Engineering, Kyushu Institute of Technology, Tobata, Kitakyushu-shi, 804 Japan
Tribollet, B.	Université Pierre et Marie Curie, Paris, France

Usui, H.	Dept. Chemical Engineering, Yamaguchi University Ube, Yamaguchi-ken, 755 Japan
Wagner, P.	Universität-GH-Essen, FB 12, Strömungsmechanik, Schützenbahn 70, 4300 Essen 1, West Germany
Walters, K.	University College of Wales, Department of Applied Mathematics, Penglais, Aberystwyth SY23 3BZ, United Kingdom
Wolff, C.	Laboratoire de Mécanique et Rhéologie, École Nationale Supéri- eure des Industries Textiles de Mulhouse, 11 rue Alfred Werner, 68093 Mulhouse Cedex, France
Zakin, J.L.	The Ohio State University, Department of Chemical Engineering, 121 Koffolt Laboratories, 140 West 19th Avenue, Columbus Ohio 43210, USA
Zimmermann, R.U.	Insitut für Mechanik, Universität Hannover, Appelstr. 11, 3000 Hannover 1, West Germany

Contents

Session 1o Injection Experiments

Chairman: A. Gyr

Session 11

Chairman: B. Gampert

Session 12 Pressure Fields, Cavitation

Chairman: S.L. Zakin

---

* Invited Lecture

\+ Paper not available at time of printing

° Presented by C. Elata

No oral presentation

Session 1

Turbulent Drag Reduction by Polymer Additives: A Survey *

JOHN L. LUMLEY, ISOROKU KUBO**

Sibley School of Mechanical and Aerospace Engineering
Cornell University, Ithaca NY 14853 USA

Summary

We first give a brief history of turbulent polymer drag reduction, and the position it has played relative to other drag reduction schemes, notably suction, compliant surfaces, form modification and heated bodies. We then survey briefly all but the most recent literature from the last twenty years, concentrating on measurements of: mean velocity profiles, turbulent intensities, sublayer streak spacing and bursting period, and flows other than pipe and channel flows. We emphasize non-intrusive measurements. From an evaluation of the various experiments, and the consensus within each group, we draw conclusions regarding the various structural hypotheses that have been put forth, and experiments which may prove fruitful in the future.

Introduction

The possibility of reducing the turbulent skin-friction drag of military and commercial vehicles has always been exciting. For surface vessels operating near the hull design speed, or for aircraft operating at transonic (or higher) speeds, it is perhaps not so interesting, since a substantial fraction of their drag is wave drag; even so, there has been interest among competitive sailors eager to reduce the net drag a percentage point or two. The possibility becomes quite interest-

---

*IUTAM Symposium, Essen 1984: Influence of Polymer Additives on Velocity and Temperature Fields, ed. B. Gampert. Berlin: Springer. Supported in part by the U. S. National Aeronautics and Space Administration, Ames Research Center, under Grant No. NSG-2382; in part by the U.S. Office of Naval Research under the following programs: Physical Oceanography (Code 422PO), Power (Code 473); in part by the U.S. National Science Foundation under grant no. ATM 79-22006; and in part by the U.S. Air Force Geophysics Laboratory.

**Present address: Mail Code 50165, Cummings Engineering Co., Columbus, IN 47201 USA.

The Influence of Polymer Additives
on Velocity and Temperature Fields
IUTAM Symposium Essen 1984
Editor: B. Gampert

ing for super-tankers, which operate well below the hull design speed, for totally submerged vehicles, or for subsonic aircraft. The turbulent skin friction drag, after all, represents the majority of the power consumption for a streamlined vehicle operating in this regime, and the possibility of reducing the drag by 50-80%, which is realistic, offers visions of increases in speed, range, payload, or profitability (depending on one's motivation) that are enthralling.

In all applications, it is as well to remember that there are other considerations. In the case of the supertanker, for example, one of the major oil companies made a careful calculation of the possible economic advantage to ejecting polymers in the boundary layer to reduce the drag, assuming that the polymer was manufactured by the company in question, taking account of the value of the crude oil displaced by the polymer which must be carried, the changes in time at sea versus time in port, and the relative costs of each, and other factors. They concluded that they would just break even, and decided not to pursue the matter.

In addition to these applications, there are many others. In the Alaska pipeline, for example, the pressure drop resulting from cooling of the crude oil during the Alaskan winter was underestimated, so that the frequency and power of the pumping stations was insufficient. This was cheaply (relative to other solutions) rectified by polymer additives to reduce the drag.

Historically, many possibilities have been considered. The possibility of a gas film in water was very attractive, and injection through porous and slotted skins, as well as ablating bodies were considered. Unfortunately, buoyancy is a problem, the air/water interface is extremely unstable (as is also the gas/solid interface of an ablating body), and the various instabilities increase the drag enormously.

Suction through porous surfaces or through discrete slots has been tried for more than twenty years. The object is to increase the stability of the boundary layer, and avoid the transition to turbulence. Unfortunately, the sucked fluid must be accelerated to the free-stream speed, and the power required to do this constitutes a drag. Hence, the amount of suction possible is distinctly limited, if the net drag is to be reduced; the maximum possible suction makes the layer more stable, but does not completely stabilize it. It can still undergo transition in response to nearly any disturbance. In air, impacted insect bodies were found to be very troublesome, and early experimental wings were covered with plastic-wrap on takeoff. In water, the holes and slots in the porous surfaces were just the right size to trap plankton and other debris, blocking the holes, and creating downstream a particularly unstable velocity profile.

In water, compliant surfaces have attracted much attention; theoretical and experimental work has been done on them for nearly twenty years. Numerous theoretical analyses of stability of laminar boundary layers over compliant boundaries have indicated that stabilization is possible by a judicious choice of material properties. Although isolated field measurements have suggested stabilization, it has never been possible to demonstrate this effect under controlled laboratory conditions, due often to the appearance of instabilities that were not considered in the analyses, which had the effect of removing the compliant layer. It has also frequently been suggested that a suitable compliant boundary might reduce the drag under a turbulent boundary layer, without stabilizing the flow. Unfortunately, not enough is known about the mechanism of a turbulent boundary layer to explain how this might come about, or to make useful suggestions regarding suitable properties for the compliant boundary.

Interest in compliant boundaries arose from anecdotal observations of the Pacific bottle-nosed Dolphin evidently making great speed near passenger ships. Calculations indicated that the apparent speed required a power output greater than mammalian muscle could produce. Hence, it was concluded that the Dolphins must be exercising some form of drag reduction.

This was reinforced by examination of sections of Dolphin skin, which indeed showed a complex structure of unexplained function. This was the birth of compliant boundary drag reduction research. Years later, tests on a trained Dolphin indicated nothing unusual in the way of drag reduction, and more realistic evaluations of the anecdotal evidence suggested that the Dolphins were riding the bow waves of the vessels. Although the matter is not settled definitively (the trained Dolphin might not have understood that drag reduction was desired during the test, if, for example, drag reduction were ordinarily applied only during emergencies), it now seems likely that the whole compliant surface drag reduction research effort was based on a misapprehension.

The addition of polymers has been investigated also for twenty years. Polymers reduce the drag of the turbulent boundary layer, without avoiding transition. Since the layer remains turbulent, there is no stringent requirement for surface smoothness, as there is for many of the other techniques. Although much is not known about the detailed mechanism, practical success can be achieved more or less routinely. As there would be in the application of any drag reduction technique, there are practical problems that must be solved to make polymers operational; these include such problems as how and with what and when to mix the polymer and how to eject it, and how to mate the propulsion system to the altered body characteristics.

Several years ago, the attention of the research community was distracted from polymers, a workable scheme with few surprises, but with engineering problems to be worked out, to heated bodies, an unproved scheme which was ultimately found to present unforseen problems.

Reducing drag by heating the surface has been examined theoretically for some time also. The principle is that (in a liquid) the hotter fluid near the surface has lower viscosity, and the profile is therefore fuller, more rounded, which is a more stable shape. The effect of the viscosity gradient on the growth of the instability itself can be shown to be a much smaller effect. Thus, surface heating can be shown at first order to be equivalent to a favorable pressure gradient. In practice, it is usually combined with a favorable pressure gradient, in such a way as to avoid transition until the latest possible point, as the body begins to close toward the rear; the body is then allowed to close as rapidly as possible without separation.

In practice it is found that, due primarily to buoyancy effects, it is not practical to make the layer extremely stable - it is possible only to considerably improve the stability. Thus, the layer remains sensitive to disturbances. Background disturbances in the fluid can cause abrupt loss of stabilization. The resulting turbulent patch is usually swept downstream, off the body, and stability restored, but the abrupt appearance and disappearance of these patches causes correspondingly abrupt changes in the drag, having serious effects on other aspects of vehicle performance. Vehicles in the ocean, of course, are not operating in a perfectly quiet environment; both natural and manmade disturbances abound.

Polymers are currently in the ascendency again, as the most developed, and most trouble-free drag reduction technique. In the mean time, other techniques are under investigation. Soviet scientists pioneered the injection of very small air bubbles in the turbulent boundary layer, which has a substantial effect on the drag. We will later suggest that this technique probably works for much the same reasons that the polymers work. In addition, it has been found that small triangular ribs (riblets) alligned with the flow, if they are sized appropriately relative to the scales of the flow, are capable of reducing the drag substantially. Again, we will see below that the mechanism is probably closely related to the mechanism for both the polymers and the microbubbles.

When drag reduction first appeared as a possibility, financially interested individuals spent business lunches calculating the number of square feet underwater in the U. S. Merchant

Marine, and estimating the cost of covering each square foot with some material, compliant or otherwise. In the case of riblets, a large manufacturer of transparent tape has been contacted about fabricating continuous moulded strips of riblets with a sticky back, in rolls, for application to the entire U. S. commercial air fleet. Similar lunches are probably spent calculating the number of square feet involved, and the possible profit.

In what follows we will examine in greater technical detail the evidence and possible mechanism for polymer drag reduction, with occasional reference to other techniques when relevant.

The experimental evidence

Extensive measurements, with all sorts of instrumentation, have been made of mean and fluctuating quantities in flows of drag reducing polymer solutions since the mid-sixties. In [1] we made a general survey of the literature through 1979. For the period 1970 - 1979, we made a particularly thorough check of the five major journals in which the majority of these papers appear. In addition, more papers were collected by checking references of review papers by four major authors. We do not reproduce here these references, but refer the reader to [1]. In making this search, we have arbitrarily excluded all data obtained with hotwire or hot film anemometers, or with pitot tube, since these are subject to considerable errors in polymer solutions. Thus, we retain only measurements by LDA and flow visualization. We summarize here our qualitative findings.

Measurements of the mean velocity profile were, of course, the first to be made, and are the most numerous in the literature. In coordinates normalized with wall variables, any profile has the same asymptotic behavior at the wall. Drag reduction corresponds to the logarithmic part (further from the wall) being higher. It has been observed for some time that the slope of the logarithmic part is the same, and a number of simple arguments explain this. On the basis of the data reviewed here, we conclude that all the observed changes in the mean velocity profiles are within the buffer layer; the viscous sublayer (defined by the first significant deviation from a linear profile) is not affected. There seems to be no well-defined maximum drag-reduction asymptote (see figure 1), but one process which limits drag reduction occurs when the thickened buffer layer reaches the center of the pipe or the edge of the boundary layer.

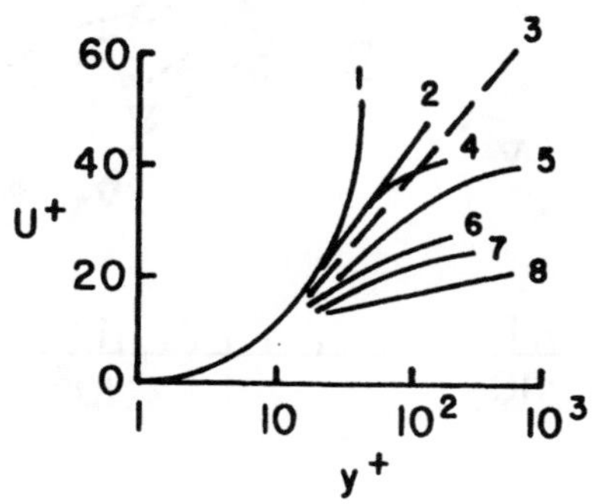

Figure 1. Sketch of representative pipe mean flow velocity profiles. 1) $U^+ = y^+$. 3) Virk's "asymptote". 8) $U^+ = 2.5\ell n y^+ + 5$. 2), 4), 5), 6) & 7), various authors. See [1] for references.

Measurements of turbulence intensity have been mainly of the streamwise component. All the measurements (figure 2) show that the peak in the intensity of the streamwise component is shifted to higher y_+. The measurements are completely in conflict on the magnitude of the peak intensity, some indicating an increase, and some indicating no change from the Newtonian case. There are indications that some of the measurements are contaminated by secondary motions. Our best estimate is that, at high Reynolds numbers, the peak intensity is close to, and probably slightly higher than, that in Newtonian fluids. Most of the data do indicate a decline in the intensity in the region $y_+ < 20$. The Reynolds number effect on the peak intensity seems significant up to at least 15,000 (based on centerline mean velocity and diameter).

Transverse intensity measurements are few, but unanimous that

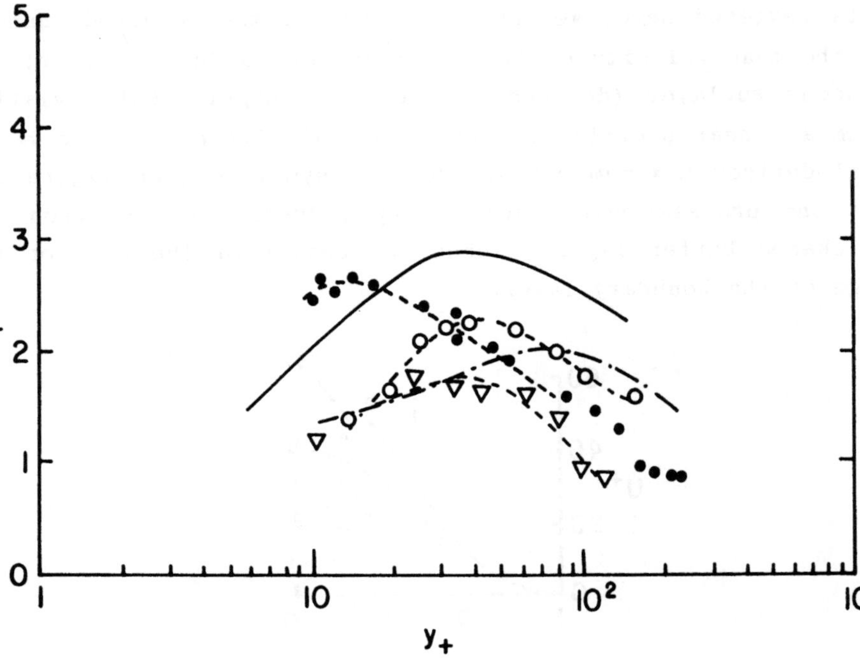

Figure 2. Representative turbulent intensity distributions. Closed circles: solvent. Triangles: Re = 7,200. Open circles: Re = 11,000. Dash-dot: Re = 9,500 - 35,000. Solid curve: Re = 20,000 - 52,000.

in the core region the intensity is the same as that of the solvent, while in the buffer layer the intensity is lower.

Frequency spectra of wall shear stress fluctuations and of velocity fluctuations indicate relative attenuation of the high frequency components (Figure 3). Measurements in boundary-free flows have also usually found that the small-scale components are suppressed relative to the large-scale.

Non-dimensional streak spacing, which is a measure of the size of the coherent eddies in the sub- and buffer layers, increases as the drag reduction increases according to all measurements. The rate of increase of the spacing is monotone and gradual, and the spacing reaches twice that of a Newtonian fluid at about 60% drag reduction (Figure 4). Minor inconsis-

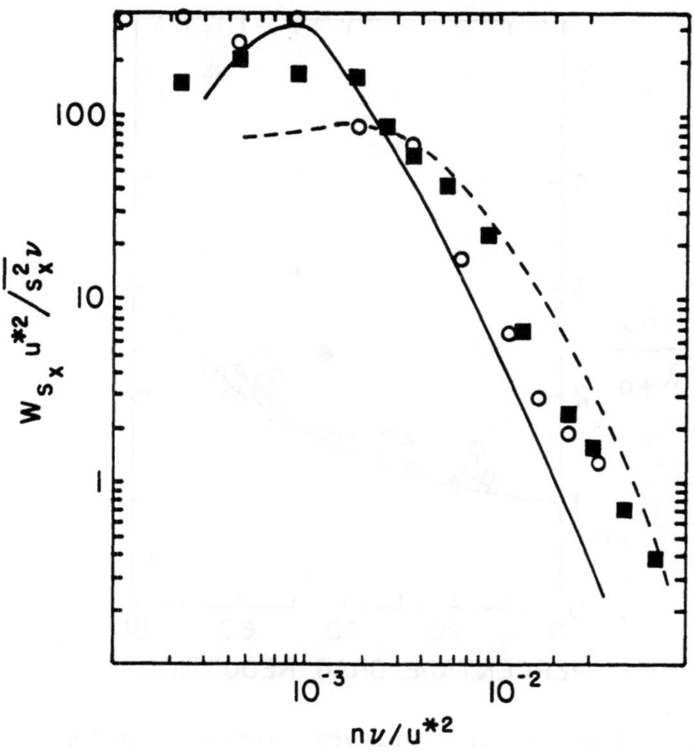

Figure 3. Spectral density function for wall shear stress fluctuations for solutions of Separan. Dashed line: solvent. Closed circles: 34.6% drag reduction, Re = 27,600. Closed squares: 64.5% drag reduction, Re = 57,600. Solid line: 62.9% drag reduction, Re = 10,000.

tencies in the data can easily be explained.

The bursting period probably corresponds to the time between occurrences of a secondary instability (see next section). Each burst is an explosive production of Reynolds stress. Measurements of the bursting period in polymer flows are divided into two groups, depending on the measurement technique used. Flow visualization indicates that the bursting period does not change from that in Newtonian fluids at the same shear velocity. The autocorrelation method suggests an increase in the bursting period in polymer flows at the same

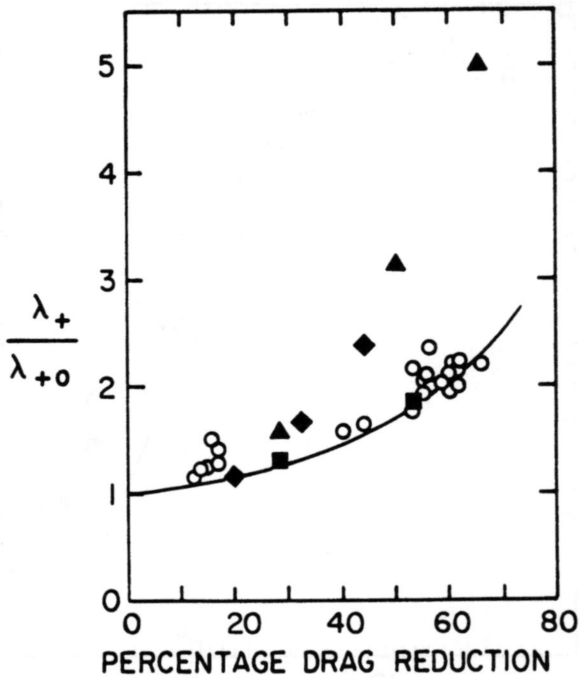

Figure 4. Changes in sublayer streak spacing, various authors.

shear velocity (Figure 5).

Drag reducing polymer flows undergo transition at a lower Reynolds number than Newtonian flows.

Structural hypotheses

A number of mechanisms have been proposed to date to explain the observations. Many of these have been highly speculative, and have not attracted the support of the scientific community, nor have they survived a brief notoriety. There are, however, two hypotheses which are more comprehensive than the others, have survived better, and have a great deal in common. These are the hypotheses of Lumley [2, 3, 4] and Landahl [5, 6].

Lumley's approach is statistical, and does not require a detailed mathematical theory for turbulent boundary layers.

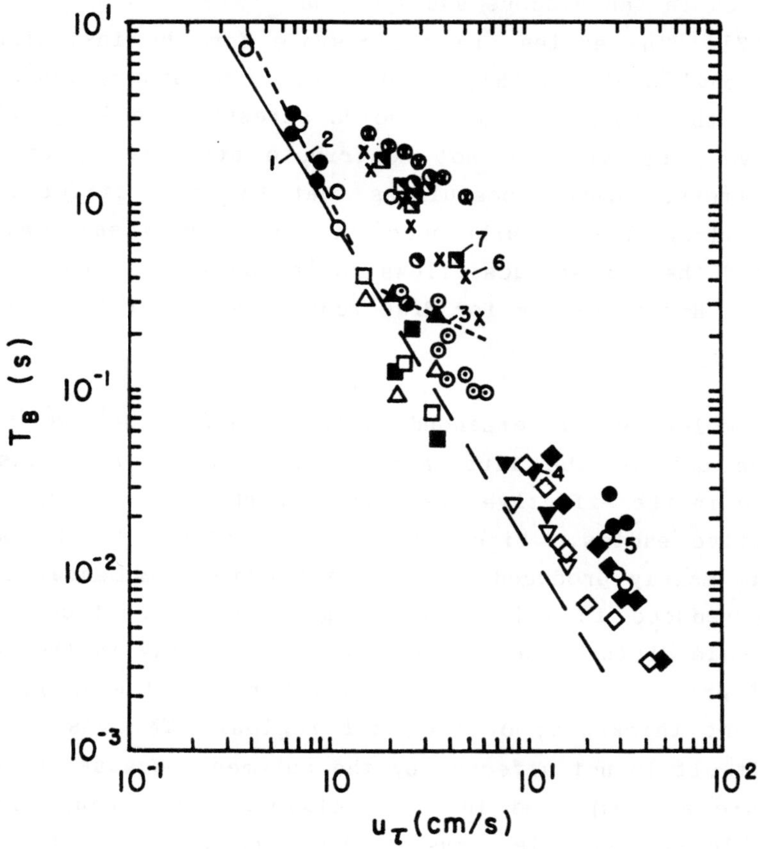

Figure 5. Sublayer burst period data in Newtonian and drag-reducing flows. The solid line represents both visual and hot-wire data in a water channel flow. Open and closed circles in the upper part of the plot represent two-dimensional visual water and polymer data, with the dashed line a pressure gradient correction. All other symbols are autocorrelation data, open for solvent, closed for polymer solutions (including open and closed circles in the lower part of the plot). Ekses are water, circled ekses are polymer; circled dots are water, half-black squares and circles are polymer.

The basic turbulence structure is assumed to be similar to that in Newtonian boundary layers, in that there exist three regions: the viscous sublayer, which is the closest to the wall, and where the dissipative forces dominate; a logarithmic region; and the buffer layer between the two. Large eddies are formed in the viscous sublayer, and extend into the buffer layer [7]. The eddies are responsible for the instantaneous inflectional profiles [8], which lead to secondary instabilities and the production of Reynolds stress. The large eddies themselves, however, do not contribute strongly to the Reynolds stress. Lumley postulates that the only difference in the boundary layer turbulence structure between Newtonian flows and the drag-reduced flows is the scaling of the various regions. His mechanism for drag reduction is outlined as follows:

Polymer molecules are expanded in the flow outside the viscous sublayer due to the fluctuating strain rate, and cause an increase in the effective viscosity, which in turn damps small dissipative eddies. Since the Reynolds stress in the buffer layer is mainly produced by the small eddies, momentum transfer is reduced in this zone owing to the reduction in the small scale disturbances. The result is a delay in the reduction of profile slope to a point farther from the wall, and a consequent thickening of the wall region. The viscous sublayer itself is not affected by the polymer, because the molecules are not expanded in the sublayer. The change in the mean velocity profile causes the expansion of the large eddies, which are determined primarily by inertial forces and kinematics [9]. Although in Lumley's hypothesis it is not necessary to be explicit, it may be pointed out that since the eddies in the buffer zone which are primarily responsible for Reynolds stress production are those associated with the secondary instability on the inflectional profile, it is presumably these (among others) which are damped.

The mechanism described above is consistent with essentially all of the existing experimental evidence which we have evaluated. Evidence for the molecular expansion on which this theory is based is summarized in [4]. Furthermore, the existence of high strain rates in shear flows was shown by water jet experiments [10]. Fluid particle elongations of sufficient magnitude are indeed apparent in near-wall flow visualization [11]. The assumption that polymers affect the small scale turbulent motion is also supported by experimental stability studies in pipe flows [12, 13], and by water jet studies [10].

This framework is too simple, however, to explain the detailed dynamics occurring in the boundary layer. For example, it does not explain during which part of the turbulent motion outside the sublayer the molecules expand, and how the expanded molecules are distributed. Prediction of the bursting period and large eddy sizes have not yet been made.

While the above approach is statistical, Landahl's is mechanistic. He applied the classical hydrodynamic stability concept to the turbulent boundary layer, and developed a so-called two-scale model. Landahl [5, 6] and Landahl & Bark [14] have applied their model to drag reduction. They considered the turbulence to consist of a coupled motion at two disparate scales: a primary one of large scale (of, say, the order of the boundary layer thickness) and a much smaller secondary one (of the order of the viscous sublayer thickness). The small-scale secondary motions ride on the primary waves, and under certain conditions grow to very large amplitude, leading to the breakdown of the primary waves. They hypothesize that the turbulent bursts are the result of the attainment of local criticality in the primary field, and that the subsequent violent motion will cause excitation of new primary waves [5]. The mechanism of drag reduction due to the additives is that the elongation of added polymer molecules stabilizes the small scale field, i.e.- stabilization of secondary inflectional instability, and thereby inhibits turbulence production. The inflectional stability was calculated using a hyperbolic tangent type of velocity profile, with several rheological models, including a convected Maxwell model and a suspension of rigid aligned rods. Their conclusion is that there is a strong secondary stabilization for all cases where there is clear experimental evidence of drag reduction, and that for drag reduction, viscoelasticity as such does not seem to be the key fluid property; rather, the anisotropic stress caused by the extension of the molecular coils (or molecular aggregates) in a turbulent flow of sufficiently high wall stress is a more likely candidate.

There is no real contradiction between Landahl's and Lumley's hypotheses. Both agree that it is the change in the stress-deformation relation for the solution due to molecular stretching that is responsible for damping smaller disturbances in the buffer region; in fact, (by inference in Lumley's case) both agree that it is the secondary instability which grows on the inflectionary profile that is damped. The only difference is that Landahl is much more specific regarding the mechanism of the primary and secondary instability, while Lumley's hypothesis does not require such specificity. In fact, Landahl postulates a mechanism which in detail goes considerably beyond what can be supported by experimental evidence in either the normal or the drag reducing boundary layer, although of course it is consistent with existing evidence. Most simply stated, Landahl's hypothesis might be characterized as a specific, rather speculative, mechanism consistent with Lumley's more conservative general hypothesis.

Relation to other schemes

Drag reduction by the injection of microbubbles through a porous surface has been investigated by workers in the Soviet Union (see, for example, [15, 16]. There are two effects here. First, at a constant shear velocity, the reduction of average density would reduce the turbulent stress. Thus, even if the structure of the boundary layer remained unaltered, so that the mean velocity remained unchanged, there would be a drag reduction. However, there is probably more going on than that. If the bubbles are in the turbulent fluid only, and not in the sublayer, which would be true on a top surface (since the bubbles are buoyant), or if the bubbles are too large to enter the sublayer, or migrate out of it, then they will substantially increase the lossiness in the turbulent part of the flow, both from the usual increase of viscosity due to particles, and from the particle inertia, causing large relative motion between the fluid and the particles (see [17]). Here we have the same effect that we saw in the case of polymers: an increase in the viscosity in the turbulent part of the flow, which will damp small scale motions everywhere, but particularly in the buffer layer, where the secondary instabilities responsible for the Reynbolds stress production will be damped. The buffer layer will thicken, the large eddies will scale up, and the mean velocity will increase, at the same shear velocity, resulting in drag reduction.

Another case of considerable interest is that of riblets. Here, the mechanism is probably somewhat different, but related. Drag reduction by riblets seems to be most effective when the riblets have a spacing about half the diameter of the coherent streamwise vortices in the sub- and buffer layers. Thus, the transverse velocities in these large eddies, as they rub against the wall, are forced to negotiate the sharp peaks of the riblets, causing increased losses. The coherent eddies gain their energy from being stretched by the mean strain rate. If their lossiness is increased, they must increase their energy gain to stay in equilibrium. The only way to do this is to grow larger, since the energy gain will go up with the cross-sectional area, while the loss will go up as the periphery in contact with the riblets. We thus expect them to become larger, again resulting in the secondary instability occurring farther from the wall, so that the sharp change in profile slope occurs farther from the wall, and hence that the mean velocity is higher for the same friction velocity.

We thus see that the rather general structural hypothesis we have formed is also able to explain (at least qualitatively) drag reduction by two apparently completely different mechanisms. We are interested here not so much in these other drag reduction schemes for their own sakes, interesting as they are, but in the reinforcement that this additional success gives to our structural hypothesis. The more disparate things we can explain with the same hypothesis, the more likely the hypothesis is to be correct.

Future Experiments

The phenomenon of drag reduction is of interest not only

because of its potential for engineering applications, but also for fundamental fluid mechanical reasons, e.g., to increase the understanding of the nature of the turbulent boundary layer. In polymer additives, one now has an effective tool to control the turbulent process, and it should be possible to make use of this tool to widen experimental and theoretical research on basic turbulent structure. Any structural hypothesis on the turbulent boundary layer should be checked for validity based on its ability to explain the phenomenon of drag reduction.

As we have seen above, a general consensus concerning the experimental evidence in polymer drag reducing flows has started to emerge. Much more work, however, is needed to establish the changes that can be observed in the turbulent structure of polymer solutions as compared with Newtonian fluids.

One of the most interesting questions in the turbulent boundary layer concerns the bursting period. What is in question here specifically is the mechanism by which the burst of Reynolds stresses (associated with the instability of the inflectionary profile produced by the vortex pairs near the wall) produces a new vortex pair. It has been speculated in the Newtonian layer, that the burst is felt all the way to the outer edge, and that the deformation of the turbulent/non-turbulent interface is reponsible for the formation of the next vortex pair. This mechanism, of course, cannot be operative in fully developed pipe flow, and it seems unlikely that there are two distinct mechanisms in boundary layers and in pipe flow. Thus, if there is a mechanism something like this, it is probably not dependent on the presence of a turbulent/ non-turbulent interface, but rather has to do with local deceleration of the outer flow, producing a pressure signal that is transmitted to the sublayer. In fact, no pressure signal may be necessary; a burst of Reynolds stress produces strong local deceleration, which might be expected to produce a necklace vortex, whose downstream trailing arms would become the next big eddies. It is not, of course, clear that the loop is closed, so that bursts necessarily cause eddies; that is, the eddies may arise as the result of an instability mechanism such as that proposed by Perry [19], or related to the mechanism that gives rise to Langmuir cells [20].

While early data was ambiguous regarding whether the bursting period should be scaled with wall variables or with outer flow variables, it has now been tentatively concluded that the scaling should be with outer variables, and we feel that this is consistent with the mechanism we propose. The findings of [18], however, are at variance with this; if their findings are correct, the bursting period scaled with wall variables is invariant, even when the scale of the outer flow is changed relative to the wall layer. This would have very serious implications for the mechanism which closes the loop in the continuous production of the vortex pairs in the wall layer. We consequently feel that any measurement that clarifies the behavior of the bursting period in drag reducing flow should have first priority.

More ambitious, but equally necessary, is greater knowledge of the structure of the large eddies in the wall region, both with and without drag reducing additives. The structure of the large eddy pairs, their growth, the formation of the inflectionary profile, and the secondary instability is certainly poorly enough understood in a Newtonian boundary layer, and much better quantitative information is needed. Comparable information obtained under drag reducing conditions would probably permit some conclusions to be drawn regarding the mechanism of these eddies, which remains somewhat speculative. While it is likely that their lateral (and probably also vertical) extent is determined by the mean velocity profile, since they scale up with the increasing thickness of the wall region, almost nothing is known about their streamwise extent.

References

1. Kubo, I. & Lumley, J. L.: A study to assess the potential for using long-chain polymers dissolved in water to study turbulence. Annual Report, NASA-Ames Grant No. NSG-2382. Ithaca, NY: Cornell (1980).

2. Lumley, J. L.: Drag reduction by additives. In Annual Review of Fluid Mechanics, 1, ed. W. R. Sears, p. 367. Palo Alto, CA: Annual Reviews Inc. 1969.

3. Lumley, J. L.: Drag reduction in turbulent flow by polymer additives. J. Polym. Sci. Macromol. Rev. 7 (1973) 263.

4. Lumley, J. L.: Drag reduction in two phase and polymer flows. Phys. Fluid 20 (1977) S64.

5. Landahl, M. T.: Drag reduction by polymer additives. In Proc. 13th Int. Cong. Theo. and Appl. Mech., Moscow, eds. Becker, E. & Mikhoilov, G. K., p. 177. Berlin: Springer 1973.

6. Landahl, M. T.: Dynamics of boundary layer turbulence and the mechanism of drag reduction. Phys. Fluids 20 (1977) S55.

7. Bakewell, H. P. Jr., & Lumley, J. L.: The viscous sublayer and adjacent wall region in turbulent pipe flows. Phys. Fluids 10 (1967) 1880.

8. Blackwelder, R. F. & Kaplan, R. E.: Turbulent Shear Flows, AGARD Conf. Proc, 931-5 (1972).

9. Lumley, J. L.: Applicability of Oldroyd constitutive equation to flow of dilute polymer solutions. Phys. Fluids 14 (1971) 2282.

10. Hoyt, J. W. & Taylor, J. J.: Turbulence structure in a water jet discharging in air. Phys. Fluids 20 (1977) 253.

11. Kim, H. T., Kline, S. J. & Reynolds, W. C.: The production of turbulence near a smooth wall in a turbulent boundary layer. J. Fluid Mech. 50 (1971) 133.

12. Paterson, R. W. & Abernathy, F. H.: Transition to turbulence in pipe flow for water and dilute solutions of polyethylene oxide. J. Fluid Mech. 51 (1972) 177.

13. Berman, N. S. & Cooper, E. E.: Stability studies in pipe flows using water and dilute polymer solutions. AIChE Journal. 18 (1972) 2.

14. Landahl, M. T. & Bark, F.: Application of a two-scale boundary layer turbulence model to drag reduction. In C. N. R. S. Colloque Int. Polymères et Lubrification, Brest.

15. Bogdevich, V. G. & Yevseyev, A. R.: Effect of gas saturation on wall turbulence. In Issledovaniya po Upravleniyu Pogranichnym Sloyem, eds. S. S. Kutateladze and §G. S. Migirenko, Thermophysics Institute Publishing House,

pp. 49-61 (1976).

16. Dubnishchev, Yu. N., Evseev, A. R., Sobolev, V. S. & Utkin, E. N.: Study of gas-saturated turbulent streams using a laser-Doppler velocity meter. Zhurnal Prikladnoi Mekhaniki i Tekhnicheskoi Fiziki (1975) 147-153.

17. Lumley, J. L.: Two phase and non-Newtonian flows. In Turbulence (Topics in Applied Physics, Vol. 12), ed. P. Bradshaw, pp. 289 - 324 (1978).

18. Donohue, G. L., Tiederman, W. G. & Reischman, M. M.: Flow visualization of the near-wall region of a drag-reducing channel flow, J. Fluid Mech. 56 (1972) 559.

19. Perry, A. E. & Chong, M. S.: On the mechanism of wall turbulence. J. Fluid Mech. 119 (1982) 243-256.

20. Leibovich, S.: The form and dynamics of Langmuir circulations. Ann. Rev. Fluid Mech. 15 (1983) 391-427.

Session 2
Extensional Viscosity

A Review on Extensional Viscosity

D.F. JAMES

Department of Mechanical Engineering
University of Toronto
Toronto, Ontario, M5S 1A4, CANADA

When minute amounts of polymer cause significant changes in a flow, the key fluid property thought to be associated with the effect is extensional viscosity. Consequently, there is considerable interest in this non-Newtonian property and current research activity is devoted to devising means of measuring the property and to understanding its origins at the microscopic level. To help focus on this property, this survey critically examines the evidence which links extensional viscosity to the observed effects, and reviews the techniques which are used to measure the property.

The evidence which points to the role of extensional viscosity is partly direct and partly circumstantial. In the case of laminar flows, the role is obvious: in flows through porous media or around small bodies, regions of high extensional deformation are easily identified and strain rates in these zones are consistent with the rates necessary for high extensional viscosity. In turbulent flows altered by polymer addition, principally boundary layers and jets, the regions most affected by the polymer are zones which are dominated by shear but periodically subjected to significant extensional motion; hence extensional viscosity is linked with the effect. However, in these turbulent flows, simultaneous measurements of the hydrodynamic effect and the fluid property have not yet been made and thus a direct relationship has not been established. Another category of polymer-influenced flows are those in which stability is altered. In laminar boundary layer separation, for example, the polymer changes the flow's stability to

The Influence of Polymer Additives
on Velocity and Temperature Fields
IUTAM Symposium Essen 1984
Editor: B. Gampert

shearing motion and thus extensional viscosity seems to play no part. But shear stability also is an essential element in the turbulent flows just considered, and so it is less certain that extensional viscosity is the operative property there. All in all, then, extensional viscosity seems to be the crucial property, even though the polymer apparently affects some flows by altering its shear stability.

Measuring extensional viscosity is not straightforward because it is difficult to create flows with well-defined fields of stress and strain rate. Furthermore, in many extensional flows inertia is important and so the experimentalist is forced to separate non-Newtonian stresses from inertial stresses. Separation is difficult because the former are generally small for dilute solutions and the latter are generally large because the strain rates necessary to induce non-Newtonian effects are high. Additional problems arise with particular techniques. For extensional flows open to the atmosphere - the free jet, falling stream or tubeless siphon - local strain rates can be determined accurately, but surface tension may be an unwanted parameter, stability may be a factor, and entrance or exit conditions are often not well defined. In internal flows, such as those through orifices or converging channels, the main problems are secondary flow and small dimensions. All of these problems have been dealt with in one way or another, but there still remains an overwhelming need for techniques which yield reliable measurements of extensional viscosity.

<u>References</u>

1. Acierno, D.; Titomanlio, G., and Nicodemo, L.: Rheol. Acta, 13, 532-537 (1974).

2. Armstrong, R.C.; Gupta, S.K.; Basaran, O.: Poly.Eng. + Sci, 20, 466-472 (1980).

3. Astarita, G, and Nicodemo, L.: Chem. Eng. Journal, 1, 57-66 (1970).

4. Baid, K.M., and Metzner, A.B.: Trans. Soc. Rheology, 21, 237-260 (1977).

5. Bakewell, H. and Lumley, J.L.: Phys. Fluids, 10, 1880-89, (1967).

6. Balakrishnan, C. and Gordon, R.J.: AIChe Journal, 21, 1225-1227 (1975).

7. Balmer, R.T.: Journ. N. Newt. Fluid Mech., 2, 307-322 (1977).

8. Balmer, R.T., and Hochschild, D.J.: Journ. Rheol., 22, 165-180 (1978).

9. Blackwelder, R.F.: Phys. Fluids, 22, 583 (1979).

10. Chao, K.K.K.; Williams, M.C.: Journal Rheol., 27, 451-474 (1983).

11. Chiou, C.S., and Gordon, R.J.: J. Appl. Poly. Sci., 21, 1359-1376 (1977).

12. Chiou, C.S., and Gordon, R.J.: Polymer Eng. Sci., 20, 456-466 (1980).

13. Cottrell, F.R.; Merrill, E.W.; Smith, K.A.: J. Poly. Sci., 7, 1415-1434 (1969).

14. Fruman, D.H., and Tulin, M.P.: Phys. Fluids, 21, 1896-1899 (1978).

15. Fuller, G.G. and Leal, L.G.: Rheol. Acta, 19, 580-600 (1980b).

16. Fuller, G.G., and Leal, L.G.: J. Non-Newt. Fluid Mech., 8, 271-310 (1981).

17. Hayes, J.W., and Hutton, J.F.: Prog. Heat + Mass Transf., 5, 195-209 (1972).

18. Hoyt, J.W.; Taylor, J.J., and Runge, C.D.: J. Fluid Mech., 63, 635 (1974).

19. Hoyt, J.W.; Taylor, J.J.: J. Fluid Mech., 83, 1, 119-127 (1977).

20. James, D.F., and Saringer, J.H.: J. Fluid Mech., 97, 655-671 (1980).

21. Kline, S.J.; Reynolds, W.C.; Schraub, F.A., and Runstadler, P.W.: J. Fluid Mech., 30, 741-773 (1967).

22. Mac Sporran, W.C.: Journ. N. Newt. Fluid Mech., 8, 119-138 (1981).

23. Martischius, F.D.: Rheol. Acta, 21, 288-310 (1982).

24. Metzner, A.B., and Metzner, A.P.: Rheologica Acta, 9, 174-181 (1970).

25. Oliver, D.R., and Bragg, R.: Rheol. Acta, 13, 830-835 (1974).

26. Oliver, D.R., and Ashton, R.C., Journal N.Newt. Fluid Mech., I, 93-104 (1976).

27. Ouibrahim, A., and Fruman, D.H.: Journal N. Newt. Fluid Mech., 7, 315-331 (1980).

28. Ouibrahim, A.: Journal N. Newtonian Fluid Mech., 11, 341-347 (1982).

29. Peng, S.T.J., and Landel, R.F.: Journ. Appl. Phy., 47, 4255-4260 (1976).

30. Rubin, H., and Elata, C.: Phys. Fluids, 9, 1929-1933 (1966).

31. Schümmer, P., and Tebel, K.H., Journal N. Newt. Fluid Mech., 12, 331-347 (1983).

32. Townsend, A.A.: The Structure of Turbulent Shear Flow, Cambridge University Press (1976).

33. Tsang, H.Y., and James, D.F.: Journ. Rheol., 24, 589-601 (1980).

34. Usui, H., and Sano, Y.: Phys. Fluids, 24, 214-219 (1981).

35. Waters, N.D.; King, M.J., and Oliver, D.R.: Jour. N. Newt. Fluid Mech., 2, 385-391 (1977).

36. Yomita, Y., and Jotaki, T., Phys. Fluids, 20, 75-77 (1977).

An Experimental Study of Polymer Induced Drag for Flows Through Porous Medium

Torgny Lagerstedt

The Royal Institute of Technology,
Department of Mechanics,
Stockholm, Sweden*

*Present address, Alfa-Laval AB. Tumba, Sweden

Summary

Measurements of pressure drop and flow rate for dilute solutions of high molecular weight polymers flowing through a porous medium show an unusually high pressure drop. The excess pressure drop due to the presence of polymers was found to be very high with a marked non-Newtonian behaviour. The intrinsic viscosity of the polymers attained a value two orders of magnitude greater than that measured in pure shear flow. After an abrupt onset the intrinsic viscosity increased rapidly with increasing strain rate to a peak value, whereafter it decreased. The increase in intrinsic viscosity correlates with a permanent lowering of the apparent molecular weight, i.e. a polymer degradation.

A hypothesis for the high intrinsic viscosity is that the polymers stretch when passing the porous medium. This is substantiated with a experimental procedure. Using a second porous medium downstream the first one it was found that the polymer entering the second medium gave a pressure drop that depends on the time of flight between the two porous media. The relaxation time obtained in this manner is about 0.1 seconds for polyethyleneoxide and 0.4 seconds for polyacrylamide. Measurements at different solvent viscosities gave that the relaxation time scales about linearly with the solvent viscosity.

The turbulent drag reduction and porous medium drag increase appear to be related properties of the polymers. This relation between drag reduction and drag increase is further enhanced by showing that an additive which magnifies the drag reducing ability of polyethyleneoxide also magnifies its drag increasing properties in a porous medium. It is suggested that the phenomenon of drag increase could be used as method to measure on and characterize drag reducing polymers. An apparatus designed to use this method was employed in experiments on cavitation in flows with polymer additives.

The Influence of Polymer Additives
on Velocity and Temperature Fields
IUTAM Symposium Essen 1984
Editor: B. Gampert

Introduction

Polymer macromolecules in a solvent may give rather remarkable effects on the fluid behaviour in different flow situations both in the dilute and the non-dilute range.

In the dilute range the most striking effects are the reduction in turbulent skin friction, Granville [1] and the increase in drag for flows through a porous medium (a filter), Savins [2]. The drag reduction and drag increase appear to be related properties of polymer macromolecules with the following similarities.

- A minute amount of the polymer is needed. 10 ppm may give 80% reduction of the turbulent skin friction. 10 ppm may also double the apparent viscosity of a solution when flowing through a filter.

- The polymers which give drag reduction and drag increase have all large unbranched or moderately branched macromolecules with extremely high molecular weights.

- A certain flow condition must be reached before drag reduction and drag increase is experienced.

- The polymers are all easily degraded in a turbulent or drag increasing flow situation resulting in a lowered drag reducing and drag increasing efficiency.

The aim of the present report is to document the main experimental results from a series of experiments on flows through porous filters. The measurements are reported in full by Lagerstedt [3], [4].

A mechanism behind the drag reduction and drag increase may be that polymer molecules or possibly entanglements of polymer molecules stretch in the flow field and become elongated. This stretching is quantified by measuring on a relaxation time behaviour for the polymer solution leaving the filter.

Experimental procedure

A schematic layout of the experimental set up is shown in figure 1. The apparatus was designed to measure the pressure drop in dilute polymer solutions flowing through a porous filter. Ordinary tap water was used as solvent.

The filtered hot and cold tap water was mixed to the desired temperature in a thermostat. For the measurements in the present report the temperature was 20°C. The pure water was led to a constant head overflow tank, which ensures a steady head. The flow rate was measured with a turbine flow rate meter.

The high concentration (0.1 - 1%) parent polymer solution was injected into an active in-line mixer to ensure a homogeneous mixing. Earlier experiments showed that a highly concentrated parent solution of polyethyleneoxide (Polyox) mixes badly in the water, and some kind of active mixing must be used to obtain a homogeneous final solution. The speed of the mixer was kept at a level that gave good mixing but low enough to avoid degradation. After mixing, the dilute polymer solution flowed into the measuring section where the porous filters were fitted.

The apparatus could also be run with a bulk solution at the final concentration. The bulk solution was fed to the test section from a vessel that could be set under pressure. Some measurements were conducted with a bulk solution of Polyox to ensure that the measurements were independent of the mixing method. A bulk solution was used when having Congo Red as a drag increasing additive to Polyox. The measurements with the polyacrylamide were made also with a bulk solution.

The filters were sintered plates about 3 mm thick made of spherical beads (0.6-0.15 mm). The experiments presented with figures in the present report were made with the 0.6 bead filter.

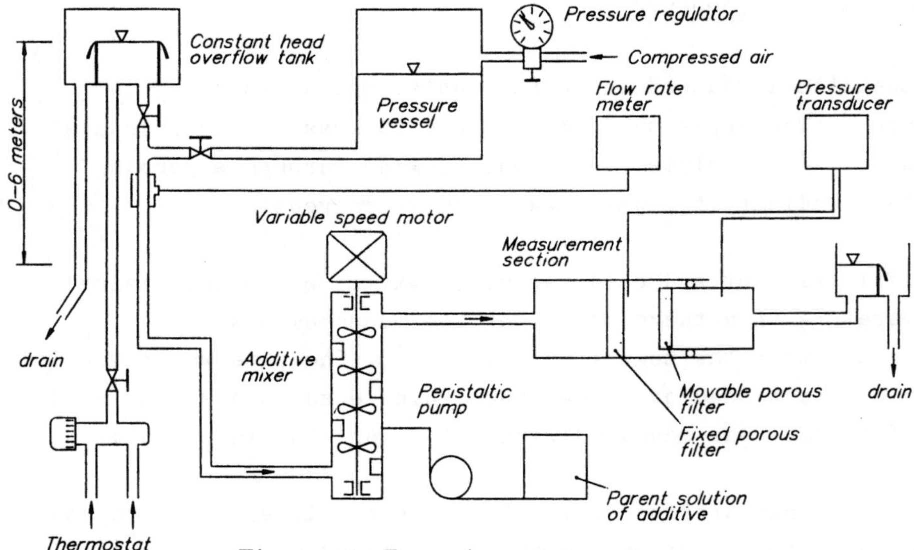

Figure 1. Experimental set up.

The measurements of the pressure drop across a porous filter due to polymers are intended to result in a phenomenological description of the polymer behaviour. Therefore, the most direct and simple scalings are used and the pressure drop is not turned into a nondimensional form. The pressure drop across the porous filter is scaled as

$$P = \frac{\Delta P}{c \cdot t}$$

where ΔP is the excess pressure drop due to polymer additives (the pressure drop due to pure water is subtracted from the total pressure drop) [kg/ms^2]

c is the polymer concentration. [kg/m^3]

t is the filter thickness [m]

The scaling with filter thickness is a conventional scaling used in flow experiment with porous media.

The pressure drop is also scaled linearly with the concentration of polymers which is consistent with the experimental data.

The flow rate through the porous filter is represented by a Reynolds number given by, Bird et al. [5].

$$RE = \frac{V \cdot D}{\nu(1-\varepsilon)}$$

where ε is the void volume defined as volume of voids/volume of filter

D is the diameter of the beads in the filter [m]

ν is the kinematic viscosity of the solvent [m^2/s]

V is the mean velocity of the fluid entering the filter [m/s]

The polymers used were mainly different grades of polyethylene-oxide (Polyox) with molecular weights ranging from $5 \cdot 10^6$ to $0.1 \cdot 10^6$ and polyacrylamide (Separan) with a molecular weight of $5 \cdot 10^6$. The measurements presented with figures in this report were made with Polyox WSR 301, $M = 4 \cdot 10^6$.

Experimental observations

Pressure drop - Reynolds number: A typical pressure drop behaviour as function of Reynolds number is ıown in figure 2. Here it may be noted that pressure drop has an "onset" at a specific Reynolds number. At higher Reynolds numbers the slope of the curve decreases with increasing Reynolds number. Measurements across an identical second filter situated downstream the first filter shows a similar pressure drop behaviour, but at a reduced level which may be attributed to degradation of the polymer. A plot of the pressure drop difference between the two filters as a function of Reynolds number showed that the onset of degradation (difference in pressure drop between the two filters) correlates with the onset of pressure drop.

The intrinsic viscosity of the Polyox as measured with porous filters is shown in figure 3. The intrinsic viscosity $[\eta]$ is defined as.

$$\nu_m = \nu(1+[\eta]c)$$

where ν_m is the measured viscosity of the solution $[m^2/s]$

Three measurements are displayed one with the uppstream filter and two with the downstream filter at different positions. At a large distance of separation the intrinsic viscosity for the downstream filter is similar to that experienced with the upstream filter, but at a lower level. For the downstream filter close to the upstream filter, however, the intrinsic viscosity is larger than that experienced with the upstream filter. This phenomenon will be investigated later on in the report.

Pressure drop - Molecular weight : For high molecular weights of Polyox, ($\geq 0.4\cdot10^6$), the pressure drop was found to be proportional to the molecular weight. For molecular weights below $0.4\cdot10^6$, however, the pressure drop was proportional to the molecular weight raised to the 1/4 power.

Degradation: To quantify degradation a second porous filter was fitted downstream the first one. Figure 2 shows measurements with two filters 35 mm apart. The pressure drop across the second filter is lowered by 40% due to degradation. It is seen that degradation starts at the onset of pressure drop. Experiments with three molecular weights of Polyox ($4\cdot10^6$, $0.6\cdot10^6$, $0.4\cdot10^6$) did not show any difference in degradation behaviour, also, the onset of pressure drop and degradation took place at about the same Reynolds number for all three molecular weights when using the same filter.

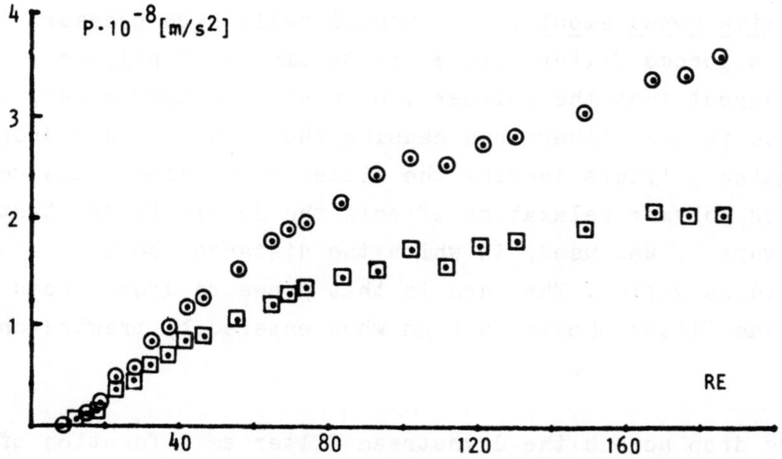

Figure 2. Polymer pressure drop versus Reynolds number. 6.7 ppm of Polyox WSR 301. ○ Upstream filter; □ Second filter 35 mm downstream the first filter.

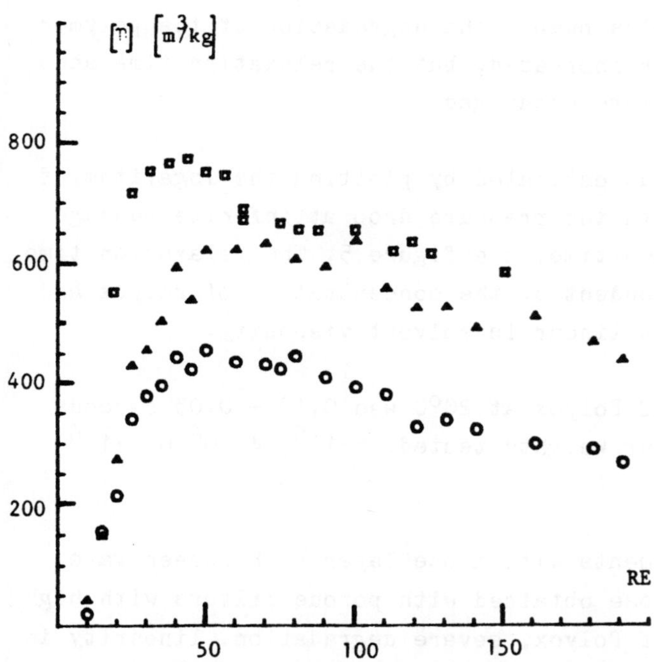

Figure 3. Intrinsic viscosity of Polyox WSR 301 versus Reynolds number measured with two filters. △ Upstream filter; ○ Downstream filter, 35 mm downstream; □ Downstream filter close to upstream filter

Relaxation time measurements: The exceptionally high pressure drop across a porous filter with a minute amount of polymer additives suggest that the polymers, molecules or aggregates, are elongated in the filter thus causing the high pressure drop. These elongated polymers leaving the filter relax after passage. To measure on polymer relaxation effects the double filter test section, figure 1, was used, in which the distance between filters could be varied. The idea is that pressure drop across the downstream filter should be high when entered by prestretched polymers.

The pressure drop across the downstream filter as a function of the mean passage time of fluid element flowing from the upstream to the downstream filter displays a relaxation behaviour, figure 4a-4c. The measurements were made at several Reynolds numbers. With increasing Reynolds number the degradation of the polymer in the upstream filter increased, but the relaxation time and relaxation behaviour were unchanged.

The relaxation time was estimated by plotting the logarithm of the pressure drop (with the pressure drop at infinite passage time subtracted) versus time, see figure 5. The relaxation time was found to be independent of the concentration of Polyox and of Reynolds number but linear in solvent viscosity.

The relaxation time of Polyox at 20°C was 0.12 ± 0.03 seconds for the three molecular weights tested, $5 \cdot 10^6$, $4 \cdot 10^6$ $0.6 \cdot 10^6$ respectively.

Mesh screens: Measurements with a one layer mesh screen gave results similar to those obtained with porous filters with high intrinsic viscosity of Polyox, severe degradation, linearity in concentration and onset of pressure drop. Also a relaxation time of about 0.1 seconds was obtained when using screens. Any configuration of screens and porous filters in the test section gave the 0.1 seconds relaxation time. The screen was a wire mesh screen with a wire diameter of 0.05 and a separation between wires of 0.125 mm.

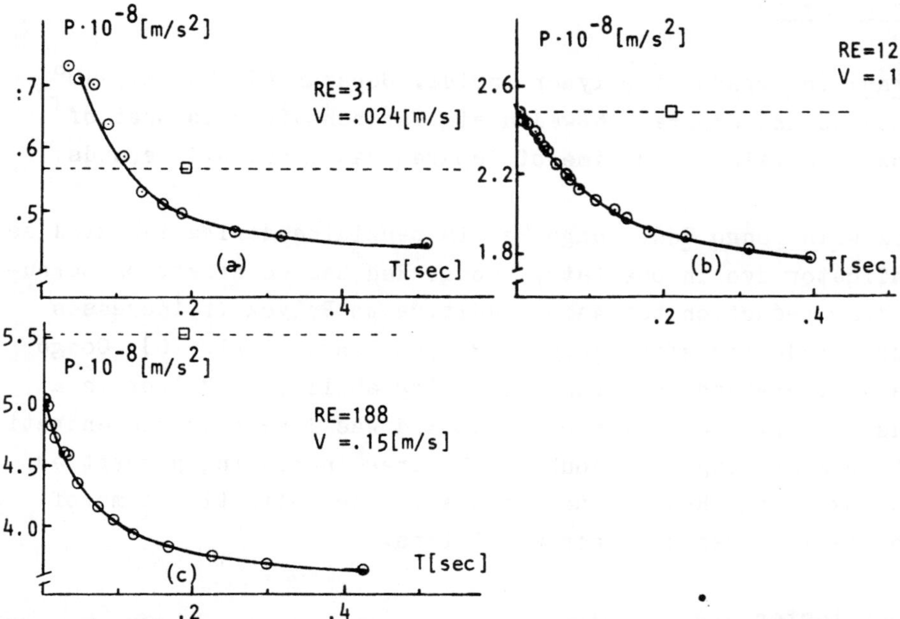

Figure 4. Polymer pressure drop versus passage time between filters. 15 ppm Polyox WSR 301. ⊙ Downstream filter; ⊡ The constant pressure drop across the upstream filter.

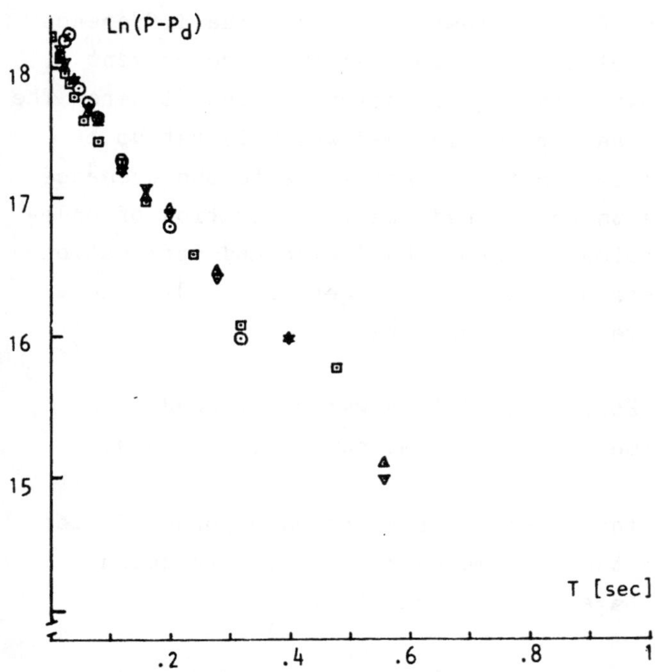

Figure 5. Relaxation time measurements with Polyox WSR 301 at different concentrations. □ 50 ppm; △ 25 ppm, ○ 15 ppm; ▽ A repeated measurement at 25 ppm. RE = 132, V = 0.13 [m/s] for all measurements. P_d pressure drop at infinite passage time

Other polymers

Separan: One grade of polyacrylamide, Separan AP 273, $M=5\cdot 10^6$ was tested and Separan showed a similar behaviour to that of Polyox. The relaxation time of Separan was about 0.4 seconds.

Polyox with Congo Red: Congo Red is benzidine derivative used as an indicator dye in chemistry. Congo Red has no effect on turbulent drag reduction but as an additive to Polyox it increases the drag reducing efficiency of Polyox, Inge et al. [6]. Congo Red also increased the drag increasing ability of Polyox in a porous filter. The effect of Congo Red was linear in concentration and 10 ppm of Congo Red doubled the drag increasing properties of Polyox. Congo Red had no effects on the relaxation time of Polyox as measured with porous filters.

Polymer tester

Based on the knowledge gained in the porous filter work a Polymer Tester was designed. The Polymer Tester measures the difference in flow rate between a polymer solution and the pure solvent when simultaneously flowing through identical porous filters. The flow is driven with the same pressure head which is set up by compressed air. The Polymer Tester is sensitive to the presence of Polyox and measurements can be made on concentration of undegraded Polyox WSR 301 below 0.1 ppm. The design and application of the Polymer Tester are documented in Lagerstedt [7]. Some applications will, however, be cited here.

- A 10 ppm solution of Polyox WSR 301 in water degraded exponentially with time, with the time constant 50 hours.

- Fish mucus increased the pressure drop across a porous filter and it was also shown that the mucus reduced the turbulent drag in a pipe flow, Lagerstedt et al. [8].

- The Polymer Tester was indispensable in experiments on cavitation noise and cavitation erosion because of the heavy degradation of polymers in those experiments, Lagerstedt [9],[10].

References

1. Granville, P.S Progress in frictional drag reduction, 1975 to 1976. David Taylor Naval Ship Research and Development Centre, December 1976.

2. Savins, I.G Non-Newtonian Flow Through Porous Media. Indust. Eng. Chem. 61, 18-47, 1969

3. Lagerstedt, T. On the Behaviour of Polymer Solutions in a Porous Medium, part I, Ph.D. Theses. TRITA-MEK-79-4. Royal Institute of Technology, Dept. of Mechanics. 1979

4. Lagerstedt, T. On the Behaviour of Polymer Solutions in a Porous Medium, part II Ph.D Thesis. TRITA-MEK-79-4. Royal Institute of Technology, Dept. of Mechanics. 1979

5. Bird, R.B., Steward, W.E., Lightfoot, E.N. Transport phenomena. New York, John Wiley & Sons, 1960

6. Inge, C., Johansson, A.V. Effects of Congo Red on the drag Reduction properies of polyethylene oxide. Phys. Fluids 22 (5), May 1979.

7. Lagerstedt, T. An Apparatus for Detecting and Characterizing Polymer Additives in Highly Dilute Solutions, TRITA-MEK-78-10. Royal Institute of Technology, Dept. of Mechanics, Stockholm, Sweden. 1978

8. Lagerstedt, T., Alfredsson, P.H. A Note on the Drag Reducing Properties of Fish Mucus. TRITA-MEK-83-11: Royal Institute of Technology, Dept. of Mechanics, Stockholm, Sweden. 1983

9. Lagerstedt, T., Bark, G. The influence of Polymers on Propeller Cavitation. An Experimental Investigation in a Cavitation Tunnel. Swedish State Shipbuilding Experimental Tank (SSPA) and The Royal Institute of Technology, Dept. of Mechanics, Stockholm, Sweden. 1978

10. Lagerstedt, T. An Experimental Investigation of the Influence of Polymers on Cavitation Erosion. TRITA-MEK-78-03. Royal Institute of Technology, Dept. of Mechanics, Stockholm, Sweden. 1978

Session 3
Molecular Parameters I

Molecular Parameters as a Basis of Structure-Property-Relations in the Flow of Polymer Solutions

W.-M. Kulicke and J. Klein

Institute of Technical Chemistry
Technical University
Hans-Sommer Str.10
3300 Braunschweig, F.R.G.

Summary

The intention of this paper is to identify the molecular parameters, which are of particular relevance for the qualitative and quantitative understanding of the flow phenomena caused especially by water soluble polymer additives. The parameters can be classified with respect to primary, secondary, tertiary and quarternary structure. These structure parameters do not only refer to the average molecular weight, molecular weight distribution and co-polymer compositions, but also to the coil dimensions, flexibility and to intermolcular interactions. Some experimental examples will be used to illustrate the applicability of the various methods with special emphasis on high molecular weight water soluble polymers.

1. Introduction

Aqueous solutions of neutral polymers as well as polyelectrolytes are of great use in different technical areas, as for instance in drag reduction, polymer flooding, flocculation sludge dewatering etc. Table 1 illustrates the production-levels of different technical applications for different water soluble polymers.

Tab. 1: Production-levels (U.S.A. and Europe) (totals in 10^3 t/year for 1978 and 1983)

	H_2O-treatment	paper-industry	mineral-gain	enhanced oil recovery
polyacrylamides	33/75	20/26	8/12	12/20
polyamines	9/20	24/28	0.3/0.3	1/1.8
quarternary polybases	6/14	-	8/10	0.5/0.8
polyacrylates	3/ 5	1.1/0.6	-	-

The Influence of Polymer Additives on Velocity and Temperature Fields
IUTAM Symposium Essen 1984
Editor: B. Gampert

A comparison of the production levels of the years 1978 and 1983 demonstrates that in most cases the levels increased. This clearly indicates the technical importance of water soluble polymers. However, since in most cases the mechanisms are still unknown, the extent of their applicability and their effectiveness is thus limited. A reason for this can be seen in the fact that the purification of aqueous solutions is far more difficult than in the case of organic solvents[1]. This is due to the polar character of water, which attracts dust, fibres and particles more readily. In such cases the determination of the microstructures, using light scattering parameters etc., is hindered and very often impossible. Although this is a well-known fact, it is still of considerable importance to improve the purification procedure, to continue further research into flow phenomena and to correlate the results to molecular structure parameters[2]. The target of this paper is to transfer flow phenomena to its primary, secondary, tertiary and quarternary structure parameters, which are to be described as follows. The primary molecular structure parameters: chemical nature, ionic - nonionic, $M = 10^3 - 10^8$ g/mol, the secondary molecular structure parameters: molecular weight distribution, branching, copolymer composition, the tertiary molecular structure parameters: molecular shape, radius of gyration, critical concentration, second virial coefficient and the quarternary ones: dynamic parameters, interactions (segment mobility), entanglements, energetic interactions, (associations, aggregations and hydrogen bonds).

2. Determination of the molecular weight

The determination of the absolute average molecular weight, which refers to the primary structure parameters, should be discussed first, because it is mainly responsible for the viscoelastic flow phenomena observed with polymer solutions in shear and elongational flow[3]. The absolute average molecular weight may be determined, either by light scattering (M_w) or by osmotic pressure measurements (M_n). It is a fact, that with increasing molecular weight, polymer additives show pronounced flow phenomena (e.g. see papers of this symposium by Gampert-

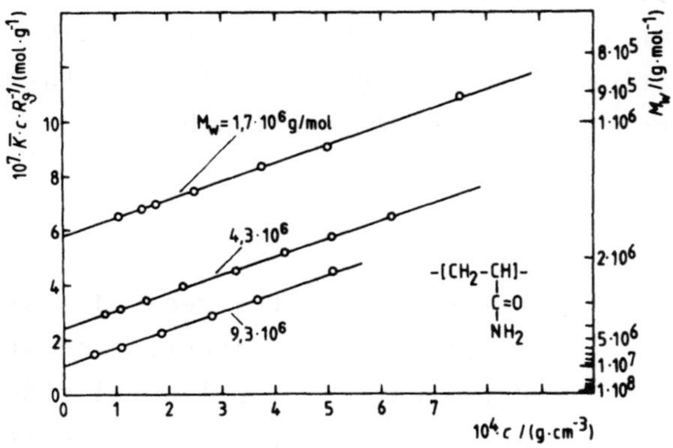

Fig.1
Light scattering results of polyacrylamide in 0.1 M Na_2SO_4 - solution at 25°C. Note: at $K \cdot c/R_\theta = 0$ the molecular weight should be $M_w = \infty$ by definition.

Wagner and Haas - Kulicke)[4]. Ultra-high molecular weights can only be measured by using the light-scattering technique. Osmotic pressure measurements are restricted to molecular weights $M_n \leqslant 1 \cdot 10^6$ g/mol. As recently shown, the contamination by dust and fibres of polymer solutions is a major problem in light scattering[1]. Therefore one has to purify the solutions very carefully, which is often not easy, especially when commercial polymer samples are used. Fig. 1 shows some light scattering results of polyacrylamides in salt solution[2]. The results are obtained using a low angle laser-light-scattering-photometer with a wave-length of 633 nm. From such measurements one can take the absolute average molecular weight and the second virial coefficient A_2. It is possible to measure molecular weight values up to infinity with the use of light scattering. The intercept of the straight line of the experimental data with the ordinate at c = 0 gives $1/M_w$. At $K \cdot c/R_\theta = 0$, the value of M_w goes to infinity. The corresponding molecular values are plotted on the right side of this figure. Therefore, above a molecular weight of $5 \cdot 10^6$ g/mol, the possibility of error becomes very significant. The reason is that the molecular weight M_w increases hyperbolically as $K \cdot c/R_\theta$ decreases, as can be seen by the right ordinate of figure 1. Therefore, the determination of ultra-high molecular weight polymers is, in general, not easy to perform (purification problem), and in addition the determination of increasing molecular weight becomes increasingly limited (see Fig 1).

Other absolute determination methods for detecting high molecular weights lead to similar problems. Therefore the normal method for the determination of these high molecular weights is by extrapolation of the MARK-HOUWINK-relationship

$$[\eta] = K\ M^a$$

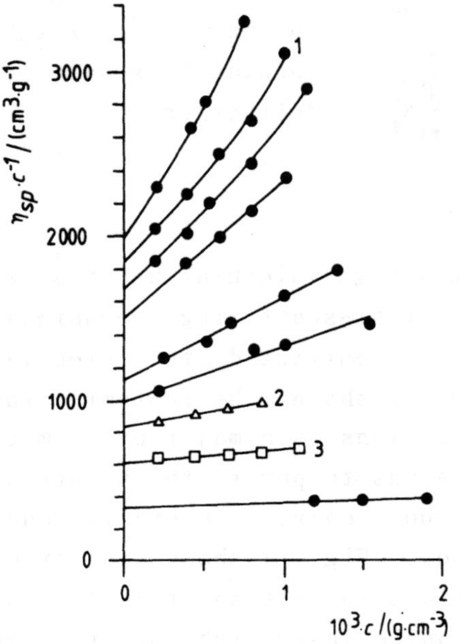

Fig. 2
Dependence of the reduced viscosity versus concentration of different polyacrylamide samples in different solvents at 25°C. (• water, ▵ formamide, ▫ ethylene glycole. All samples have the same ratio of $M_w/M_n \cong 2.5$. Same sample for curve 1,2 and 3)

where K and a depend on the selected polymer-solvent-temperature system. The MARK-HOUWINK relationship, (found in polymer handbooks[5]), is therefore limited in most cases to the determination of absolute molecular weights lower than $2 \cdot 10^6$ g/mol. The advantage of this method is, that the reduced viscosity can be determined quickly, which moreover is much cheaper and less time consuming than absolute determination methods such as light scattering measurements. η_{sp}/c at $c = 0$ gives the intrinsic viscosity $[\eta]$, from which the molecular weight can be easily calculated. Fig. 2 shows some experimental results of such measurements where a homologous series of polyacrylamide samples has been used. From this figure, one can see that the thermodynamic quality decreases from water to formamide down to ethylene glycole. In addition, the black circles, representing aqueous solutions of polyacrylamide, have different molecular weights. An increase in the intrinsic viscosity of course reflects a higher molecular weight of the polymer sample. With this the dependence between the reduced viscosity versus concentration is no longer a linear relationship (see figure 2.) which is due to intermolecular interactions between polymer molecules and will be discussed later. The determination of the intrinsic viscosity should of course be performed in the dilute state of the solution. However, in the dilute concentration range the increase in the polymer solution viscosity compared to the solvent is too small and therefore most of the points of measurement are performed using a moderately concentrated solution. This is one of the disadvantages of using relative determination methods, especially with increasing molecular weights which has already been discussed in detail[2].

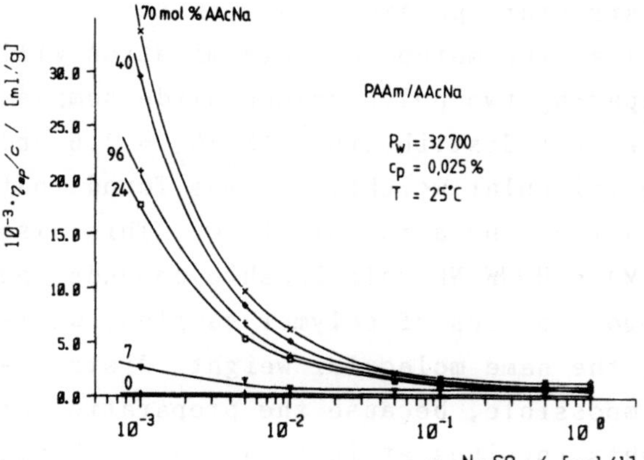

Fig. 3
The influence of salt concentration on the reduced viscosity of different polyacrylamide-co-acrylates which have the same degree of polymerization and chain length distribution.

In the case of polyelectrolytes another parameter has to be considered: the degree of dissociation. Molecular weight determination in pure water can not be performed. The reason for this is that the viscosity is also very high in the dilute state and therefore the purification procedure (filtration, centrifugation) is difficult to perform. In addition, the difference between the scattering intensity of the light in the solvent and in the polymer solution is very small at this low concentration, which makes such measurements impossible. Therefore, all measurements have to be performed in a low molecular weight electrolyte. The influence of a low molecular weight electrolyte on the viscosity level is shown in figure 3. One can see that, by increasing the content of the carboxylate side groups, the viscosity of a given salt concentration increases. On the other hand, the sample with 96% acrylate side groups exhibits a significantly lower viscosity level than expected. This unusual behaviour will be discussed in detail later. However, the remaining problem is evident in the fact that the COULOMB repulsion is continuously suppressed with increasing salt concentration. Therefore, in the case of polyelectrolytes countless MARK-HOUWINK-relationships can be in general established, which shows the increasing problems of simple molecular weight determinations on charged systems.

3.Molecular weight distribution

The influence of molecular weight distribution, (which also refers to the secondary structure parameters), on the behaviour of solutions as well as the method of determination will now be discussed. By preparing two polyethylene oxide samples with different molecular weight distributions (M_w/M_n = 1.6 and 4.1) but the same average molecular weight, it was found that the zero-shear viscosity differs by a factor of 10. This also shows clearly that every MARK-HOUWINK-relationship depends on the quality of the homologous series of polymer samples, where every sample has exactly the same molecular weight distribution. This is obviously impossible, because the preparation of different molecular weights involves at least a small change in the molecular weight distribution. Therefore, in the last ten years increasing interest has been focused on the determination of molecular weight distribution. The normal procedure in the determination of the number of particles involves use of the refractive index, where the separation of the polydisperse sample particles is caused by using the size-exclusion-chromatographic method. To calculate the molecular weight distribution, one has to know the calibration curve for this polymer-solvent-system. This is in fact an indirect method and in addition, as in the case of aqueous polymer solutions, some peculiarities may be observed[6]. The molecular weight distribution, including the calculation of M_w and M_n of an aqueous polymer solution system, was recently determined by simultaneously and continuously monitoring the effluent from a size-exclusion-chromatographic column (SEC) filled with porous glasses with a low-angle-laser-light-scattering (LALLS) photometer, see fig. 4.The recorder traces of the scattering intensity and the refractive index difference have been digitalized in about 400 steps corresponding to 200 μl fractions. Each fraction has been assumed to be monodisperse. This method gives absolute data rather than relative data obtained from any other calibration method. The investigations clearly show a flow-rate dependent degradation of samples with relatively low molar masses (M_w = 500 000 g/mol at $M_w/M_n \cong 2.5$) at typical flow rates Q = 1 ml/min[7]. Consequently, such a determination of the absolute molecular weight distribution is not an easy task. The characteristic ratio M_w/M_n has often been used to indicate the width of the molecular weight distribution, which is, of course, the initial stage of approximation and can also only be used up to an average molecular weight of $1 \cdot 10^6$. On the other hand, the high molecular weight tail of

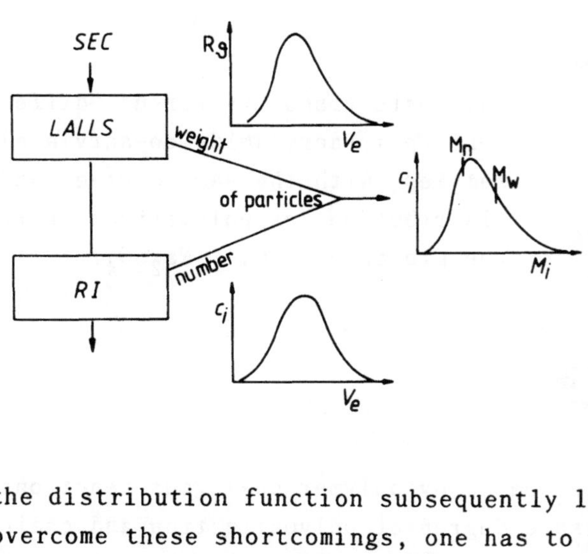

Fig.4
Schematic representation of the absolute determination of the molecular weight distribution of aqueous polymer solutions, using a combined size-exclusion chromatographic and low-angle-light-scattering apparatus.

the distribution function subsequently limits this method. To overcome these shortcomings, one has to improve this method at least for higher molecular weight samples by changing matrix, volume speed, thermodynamic quality of the solvents etc[7].

4.Specific effects

The characterization of the covalent arrangement of the monomers within one polymer chain and within the mixture of polymer chains refers to the secondary structure parameters as well as branching (long and short chain branches).

In paragraph 2 the influence of average molecular weight on solution properties was discussed and in the third paragraph the influence of molecular weight distribution. The influence of the ultimate chemical composition can be investigated by comparing two different polymer samples with the same average molecular weight. However, the distribution is different and may strongly influence the results. In the case of co-polymers the chemical composition is based on two monomer units which makes a comparison more complicated. By using a basis polymer and changing the chemical nature by polymer analogous reactions, one will have different chemical compositions but still the same degree of polymerization and chain length distribution. The influence of chemistry on solution properties can only be investigated under such circumstances. This should be demonstrated by a series of linear poly(acrylamide-co-acrylates)

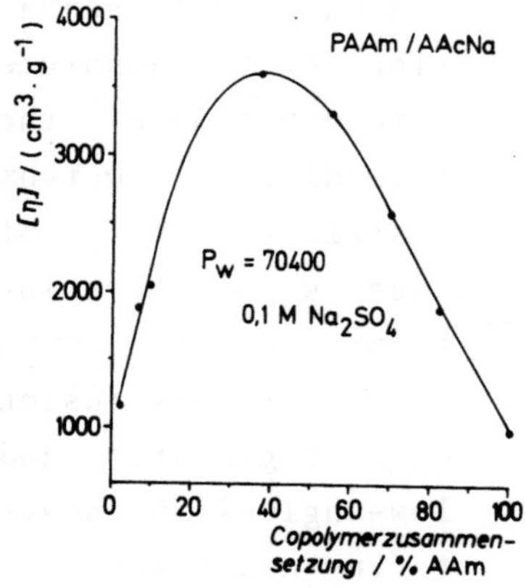

Fig. 5
Intrinsic viscosity versus different poly(acrylamide-co-acrylate) samples with the same degree and distribution of polymerization in the presence of 0.1M Na_2SO_4.

(PAAm/AAcNa) which are prepared by polymer analogous reactions and therefore have the same degree of polymerization and chain length distribution. Figure 5 shows the viscosity behaviour at zero-concentration for co-polymer compositions between 0 and

96.3 mole% acrylate units. All measurements were performed in a low molecular weight electrolyte. The intrinsic viscosities exhibit maximum behaviour, whereas mixtures of poyacrylamide and polyacrylic acid will not show such maximum behaviour (not shown in fig. 5)

Further investigations show that the degree of the maximum behaviour depends strongly on the concentration of the added low molecular weight electrolyte, see fig. 6.

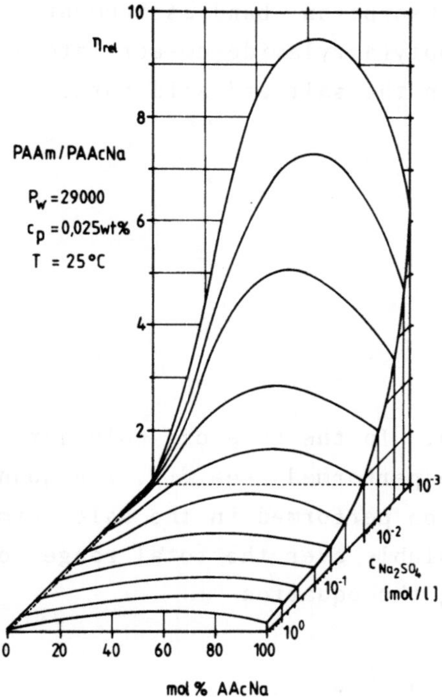

Fig.6
The relative viscosity of polyacrylamide-co-acrylates versus different copolymer compositions depending on the presence of a low molecular weight electrolyte ($c(Na_2SO_4)$). All samples have the same degree and distribution of polymerization.

Measurements of viscosity and light scattering have proven that maximum coil dimensions of PAAm/AAcNa in salt solutions are achieved at about 67 mole% AAcNa. The maximum behaviour was researched by measuring the radius of gyration, second virial coefficient, molecular weight, preferential solvation, viscosity and absorption bands of the copolymer series. Thus, an increase in the possible ways of arranging AAcNa-AAm-AAcNa units (via H-bonds) will lead to extended coil dimensions and therefore influence the viscosity level. In addition, it can therefore be concluded from the results that the maximum behaviour is real and not influenced by the different distributions of the samples, for details see[8]. The content of ionic side groups in such charged systems is not easily determined by titration. Nevertheless, an infra-red (IR)-determination would be very practical in determinating copolymer compositions quickly, which is of course only possible if the

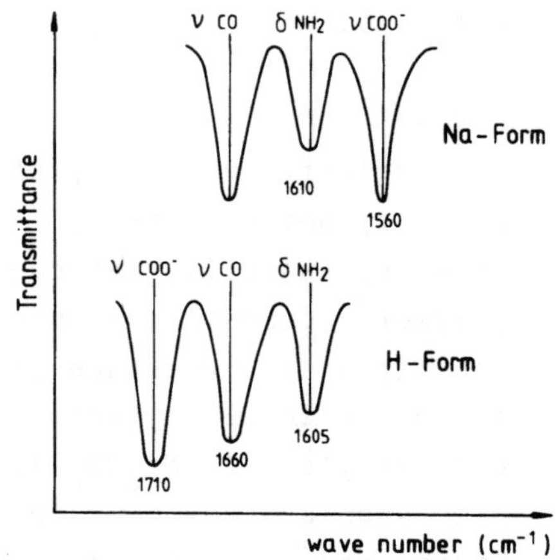

Fig.7
Schematic representation of the absorption band assignment of poly(acrylamide-co-acrylates) in the salt and acid form.

absorption bands do not overlap. In the case of poly(acrylamide-co-acrylate) this could be performed, see fig. 7.A quantitative determination can only be performed in the salt form, because the acid form is not soluble over the total range of co-polymer composition. By using the equation

$$\text{mole\% AAm} = 96.78 \cdot e^{-0.586 \cdot E_{coo^-}/E_{co}}$$

the co-polymer composition in the salt form can now be determined quickly[8]. Long chain branched polymers have different properties compared with linear ones of the same total molecular weight. This was examined in theory by ZIMM and STOCKMAYER. They used a characteristic g-factor, which is a ratio of the average radii of gyration between branched and linear molecules.

$g = R_v^2/R_l^2$

From viscometric measurements one can determine the intrinsic viscosities of branched and linear polymer molecules which also gives an indication of the coil dimensions between them. Therefore is $g' = [\eta]_v/[\eta]_l$ and $g' = g^b$, which shows the correlation between light scattering and viscometric investigations. By preparing long chain branched polyacrylamides, we were able to show similarities between polystyrene in benzene and polyacrylamide in water. Investigations into branched polyacrylamide, (backbone molecular weight: 794 000 g/mole, molecular weight of the branches: 48 000 g/mole, number of branches: 20), using light scattering and viscometric measurements, show that the branched polymers have a smaller hydrodynamic volume and a smaller HUGGINS constant than the linear ones with the same total molecular weight. By analysing a few technical applications we found that the flocculation speed of linear structured poly(acrylamides) is better than that of branched ones.

5. The size and shape of polymer molecules

All factors related to the arrangement of the polymer chain in space are classified as tertiary structure. Measureable parameters are directly, (R_G: radius of gyration, $\bar{h}$: end to end distance, R_H: hydrodynamic radius, z: asymmetry in light scattering intensity), or indirectly, (χ: interaction parameters, A_2: second virial coefficient) related to the dimensions, e.g. size and shape of the polymer chain in a specific solvent under given conditions of temperature and pressure. For the exact determination of the coil size of macromolecules, it is necessary to ensure that measurements are carried out in dilute solutions, because in moderately concentrated solutions intermolecular interactions will lead to errors in the results. In general, the range of particle (dilute) and concentrated solutions may be definitely separated by a critical concentration c^*. c^* can be easily determined by using viscometric measurements, which yield a relative number or by performing, for example, light scattering measurements, which yield to absolute values of c^*. The viscometric measurements are most commonly used, because results are quickly obtained. However, this relative method is not equivalent to the absolute value of c^*_{LS} and only absolute determination methods like light scattering result in correct values of c^*. Figure 8

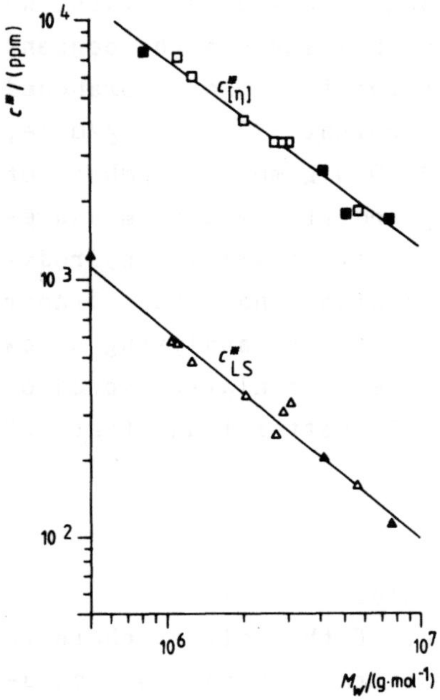

Fig. 8
Critical concentrations of polyacrylamide in water at 25°C versus molecular weight.

$$c^*_{[\eta]} = 2.500.000/[\eta] \qquad \text{(ppm)}$$

$$c^*_{LS} = 1.31\ 10^{-19} \frac{M}{(R_G^2)^{3/2}} \qquad \text{(ppm)}$$

demonstrates the results of some such investigations.One can see that both methods lead to a linear dependence between c^* and M_w but differ by a factor of ten. Similar results were found for polystyrene in toluene and polystyrene in cis-decalin, which shows that this was not an exception. The reason for this is seen in the fact that $c^*_{[\eta]}$ depends on a model (EINSTEIN's law), whereas c^*_{LS} gives absolute results. In both cases the geometric shape of the polymer coils are assumed to be spherical, but in accordance with the findings of KUHN and others we know that the most probable form can be best represented as a bean-like (irregularly ellipsoidal) structure. The reason is that for such an irregular overall shape this is the largest number of possible formations. A polymer chain in dilute solution can be seen as a coil which changes its form continuously with time under the influence of statistical thermal movements. Over an average time interval a conformation can be determined on the basis of the end to end distance $\bar{h}$ and a radius of gyration R_G.

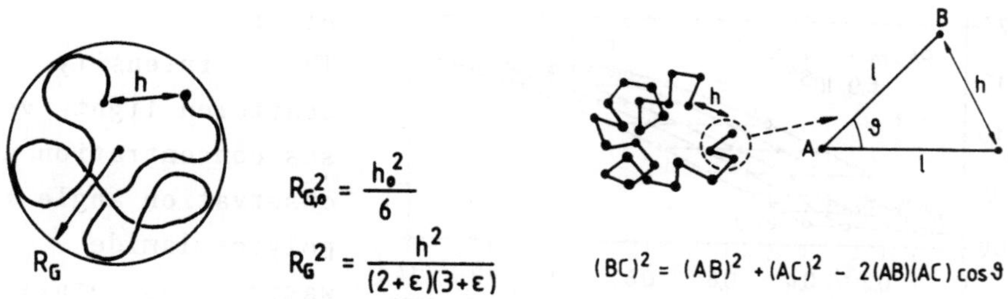

$$R_{G,\theta}^2 = \frac{h_\theta^2}{6}$$

$$R_G^2 = \frac{h^2}{(2+\varepsilon)(3+\varepsilon)}$$

$$\varepsilon = 2/3a - 1/3$$

$$(BC)^2 = (AB)^2 + (AC)^2 - 2(AB)(AC)\cos\vartheta$$

$$h^2 = 2l^2 - 2l^2\cos\vartheta$$

$$h^2 = nl^2$$

Fig.9
Correlation between end to end distance and radius of gyration in pseudo-ideal conditions and for good solvent qualities.

As can be seen from figure 9 the radius of gyration can be correlated to the end to end distance for different solvent qualities. The end to end distance can be calculated theoretically as is described in the "random walk model". The end to end distance for the simplest case containing only two monomer units is given by the cosinus sentence. If the polymer chain contains a great number of monomer units allowing the possibility of any angle θ so that the sum of all angle terms will be zero, it leads to the simple equation of $h^2 = nl^2$. Under pseudo ideal conditions we will have the simple equation see fig. 9. And for good solvent qualities we have to know ε which can be calculated easily from the exponent a of the MARK-HOUWINK equation. The problem in using this model is that long and short range interactions will occur and in addition excluded volume effects will limit the precision of calculation. The radius of gyration can be measured by using the light scattering method. Figure 10 shows the result of such investigations for polyacrylamide in two different solvents. The intensity of the scattered light can not only be measured as a function of concentration (see fig. 1, from the slope one can

can calculate A_2) but also as the function of the angle of observation. In this case parts of the scattered light are diminished by interference depending on the angle of observation. Therefore from the dependence of the angle of observa-

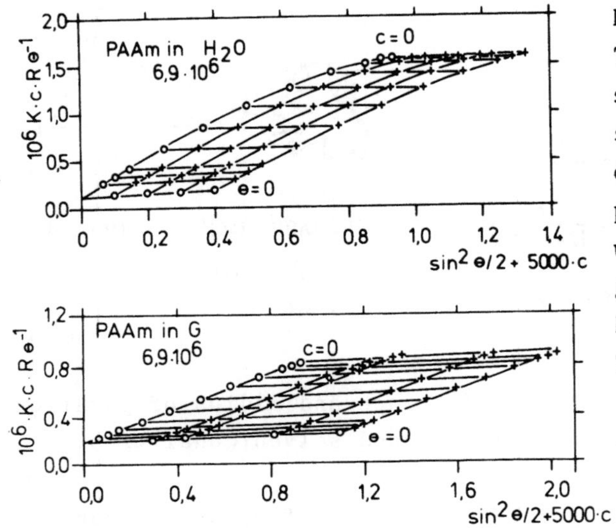

Figure 10
The intensity of scattered light versus concentration and observation angle for polyacrylamide in water and ethylene glycole.
$M_w = 6.3 \cdot 10^6$,
T = 25°C.

tion one can calculate the radius of gyration. From one investigation one can therefore calculate the absolute average molecular weight, the coil size (R_G) and the thermodynamic quality of the solvent[2]. By measuring the scattering functions at eg. 45° and 105°, one can calculate their dissymmetry z from their ratio. The dissymmetry, which is a result of interference, becomes larger with increasing difference of the angle. However, from such measurements one can get an idea of the nonspherical shape of polymer molecules. Other cross correlations like the dependence of the end-to-end distance versus molecular weight were also investigated, see figure 11. The corresponding equation from figure 11 of end-to-end distance versus molcular weight is then given as (assuming SCHULZ distribution) $(h^2)_w^{0.5} = 0.39\ M_w^{0.59}$. The experimental results for the end-to-end results of light scattering are z-averages, whereas the moleccular weights are weight averages. By assuming a SCHULZdistribution of the molar mass, the z-average can be calculated if M_w/M_n is known.

$U = M_w/M_n - 1$

$M_z = (1 + 2U)/(1 + U)\, M_w$

By means of this assumption the weight average end-to-end distances canbe calculated by the following equation,

$$\left(h^2\right)_w^{0.5} = \left(h^2\right)_z^{0.5}/(M_z/M_w)^{0.5}$$

from which the above equation was calculated.

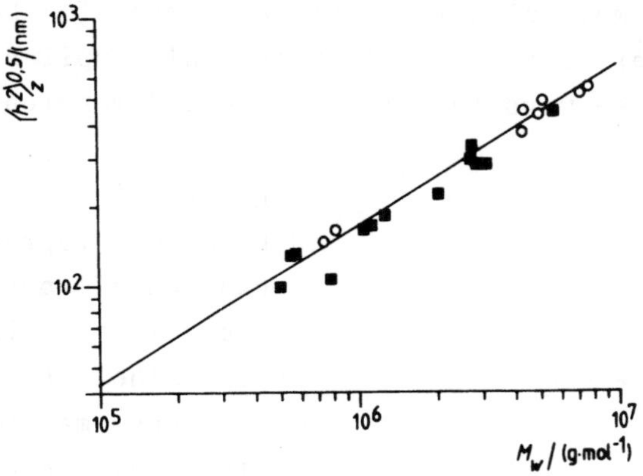

Fig. 11
Double logarithmic plot of end-to-end distance versus molecular weight for polyacrylamide in water at 25°C[2].

6.Conclusion

The influence of polymer additives on velocity fields may be described by using molecular structure parameters. The aim of this contribution is to show how the molecular structure parameter can be determined. Especially in aqueous solutions the measurements are not as easy to perform as in organic solvents. Measurements of polyelectrolytes are however more difficult to perform than measurements of non-ionic polymers, because polyelectrolytes have the properties of both i.e. charge and chain-structure. However, as shown before, by recognising these problems and by working very carefully it is possible to determine such important structure properties as molecular weight, molecular weight distribution, radius of gyration, end-to-end distance etc..

By establishing structure property relationships one should be able to select polymers for specific technical applications. To give an example, we established a zero-shear-viscosity - molecular weight - concentration relationship which is valid over a great range of concentration and molecular weight[2]. On the other hand we were able to show how to determine the molecular weight of ultra-high molecular weight polymers by using a porous media flow device and applying some theoretically derived elongational flow equations[3]. In addition the strong influence of co-polymer composition on solution properties was shown[8]. From all these results we got the feeling that the size of polymer molecules in solution is one of the keys to increasing the effectiveness of technical applications. Figure 12 shows some ways in which the coil dimensions can be influenced [9].

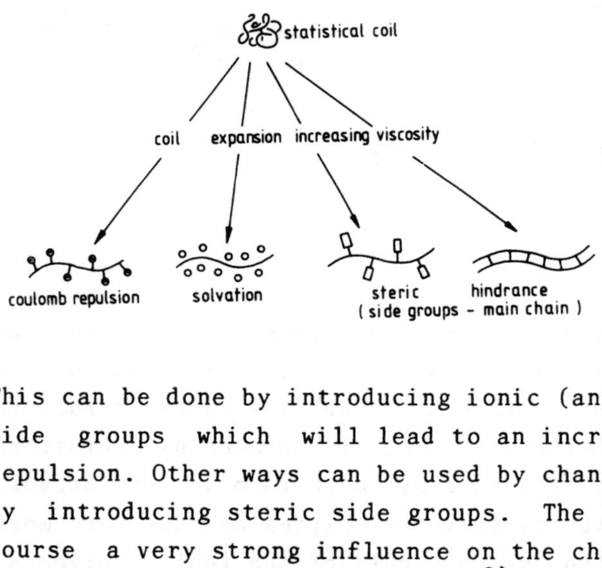

Fig. 12
Schematic representation of possible ways, in which the coil dimensions may be influenced, which can be tested by simple viscometric measurements.

This can be done by introducing ionic (anionics and cationics) side groups which will lead to an increase in the COULOMB repulsion. Other ways can be used by changing the solvation or by introducing steric side groups. The last example has of course a very strong influence on the chain flexibility which can be measured in different ways[2]. However, these examples should only show that besides the molecular weight and distribution etc., the coil dimensions can be influenced by changing the chemical structure. It is a well-known fact that with increasing molecular weight the mechanical stability decreases. Looking at figures 5 and 6 one can see that the coil dimensions can be changed up to a factor of 8 whereas the degree and distribution of polymerization is unchanged. GAMPERT and WAGNER[4] were able to show that if the degree of polymerization is increased, this will also lead to an increase in drag reduction, which serves as an example of technical application. On the other hand with an increasing degree of polymerization, there will also be an increase in the coil dimensions. Therefore, the question arises whether the controlling factor for the improvement of the effectiveness of technical applications is the coil dimension or the chain length. The influence of chain flexibility and chemical nature on the possible improved effectiveness on technical applications should also be researched in the future. These questions can only be answered by researching the molecular structure parameters and their influence on velocity fields.

<u>References</u>

1) Kulicke, W.-M.; Kniewske, R.: Purification of aqueous and non-aqueous polymer solutions for light scattering measurements and a comparison of molecular weights determined with a low-angle and a wide-angle-photometer. Makromol. Chemie 1 (1980) 719 - 721

2) Kulicke, W.-M.; Klein, J.; Kniewske, R.: Preparation, characterization, solution properties and rheological behaviour of polyacrylamides. Progress in Polymer Science 8 (1982) 373 - 468

3) Kulicke, W.-M.; Haas, R.: Flow behaviour of dilute polyacrylamide solutions through porous media, part one and two. Ind. & Eng. Chemistry, Fundamentals 23/3 (1984) 308 - 319

4) Gampert, B.; Wagner, P.; Haas, R. and Kulicke, W.-M., this symposium.

5) Kurata, M.; Tsunashima, Y.; Iwama, M.; Kamada, K.: Viscosity-molecular weight relationships and unperturbed dimensions of linear chains molecules. Polymer Handbook edited by Brandrup, J. and Immergut, E.H., 2nd edition, Wiley-Interscience, N.Y. (1975) IV-1ff

6) Grüneberg, M.; Klein, J.: Determination and use of cumulants of the peak broadening function in steric exclusion chromatography. K- Öoqiod Chromat. 3(10) (1980) 1593 - 1615
Klein, J.; Grüneberg, M.: Mass transfer of macromolecules in steric exclusion chromatography, part I, II and III. Macromolecules 14/5 (1981) 1419 - 1422
Klein, J.; Westerkamp, A.: Peculiarities of polyacrylamide analysis by aqueous GPC. J. Polym. Sci. 19 (1981) 707 - 718

7) Kulicke, W.-M.; Böse, N.: Bestimmung der Molmassenverteilung sowie der Stabilitätsgrenze von Polyacrylamiden unter Benutzung einer kombinierten Ausschlußchromatographie/Kleinwinkel-Laser-Streulichtphotometer Anlage. Coll. & Polym. Sci. 262 (1984) 197 - 207

8) Klein, J.; Heitzmann, R.: Preparation and characterization of Co-poly(acrylamide/acrylic acid). Makromol. Chemie 179 (1978) 1895 - 1904
Klein, J.; Conrad, K.-D.: Molecular weight determination of poly(acrylamide) and poly(acrylamide-co-sodium acrylate). Makromol. Chemie 179 (1978) 1635 - 1638
Kulicke, W.-M.; Hörl, H.-H.: Preparation and characterization of a series of poly(acrylamide-co-acrylates), with a copolymer composition between 0-96.3 mol% acrylate units with the same degree and distribution of polymerization. Coll. & Polym. Sci. (submitted) 1 - 30

9) Klein, J. et al.: Entwicklung neuer Polymertypen zur Tertiärförderung von Erdöl aus Lagerstätten hoher Salinität. BMFT Forschungsbericht ET 3159 A, (Okt. 1978)

Drag Reduction of Non-Ionic Surfactant Mixtures

R. C. Chang and J. L. Zakin

Department of Chemical Engineering
The Ohio State University
Columbus, Ohio 43210 U.S.A.

SUMMARY

An investigation of mixed nonionic surfactant systems was made to find conditions in which high molecular weight aggregates suitable for use as drag reducers would be formed. Viscosities of a series of mixtures of Brij 96 ($C_{18}H_{35}$-$(OCH_2$-$CH_2)_{10}$-OH) and Brij 92 ($C_{18}H_{35}$-$(OCH_2$-$CH_2)_2OH$) were measured at various temperatures to find conditions for maximum relative viscosity. It had been shown earlier that this is a convenient screening technique for finding solutions exhibiting high drag reduction. A 1% (90:10 by weight) mixture of Brij 96 and Brij 92 gave the highest relative viscosity, 6.9 at 46°C, just below the cloud point.

Addition of varying amounts of Na_2SO_4 or of phenol lowered the cloud points and gave maximum relative viscosities of 9.4 and 11.0 for 1% 90/10 mixtures at temperatures close to their cloud points. Cloud points can be adjusted to convenient temperatures by controlling the concentration of these additives. Light scattering measurements indicated molecular weights of the surfactant aggregates with sodium sulfate or phenol near their cloud points may be as high as 10 million or more.

Drag reduction results for a 1% 90:10 Brij 96/92 mixture with 0.5 Normal sodium sulfate showed that the greatest drag reduction occurred above the cloud point and no mechanical degradation was observed at wall shear stresses close to 2500 dynes/cm^2.

INTRODUCTION

The addition of small amounts of certain materials to fluids undergoing turbulent flow causes a reduction in pressure loss called drag reduction. Polymer solutions, solid suspensions in liquids and in gases, soap and surfactant solutions have all demonstrated this phenomenon.

Polymer and hydrocarbon-soluble soap additives studied so far all have inherent deficiencies in mechanical or chemical stability which limit their

The Influence of Polymer Additives
on Velocity and Temperature Fields
IUTAM Symposium Essen 1984
Editor: B. Gampert

use in practical applications. In addition, solution preparation is a serious problem with many polymer additives. Flows of liquids with solid suspension additives have settling and plugging problems in stagnant regions.

Aqueous solutions of complex soaps (1-3) are drag reducers, as are certain conventional soaps (4-6) and nonionic surfactants (7-11), and they do not have some of these deficiencies. They have the advantage of regaining their drag reducing effectiveness after subjection to high shear fields. The latter two are effective near their coacervation temperature or cloud point (upper consolute temperature). The addition of electrolyte lowers the cloud point and therefore the temperature at which effective drag reduction occurs. Cloud points can be adjusted to convenient temperatures in this manner.

The purpose of this investigation was to obtain information about the solution properties of combinations of nonionic surfactants with other additives which form high molecular weight aggregates which can be used as drag reducers.

RELEVANT PREVIOUS WORK

Two types of drag reducing aqueous soap systems have been reported. Savins (4-6) studied 0.2% sodium oleate in water solutions containing varying amounts of KCl. Increasing KCl concentration from 3.5 to 10% increased drag reduction from 45 to 82% at a given temperature and fixed wall shear stress and tube diameter. Diameter effects were similar to those for polymer solutions. Savins suggested that with increasing electrolyte concentration,initially spherical micelles were probably rearranged into cylindrical micelles which formed a network of interlaced rod-like elements.

Savins noted that at a critical wall shear stress, independent of tube diameter, the solutions suddenly lost their drag reducing ability. This loss was presumably caused by break up of micelles due to high shear stresses in the critical wall region. The rate of break up was faster than the rate of reformation. Savins also observed that drag reducing ability was immediately regained by lowering the flow rate so that the wall shear stress was below the critical value. Here the rate of reformation exceeded the rate of break up. No permanent degradation was noticed after 88 hours of continuous shear above the critical wall shear stress as fresh solution low flow rate pressure drops could be duplicated.

Despite the relatively low cost of conventional soaps and their mechanical stability, they have limited applicability because the presence of calcium and other cations present in most tap waters and in sea water causes precipitation of calcium and other insoluble soaps.

White and others (1-3) studied the complex soap, cetyl trimethyl ammonium bromide complexed with 1-naphthol in water. Up to 70% drag reduction was found and shear stress behavior the same as that reported by Savins was observed. This system also proved useful in reducing drag in 20% (by volume) slurries of sand and of plastic particles in water (2,3). Not only was drag reduced, but the critical settling velocity was also significantly reduced. In ordinary use, however, these systems lost their effectiveness in one or two days, and fresh solutions had to be used for maximum effectiveness.

Chang and Zakin (7-9) studied several straight chain aliphatic ethylene oxide nonionic surfactants, and observed that drag reduction increased as the surfactants formed larger aggregates, reaching a maximum near the cloud point, at which two phases formed. Relative viscosity, measured in Ubbelohde viscometers, dropped rapidly when phase separation occurred but drag reduction remained large above the cloud point. They postulated that this may be due to the existence of large micelles in the separated phase which are effective as drag reducers but have less effect on relative viscosity of the two phase system. The addition of electrolytes, particularly those containing multi-valent anions, lowered the cloud point significantly and increased drag reduction effectiveness (7-9, 11).

The behavior of the nonionic drag reducing solutions in shear fields was similar to that of soaps in aqueous systems (1,4-6). Chang and Lui (7-9, 11) also showed that the critical shear stress for mechanical degradation is dependent on both surfactant concentration, solution temperature, and on electrolyte nature and concentration. They studied several nonionic surfactants with varying oxyethylene group numbers. The results indicated that the particular chemical structure of the surfactant has a major effect on its micelle structure and size and hence on the effect the micelles have on drag reduction. One of the most effective additives was Brij 96 ($C_{18}H_{35}$-$(OCH_2$-$CH)_{10}$-OH).

These earlier reports on drag reduction with surfactants utilized single surfactants and described the effects of electrolytes and temperature changes on their drag reducing abilities. Recent work by Florence and Puisieux and coworkers (12,13) has shown that a combination of an insoluble surfactant ($C_{18}H_{35}$-$(OCH_2$-$CH_2)_2OH$) solubilized by a soluble surfactant ($C_{18}H_{35}$-$(OCH_2$-$CH_2)_{10}OH$) forms much larger micelles than the latter which is a very good drag reducer (11). The mixed surfactant micelles were rod-shaped. This mixed system was studied to determine its potential usefulness as a drag reducer.

EXPERIMENTAL

Several types of measurements were performed on solutions containing a number of combinations of Brij 96 ($C_{18}H_{35}(OCH_2-CH_2)_{10}OH$) or Polyoxyethylene (10) oleyl ether and Brij 92 ($C_{18}H_{35}(OCH_2-CH_2)_2-OH$ or Polyoxyethylene (2) oleyl ether, straight chain aliphatic nonionic surfactants supplied by ICI America, Inc. Measurements were also made on solutions of surfactant mixtures and sodium sulfate and phenol, two additives which are effective in lowering the cloud point (7-9, 11, 13,14).

Solutions were prepared at room temperature and measurements were made within seven days. The following measurements were taken (not all measurements were made on all solutions): cloud point by visual observation, cloud point by turbidity measurement, viscosity using Ubbelohde viscometer, molecular weight using a Brice-Phoenix Universal Light Scattering Photometer (2000 series), and drag reduction using 0.03247-, 0.1078- and 0.2611-inch tubes. The apparatus for the drag reducing experiments has been described earlier (15). Most runs were at ± 0.2°C of the nominal temperature.

Cloud points were initially measured by heating 1% solutions of surfactant in distilled water at a rate of 1°C per minute in a test tube immersed in a water bath. When cloudiness was observed the heat source was removed and the solution cooled until it was clear. The cloud point was taken as the midpoint between the observed temperature while heating and that during cooling. These were within 1.0°C of each other for pure surfactant.

Wider differences were observed for mixtures of Brij 96 and Brij 92. For these solutions, turbidity measurements were used to determine the cloud point. The intensity of light scattered at 90 degrees to a beam passing through a solution cell was monitored by a photomultiplier. Measurements were taken at temperatures spanning the expected cloud point value. Typical data for 1% 90:10 mixtures of Brij 96 and Brij 92 containing varying amounts of Na_2SO_4 and of phenol are shown in Figures 1 and 2. In each case the slopes are steeper above the cloud points and the intersections of the two regions are taken as the cloud points.

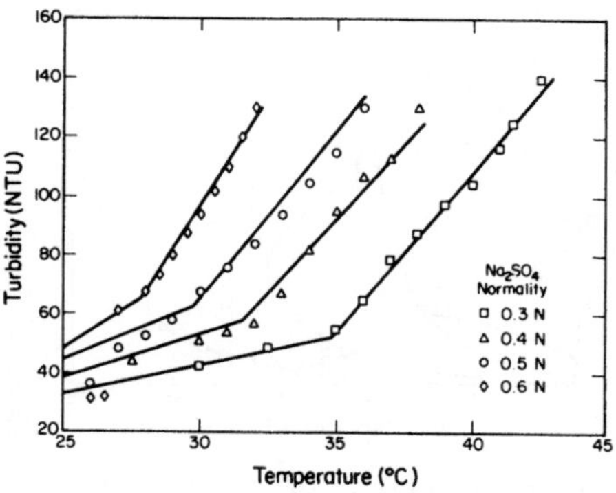

Figure 1: Effect of sodium sulfate on cloud point of 1% 90:10 Brij 96/92 mixture.

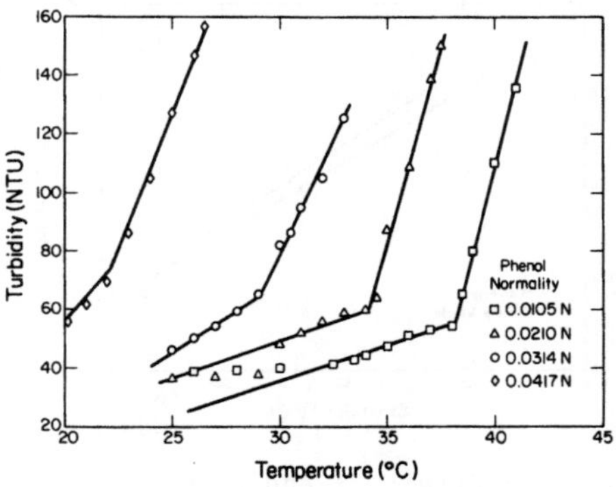

Figure 2: Effect of phenol on cloud point of 1% 90:10 Brij 96/92 mixture.

RESULTS AND DISCUSSION

Cloud point results for 1% 90:10 Brij 96/92 mixtures with varying amounts of Na_2SO_4 and phenol are shown in Figure 3. Phenol has been reported to have a large effect on the cloud point of nonionic surfactants (14) and it can be seen here to be very effective at quite low concentrations. The effect of sodium sulfate is also quite large, but at normalities more than an order of magnitude higher. The effect increases with additive concenttration. These additives act to decrease the hydration of the polyoxyethylene chains, leading to separation into two phases at the cloud point. Anions are more selective in "salting out" than cations and multivalent anions are more effective than univalent anions. The order of ion effectiveness resembles the "Hofmeister series" for proteins.

Relative viscosity measurements for 1% Brij 96/92 mixtures as a function of temperature and of 96/92 ratio are shown in Figure 4. The 100% Brij 96 sample rose to a value of 2.34 at 66.0°C which is just below its cloud point of 68.5°C. The solubilization of 5% Brij 92 lowered the cloud point to 59.5°C. Relative viscosities of the mixture were larger, reaching 5.7 at 57.5°C, two degrees below the visually-determined cloud point.

For the 10% Brij 92/90% Brij 96 mixture, difficulties were encountered in obtaining a sharp cloud point reading. The visual estimate of 53.5°C was in poor agreement with the temperature of maximum relative viscosity, 46.0°C, at which relative viscosity was 6.9. The turbidity - determined cloud point of 44.8°C is in fairly good agreement with this value. Similar discrepancies were observed for the 85/15 mixture but its maximum relative viscosity was only about 2.6. Reference 13 states that it is

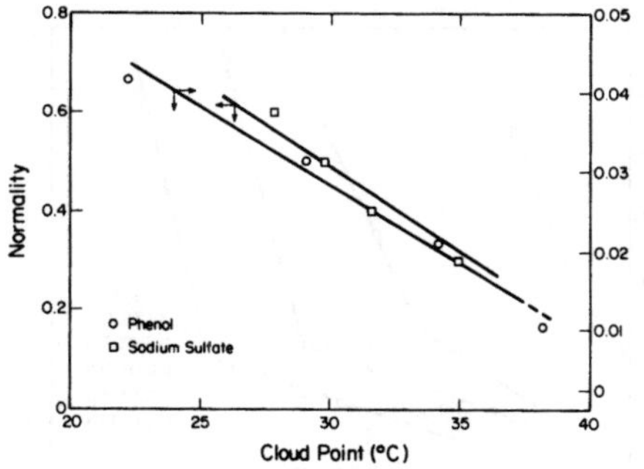

Figure 3: Effect of Additive concentration on cloud point of 1% 90:10 Brij 96/92 mixture

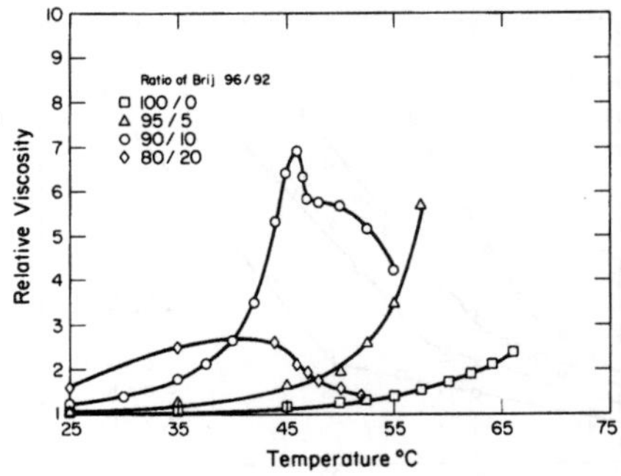

Figure 4: Effect of temperature on relative viscosity of various 1% Brij 96/92 mixtures.

possible to solubilize up to 19% Brij 92 in Brij 96. Therefore, a few measurements were also made with an 83/17 mixture but the solution was hazy so a cloud point could not be determined. The maximum relative viscosity value observed was near 2.

Since the largest relative viscosity, which should correspond to the largest aggregate size, was observed for the 90/10 mixture, further studies concentrated on this mixture.

Relative viscosity measurements were made as a function of temperature for 1% 90/10 Brij 96/92 mixtures with different concentrations of additives. These results are shown in Figure 5 (sodium sulfate) and Figure 6 (phenol).

In Figure 5, the increase in sodium sulfate concentration leads to a lowering of the temperature at which maximum relative viscosities are

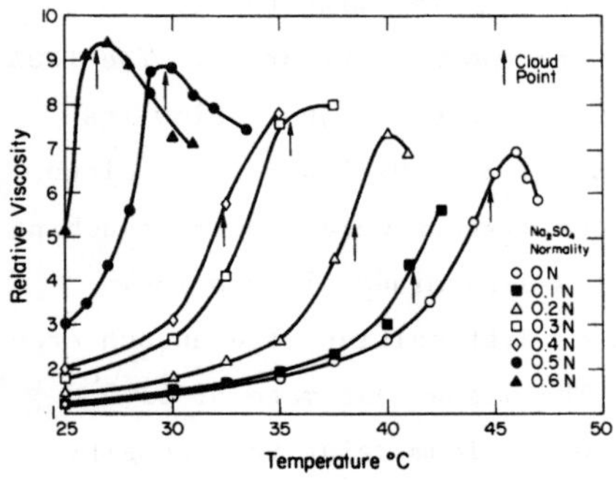

Figure 5: Effect of sodium sulfate concentration and temperature of 1% 90:10 Brij 96/92 mixtures

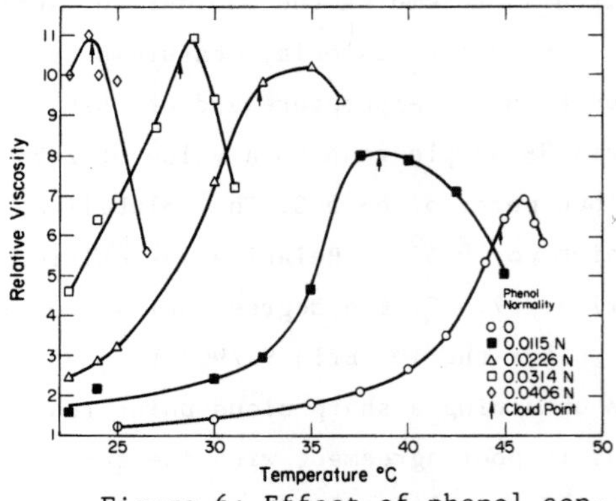

Figure 6: Effect of phenol concentration and temperature on relative viscosity of 1% 90:10 Brij 96/92 mixtures.

observed. In most cases, the turbidity-determined cloud point is close to the temperature of maximum relative viscosity. (Vertical arrows mark the cloud points). The maximum relative viscosity also increases with increase in sodium sulfate concentration. Relative viscosity drops rapidly as temperature exceeds the cloud point.

Similar effects are seen in Figure 6 where phenol is the additive. At quite low concentrations, phenol both lowers the cloud points and causes the relative viscosity to increase to values over 10. Relative viscosity drops rapidly above the cloud point.

Light scattering measurements were made to obtain estimates of the sizes of the aggregates at temperatures close to their cloud points. Since temperature control was convenient only just above room temperature, measurements with the Brice-Phoenix were made just above 30°C. For 1% 90/10 Brij 96/92 with 0.45N Na_2SO_4, molecular weight at 30.5^{o}C (cloud point $\approx$30.9^{o}C) was estimated to be over 10 million. For the same surfactant with 0.0285 N phenol, molecular weight at 31.5^{o}C (cloud point $\approx$29.4^{o}C) was estimated at 3 to 7 million. The uncertainty in both these estimates is large (factor of two), but there is no question that very high molecular weight aggregates are present. The two degree deviation from the cloud point may account for the lower estimate for the phenol system.

A test of the correlation between relative viscosity and drag reducing effectiveness was carried out using the 1% 90/10 Brij 96/92 mixture and one concentration (0.5N) of sodium sulfate. These measurements, carried out in 0.03-, 0.11- and 0.26- inch tubes at four temperatures near the cloud point are shown in Figures 7 to 10.

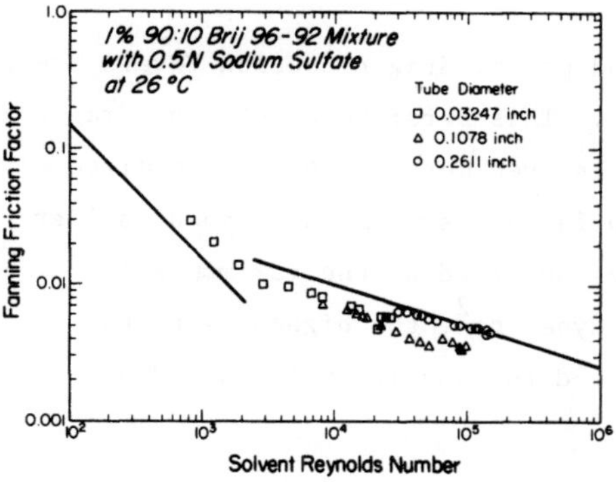

Figure 7: Friction factor vs solvent Reynolds number for 1% Brij 96/92 mixture with 0.5N sodium sulfate at 26°C

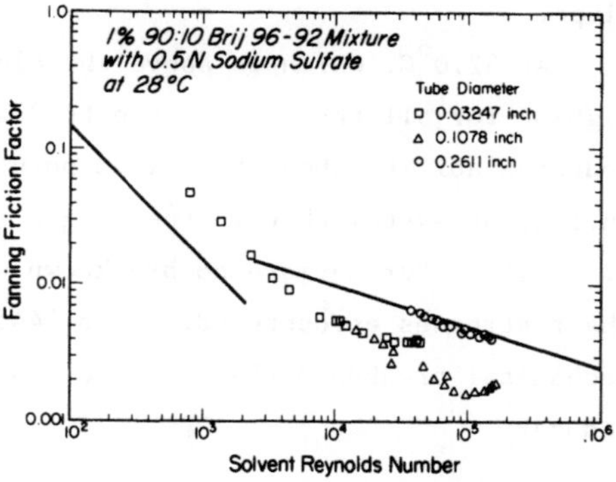

Figure 8: Friction factor vs solvent Reynolds number for 1% Brij 96/92 mixture with 0.5N sodium sulfate at 28°C

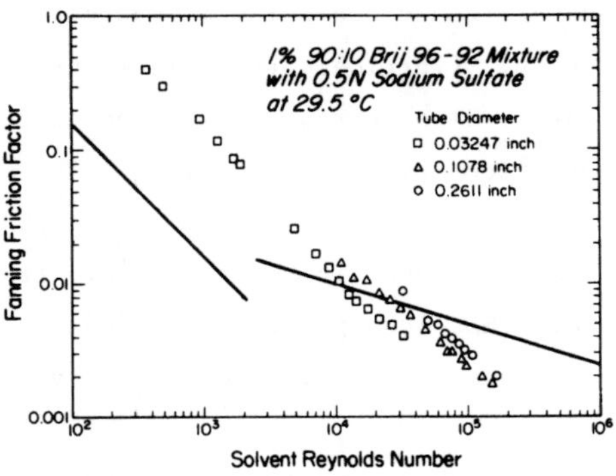

Figure 9: Fricton factor vs solvent Reynolds number for 1% Brij 96/92 mixture with 0.5N sodium sulfate at 29.5°C

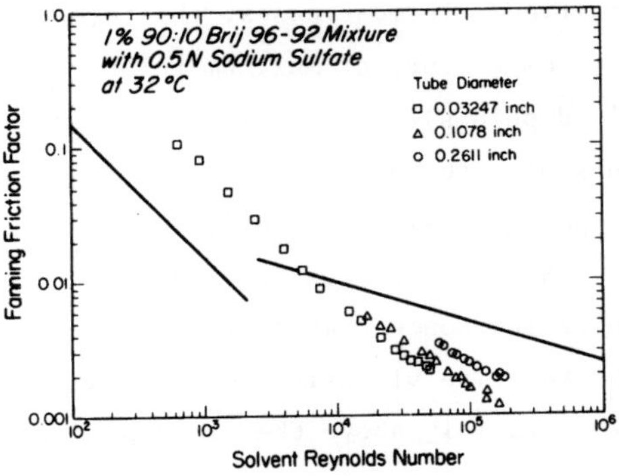

Figure 10: Fricton factor vs solvent Reynolds number for 1% Brij 96/92 mixture with 0.5N sodium sulfate at 32°C

At 26.0°C where relative viscosity in Figure 5 was 3.5, little drag reduction is seen in the 0.26-inch tube but larger amounts in the smaller tubes. At 28.0°C, where relative viscosity was 5.6, more drag reduction is observed in the smaller tubes. A maximum shear stress, above which drag reduction decreases, was observed in both the 0.03- and 0.11-inch tubes.

At 29.5°C, which is very close to the measured cloud point of 29.7°C the surfactant solution has very high relative viscosity (≈8.7) and exhibits "concentrated" drag reducing behavior (16). That is, the data show a gradual departure from a slope parallel to the laminar line. If measurements could be made at higher solvent Reynolds numbers, they would show no laminar to turbulent transition but would gradually decrease in slope.

At 32.0°C, which is above the cloud point, drag reduction levels are highest for all three tubes (up to 73%). This persistence of good drag reducing ability above the cloud point is even more marked in the mixed surfactant system than in the single surfactant systems described earlier (7-9, 11). Furthermore no breakdown was observed at the maximum wall shear stresses encountered, up to 2445 dynes cm^2, the highest with no mechanical breakdown that we have observed for any 1% surfactant drag reducer.

CONCLUSIONS:

Mixtures of Brij 96 and Brij 92 in water, in which the latter was solubilized by the former up to 15%, gave lower cloud points and higher relative viscosities than the former alone.

Relative viscosities of Brij 96/92 mixtures reached maxima at temperatures close to their cloud points. The 90/10 mixture gave the highest relative viscosity. Visual observations of cloud points gave incorrect readings with mixtures and a turbidity technique had to be used.

The addition of sodium sulfate or of phenol lowered the cloud points and increased the relative viscosities of 90/10 Brij 96/92 mixtures. Phenol was more effective than sodium sulfate at concentrations an order of magnitude lower.

Light scattering determinations near the cloud points indicated aggregate molecular weights may be as high as 10 million or more.

Drag reduction measurements on a 1% 90/10 Brij 96/92 mixture with 0.5N Na_2SO_4 showed the highest drag reduction at a temperature above the cloud point where relative viscosity had begun to decrease. No mechanical degradation was observed even at a wall shear stress of 2445 dynes/cm^2.

REFERENCES

1. White, A., "Flow Characteristics of Complex Soap Systems", Nature, 214, 585-586 (1967).
2. Poreh, M., Zakin, J.L., Brosh, A. and Warshavsky, M., "Drag Reduction in Hydraulic Transport of Solids", J. Hyd. Div., Proc. Am. Soc. Civ. Eng., HY 4, 903-909 (1970).
3. Zakin, J.L., Poreh, M., Brosh, A., and Warshavsky, M., "Exploratory Study of Friction Reduction in Slurry Flows", Chem. Eng. Prog. Symp. Series, 67, 85 (1971).
4. Savins, J.G., "A Stress-Controlled Drag Reduction Phenomenon", Rheol. Acta, 6, 323 (1967).
5. Savins, J.G., in Viscous Drag Reduction, p. 183-212, Wells, C.S., ed., Plenum Press, New York (1969).
6. Savins, J.G., "Method of Decreasing Friction Loss in Turbulent Liquids", U.S. Patent 3,361.213 (1968).
7. Zakin, J.L., and Chang, J.L., "Non-Ionic Surfactants as Drag Reducing Additives", Nature Phys. Sci., 239, 26-28 (1972).
8. Zakin, J.L., and Chang, J.L., "Polyoxethylene Alcohol Non-Ionic Surfactants as Drag Reducing Additives", in Proc. Intern. Conf. Drag Reduction, p. D1-1, Coles, N.G., ed., BHRA Fluid Engineering, Cranfield, Bedford, England (1974).

9. Chang, J.L., and Zakin, J.L., "Methods and Compositions for Reducing the Frictional Resistance to Flow of Aqueous Liquids", U.S. Patent 3,961,639 (1976).

10. Shenoy, A.V., "Drag Reduction with Surfactants at Elevated Temperatures", Rheol. Acta, 15, 658-664 (1976).

11. Zakin, J.L. and Lui, H-L., "Variables Affecting Drag Reduction by Nonionic Surfactant Additives," Chem. Eng. Commun., 23, 77-88 (1983).

12. Issa, L., Madsen, F., Florence, A.T., Treguler, J-P., Seiller, M. and Prusieux, F., "Mixed Non-Ionic Detergent Systems in Aqueous and Non-Aqueous Solvents", in Micellization, Solubilization, and Microemulsions, Vol. 1, K.L. Mittal, ed., 455-466, Plenum Press, New York, (1977).

13. Al-Saden, A.A., Florence, A.T., Whateley, T.L., Puisieux, F., and Vaution, C., "Characterization of Mixed Nonionic Surfactant Micelles by Photon Correlation Spectroscopy and Viscosity", J. Coll. Inter. Sci., 86(1),51-56 (1982).

14. Donbrow, D., and Azaz, E., "Solubilization of Phenolic Compounds in Nonionic Surface-Active Agents", J. Coll. Inter. Sci., 57, 20-27 (1976).

15. McMillan, M.L., Hershey, H.C., and Baxter, R.A., "Effects of Aging, Concentration, Temperature, Method of Preparation, and Other Variables on the Drag Reduction of Aluminum Disoaps in Toluene", Chem. Eng. Symposium Series, 67, 27-44 (1971).

16. Liaw, G.C., Zakin, J.L. and Patterson, G.K., "Effects of Molecular Characteristics of Polymers on Drag Reduction", Am. Inst. Chem. Eng. J., 17, 391-397 (1971).

The Influence of Molecular Weight and Molecular Weight Distribution on Drag Reduction and Mechanical Degradation in Turbulent Flow of Highly Dilute Polymer Solutions

B. Gampert, P. Wagner

Universität-GH-Essen, Strömungsmechanik
Schützenbahn 70, 4300 Essen 1,
West Germany

Abstract

The influence of the molecular weight and the molecular weight distribution on the drag reduction and the mechanical degradation in turbulent flow is investigated for highly pure laboratory synthesized polyacrylamides with mean molecular weights between 1.1 and $9.2 \cdot 10^6$ g/mol.

Introduction

The Toms effect is essentially dependent on the molecular parameters of the dissolved additive; the most important of these are the molecular weight and the molecular weight distribution. In this paper the influence of these parameters in the case of water-soluble polymers is investigated, whereby experiments were carried out with extremely pure laboratory polyacrylamides (PAAm).

Those polyacrylamides, which are commercially available, display as a result of various factors such as blending a wide and to a certain extent random distribution of the molecular weight. Their molecular chains are branched and generally they are present as copolymers. Such substances are not suitable for the investigation of the influence of the molecular weight on the drag reduction and the mechanical degradation. The mean molecular weight especially is not a characteristic parameter for these substances, as Patterson & Abernathy [1] have shown. In order to characterize these substances, the most appropriate method would be to determine the molecular weight distribution with the aid of suitable techniques such as gel-permeation chromatography (GPC). Berman et al. [2] have presented results of experiments with different polyethy-

The Influence of Polymer Additives
on Velocity and Temperature Fields
IUTAM Symposium Essen 1984
Editor: B. Gampert

lene oxide (PEO) fractions, which were fractionated by means of GPC. The application of this procedure is however difficult and uncertain in the case of water-soluble substances and is still a subject for further research. Polyacrylamides, which were mainly synthesized in the laboratory of the "Institut für Chemische Technologie der TU Braunschweig" (Prof. Klein, Prof. Kulicke) and partly in our own laboratory were therefore used in this work. These substances are extremely pure, free from ionic groups, consist practically of straight molecular chains and show a comparatively narrow molecular weight distribution.

The parameter molecular weight is in this manner closer defined and is adequately characterized by the mean molecular weight.

Experimental method

For characterizing the PAAm samples it was necessary first to determine the intrinsic viscosity $[\eta]$ using a Zimm-Crothers viscosimeter. Measurements with pure water resulted in a shear stress value of $\tau = 1.5 \cdot 10^{-3}$ dyn/cm^2, which meant that the viscosity measurements for the polymer solutions were in the first newtonian region; it was then not necessary to carry out the time-consuming extrapolation $\dot{\gamma} \rightarrow 0$ ($\dot{\gamma}$ = shear rate). The weight averaged mean molecular weight was thereafter calculated by means of the formula given by Kulicke, Kniewske & Klein [3]

$$[\eta] = 1.0 \cdot 10^{-2} \, \bar{M}_W^{0.755} \qquad (1)$$

Mother solutions with concentrations between 0.05 and 0.1% PAAm were prepared, stabilized by the addition of 2% by weight of isopropanol and homogenized on a jolting table. The addition of isopropanol caused a decrease of approximately 3% in the shear viscosity. A water content of 5% by weight was assumed for the weighing of the hygroscopic PAAm.

The mother solution was cautiously mixed with deionized water (with a conductivity < 1μS/cm), until the necessary degree of dilution required for the experiments was reached. The meas-

Sample	$\overline{M}_W$[g/mol]	[η] [cm^3/g]	DH
1	$1.10 \cdot 10^6$	365	0
2*	$1.43 \cdot 10^6$	443	0
3*	$1.70 \cdot 10^6$	507	0
4*	$4.26 \cdot 10^6$	1012	0
5	$4.30 \cdot 10^6$	1020	0
6	$6.67 \cdot 10^6$	1420	0
7	$6.86 \cdot 10^6$	1450	0
8*	$8.00 \cdot 10^6$	1624	0
9*	$9.20 \cdot 10^6$	1820	0

*polymerized and characterized by Kulicke & Böse [4]
DH = degree of hydrolysis

Table 1

urements were performed after the diluted solution was left to stand for several hours at room temperature. The temperature fluctuation within the period required for an experiment with a single solution was less than ± 1K. The measurement of the shear viscosity of the PAAm-solutions was carried out in an Ubbelohde viscosimeter.

To determine the influence of the polymer on the flow behaviour the friction coefficient λ ($\lambda = 2 \cdot \Delta p \cdot d / \rho\, \bar{u}_m^2\, L$) was investigated as a function of the Reynolds number ($700 \leq Re \leq 10^5$) for the different PAAm samples of varying molecular weight. The measurements were performed with a pipe flow system. The increased air pressure above the liquid in a container drove the test liquid through a 0.0061 m diameter horizontal pipe. To prevent degradation caused by pumping a single pass rather than a recirculating system was employed. The liquid flow rate was monitored by weighing the quantity collected in a second container over a measured interval of time. Two pressure taps were placed at approximately 1.5 m and 2.0 m from the upstream end of the pipe. To minimize the stresses outside of the test section, the bent pipe sections and the tapering cones were kept gradual. In addition to that, only ballcock valves instead of magnetic valves were used. Preliminary investigations support the findings of Patterson & Abernathy [1], that a sharp-edged entrance results in a higher degree of polymer degradation at higher Reynolds numbers. The pipe entrance section was therefore made gradual, as a consequence of which the laminar-turbulent transition was displaced from Re = 2300 to Re = 3500.

Results and discussion

The drag reduction DR is determined through the following relation

$$DR\ [\%] = \frac{\lambda_w - \lambda_{ps}}{\lambda_w}\Big|_{\dot{V}} \cdot 100 \qquad (2)$$

(w = water; ps = polymer solution; $\dot{V}$ = flow rate). The constant flow rate $\dot{V}$ corresponds approximately to a constant Reynolds number as the change in the shear viscosity caused by the highly pure polymer additives for a concentration of 50 ppm is only a few percent.
The λ_w-values for the corresponding Reynolds numbers of the polymer experiments were calculated with the Blasius relationship, since a very good agreement of experimental data for water with the Blasius relationship was reached, fig. 1.

Although the drag reduction is connected with a partial laminarization of the flow, the polymer addition shifts the laminar-turbulent transition towards lower Reynolds numbers. The higher the molecular weight of the polymer addition is, the lower is the Reynolds number at which the transition occurs, fig.1.

Fig. 1 shows clearly the influence of the molecular weight on the onset wall shear rate and the drag reduction. The onset value for the PAAm sample with the highest molecular weight ($\overline{M}_w = 9.2 \cdot 10^6$ g/mol) is so low, that it is already exceeded at the laminar-turbulent transition.

The influence of the concentration on the drag reduction is shown in fig.2 for four polymer samples of varying molecular weights at two different Reynolds numbers. We will first consider the results for Re = 60 000. There is a saturation concentration above which no further drag reduction is reached by

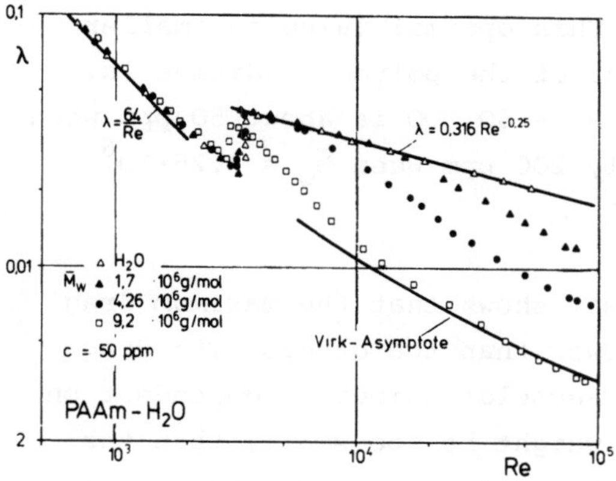

Fig. 1. The friction coefficient as function of the Reynolds number Re.

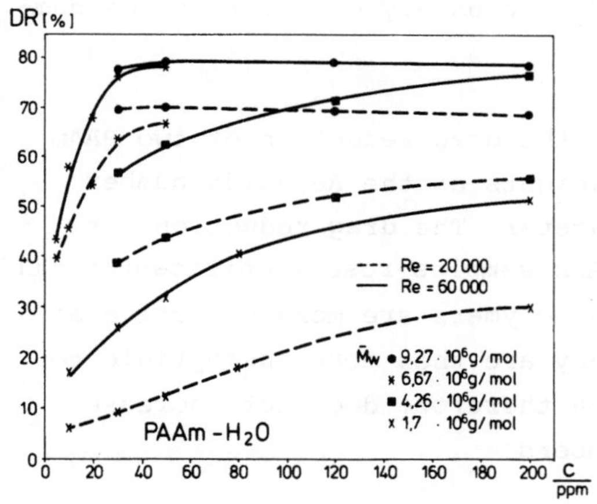

Fig. 2. Drag reduction DR as function of the polymer concentration

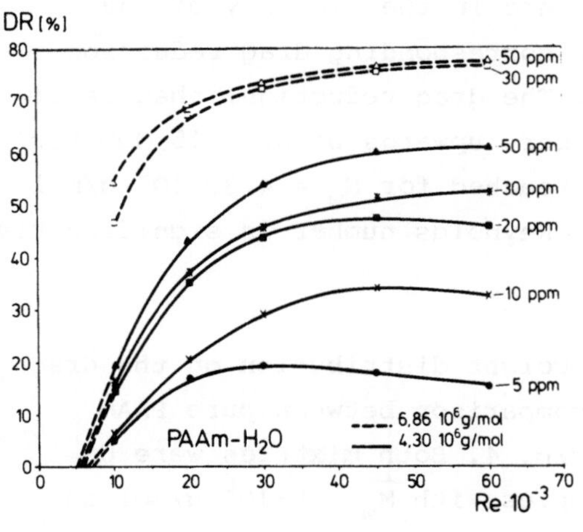

Fig. 3. Drag reduction DR as function of the Reynolds number for various polymer concentrations

increasing the concentration. This optimal value is smaller the higher the molecular weight of the polymer additive is. The optimal concentration for Re = 60 000 is about 50 ppm when $\overline{M}_w = 6.67 \cdot 10^6$ g/mol and roughly 200 ppm when $\overline{M}_w = 4.26 \cdot 10^6$ g/mol.

The curve for $\overline{M}_w = 1.7 \cdot 10^6$ g/mol shows that the maximum drag reduction of this sample is lower than the others. The drag reduction possible at a given Reynolds number is dependent on $\overline{M}_w$. Should the mean molecular weight be too small, then the drag reduction of Virk's asymptote is no longer possible. According to fig. 2 the optimal concentration is strongly dependent on the molecular weight $\overline{M}_w$ but barely on the Reynolds number.

Fig. 3 shows the dependence of the drag reduction of two PAAm samples of unequal molecular weights on the Reynolds number with the concentration as parameter. The drag reduction for the investigated laboratory PAAm samples rose significantly till approximately Re = 35 000. The polymers are more effective at higher Reynolds numbers but they are then more susceptible to degradation. The drag reduction therefore does not increase further at higher Reynolds numbers.

As the concentrations c = 30 ppm and c = 50 ppm for the PAAm sample with $\overline{M}_w = 6.86 \cdot 10^6$ g/mol are in the vicinity of the saturation concentration, their corresponding drag reduction values are approximately equal. The drag reduction, that is already observed for Reynolds numbers upwards of Re = 15 000 for $M_w = 6.86 \cdot 10^6$ g/mol, cannot be reached for $M_w = 4.30 \cdot 10^6$ g/mol polymer molecules even when the Reynolds number is significantly increased.

The influence of the molecular weight distribution on the drag reduction can be seen, when a comparison between pure PAAm samples and mixtures is made, fig. 4. Both mixtures were derived from the pure polymer samples with $\overline{M}_w = 8 \cdot 10^6$ g/mol and

$\overline{M}_w = 1.43 \cdot 10^6$ g/mol. For pure samples between Re = 20 000 and Re = 70 000 all DR-values for $\overline{M}_w = 1.7 \cdot 10^6$ g/mol are smaller than all DR-values for $\overline{M}_w = 4.26 \cdot 10^6$ g/mol and the latter smaller than all DR-values for $\overline{M}_w = 8 \cdot 10^6$ g/mol. The influence of $\overline{M}_w$ on the DR in the case of samples with comparatively low dispersities is very significant.

The difference in drag reduction between the mixtures and the pure samples is greatest for low Reynolds numbers and small mean molecular weights. The mere addition of a few long molecules can in the latter case increase the DR-value significantly. The drag reduction values of the mixtures for larger Reynolds numbers decrease with increasing Reynolds number. Substances with larger dispersities tend to display a maximum in the curve representation DR = DR(Re). The susceptibility of the long molecules of the mixture ($\overline{M}_w = 8 \cdot 10^6$ g/mol) towards degradation is apparently the reason for this. The behaviour of the mixtures demonstrates clearly the effects of the degradation. A comparison of the drag reduction behaviour of the mixtures and the pure samples of equal molecular weights shows that degradation of the fraction of larger molecules occurs already at Re = 20 000.

Fig. 5 shows the drag reduction for the pure samples as a function of the mean molecular weight for three different Reynolds numbers. As the concentration is constant (50 ppm) the number of molecules present in the solution decreases with larger $\overline{M}_w$. The largest gradient at DR = 0 is shown by the curve for Re = 60 000, as relatively short molecules are then also sufficiently stretched so as to be drag reducing effective.

The drag reduction values predicted by Virk's asymptote are approximately reached for Re = 60 000 and $\overline{M}_w = 6.8 \cdot 10^6$ g/mol and for Re = 20 000 and $\overline{M}_w = 8 \cdot 10^6$ g/mol. By extrapolating the curves DR = f($\overline{M}_w$) towards the ordinate axis (i.e. DR → 0), one obtains the cut-off molecular weight at the point of intersection. The quantitiy $\overline{M}_{wc}$ depends to a large extent on the dispersity δ. Very low values for $\overline{M}_{wc}$ can result from substances

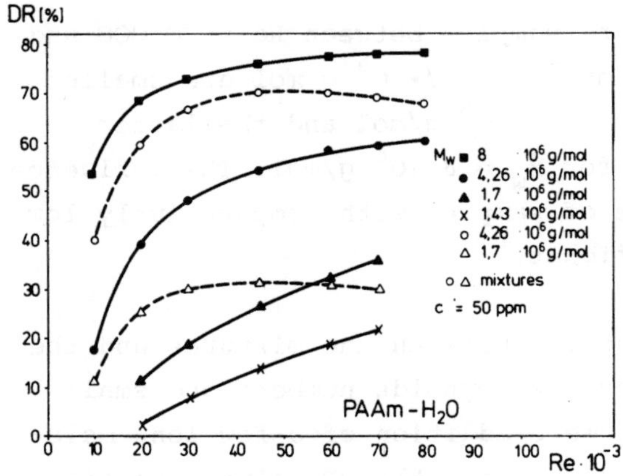

Fig. 4. Drag reduction DR as function of the Reynolds number for polymer samples with different mean molecular weights $\bar{M}_W$

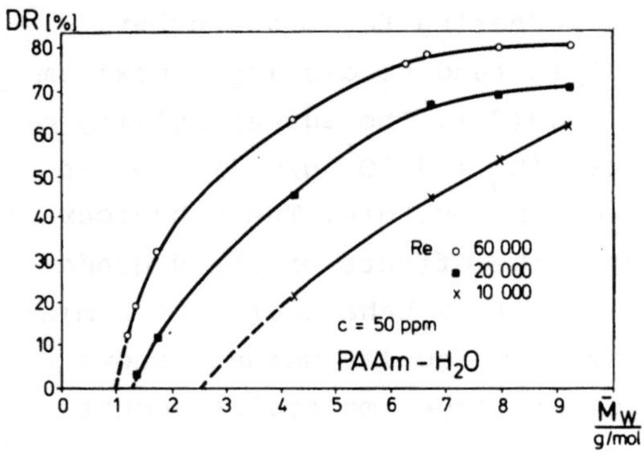

Fig. 5. Drag reduction DR as function of the mean molecular weights of the polymer samples showing cut-off values M_{WC}

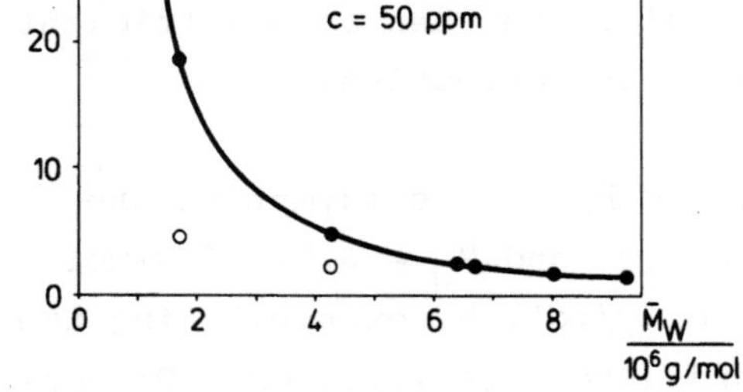

Fig. 6. Wall shear stress for the onset of drag reduction τ_{WE} as function of the mean molecular weight $\bar{M}_W$.

with a wide molecular weight distribution and high concentrations.

The dependence of the cut-off molecular weight on the concentration is small, as firstly the molecules in these highly diluted solutions (c = 50 ppm) are present as single entities, and secondly the stress in the flow field does not essentially depend on the number of polymer molecules in a volume element. However the drag reducing effectiveness of the individual polymer molecule can be better observed at higher concentrations.

It is assumed that with respect to the stress form the polymers are activated in extensional fields. As the extensional rates are not directly measurable, we characterize the flow fields by means of the wall shear stress. The onset wall shear stress is presented in fig. 6 as a function of the mean molecular weight of the PAAm sample. Apparently there is a lower limit for $\bar{M}_w$ below which no drag reduction can be attained for a realistic value of τ_w. On the other hand a certain value τ_{wE} appears to exist for large $\bar{M}_w$, below which even the longest molecules are not drag reducing effective.

A further aim of this paper was the investigation of the extent to which the drag reduction is affected by mechanical degradation. The molecular weight of the polymers was measured before and after degradation runs, so as to ascertain if and how the molecular weight was altered. As the molecular weight distribution cannot be determined in the case of water-soluble polymers with high molecular weights one is limited to experimental investigations using the mean value.

As the polyacrylamides investigated by us were free from ionic groups, the viscosity-based mean molecular weight could be measured with a great degree of accuracy. In order to avoid thermal degradation it was necessary to concentrate the polymer solutions in a rotary bulb evaporator operating at such a low pressure that boiling occured at 30° C.

The mechanical degradation was studied by subjecting all polymer solutions initially with a pressure gradient $\Delta p/l = 0.9$ bar/m during "degradation runs" in a circular pipe. Thereafter the altered viscosity-based mean value of the molecular weight was determined, and the change in the drag reduction was recorded as a function of the Reynolds number.

Fig. 7 shows the drag reduction DR as a function of the number of degradation runs n for different PAAm-solutions. The Reynolds number and the concentration for all measurements were Re = 60 000 and c = 50 ppm. The polymer solutions were made partly with pure PAAm-samples and partly with mixtures of the pure samples. The mixtures consisted of fractions from the pure samples with $\overline{M}_w = 1.43 \cdot 10^6$ g/mol and $\overline{M}_w = 8 \cdot 10^6$ g/mol.

As already reported by earlier authors the largest changes in the drag reduction occured in the case of the samples with the higher molecular weights. These changes are most significant during the first degradation runs. The drag reduction then decreases gradually in the following runs, fig. 7.

The fraction of molecules of the pure sample with $\overline{M}_w = 8 \cdot 10^6$ g/mol in the mixture with $\overline{M}_w = 1.7 \cdot 10^6$ g/mol is so small, that the drag reduction rapidly approaches that of the pure sample with $\overline{M}_{wo} = 1.43 \cdot 10^6$ g/mol (Index "o" indicates the initial $\overline{M}_w$-value for the sample). The fraction of longer molecules in the mixture with $\overline{M}_w = 4.6 \cdot 10^6$ g/mol is larger; the drag reduction therefore decreases more gradually.

The values of $\overline{M}_w$ for the mixtures also approach those values of the component with the shorter polymers with increasing n-value, fig. 8. The mixture with the initial molecular weight $\overline{M}_{wo} = 1.7 \cdot 10^6$ g/mol has with n = 2 virtually reached $\overline{M}_w \approx 1.43 \cdot 10^6$ g/mol. However the minute fraction of longer molecules causes a clear deviation of the DR-values of the mixture with $\overline{M}_{wo} = 1.7 \cdot 10^6$ g/mol from those of the pure sample with $\overline{M}_{wo} = 1.43 \cdot 10^6$ g/mol. The minute fraction of longer molecules has a stronger effect on the drag reduction than on the mean molecular weight.

Similar to the DR the $\overline{M}_w$ decreases most during the initial de-

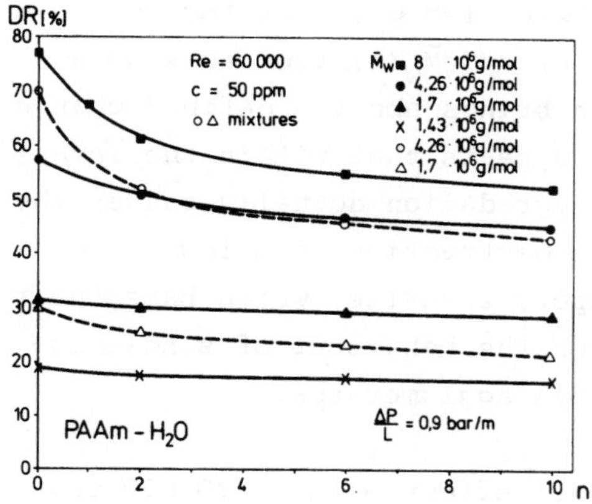

Fig. 7. Drag reduction as function of the number of degradation runs n. The Re number investigated is Re = 60 000.

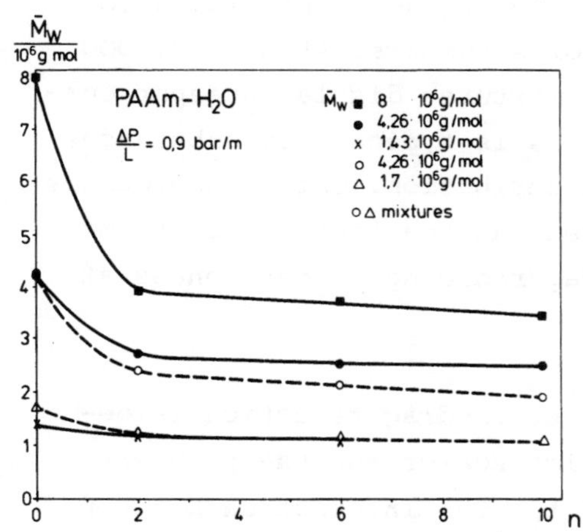

Fig. 8. Mean molecular weight of the polymer sample as function of the number of degradation runs n.

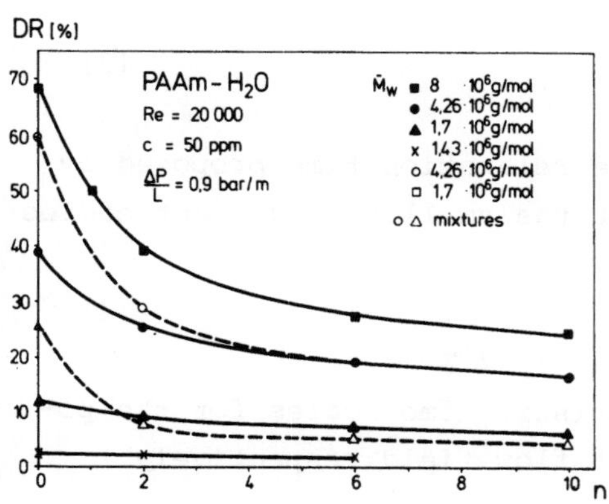

Fig. 9. Drag reduction as function of the number of degradation runs. The Re number investigated is Re = 20 000.

gradation runs. The changes are here also greatest for the longest molecules. The graphs DR(n) and $\overline{M}_w(n)$ seem to show a good correlation. The degradation brings about a defined change in the mean molecular weight. It appears that within the investigated concentration range the degradation actually causes the breakage of molecules and not the destruction of molecule agglomerates. The measurements support theories, which base their explanations of the Toms-effect on the behaviour of single molecules and not on that of molecule agglomerates.

The same change in $\overline{M}_w$ has a greater effect at Re = 20 000 than at Re = 60 000 (refer to figs. 7 and 9), as the shorter molecules (which are less susceptible to degradation) play a more significant role at higher Reynolds numbers. At Re = 20 000 the functions DR = f(n) for the mixtures display steeper curves than those for $\overline{M}_w = f(n)$; this is typical for substances with a broad molecular weight distribution. This behaviour is for Re = 60 000 not so pronounced, as the long and short molecules differ less in their drag-reducing effectiveness at higher Reynolds numbers.

As Lumley [5] has shown, the onset of drag reduction is dependent on the characteristic time scales for the polymers t_p and the turbulent flow field t_t. The interaction between the polymer and the turbulent flow field can be characterized by a Deborah number

$$De = t_p/t_t \qquad (3)$$

Elata et al. [6] have chosen the relaxation time proposed by Rouse [7] and the time scale for the smallest turbulent eddies

$$t_t = C_1 \cdot \nu/u_\tau^2 \qquad (4)$$

(C_1 = constant) as the characteristic time scales for the polymer molecule and the turbulent flow field respectively.

Experiments carried out by Fuller & Leal [8] as well as calculations by Bueche [9] indicate that polymer molecules are not sufficiently affected by shear stresses so as to substantiate the hypothesis that drag reduction and degradation are linked to shear fields. Furthermore it is shown in [8], that in extensional fields a strong interaction takes place between polymer molecules and the flow field.

Durst et al. [10], [11] considered the extensional flow between turbulent eddies and derived initially a time scale, which was dependent on the local position, and from which they tried to obtain information about the local alteration of the velocity profiles caused by the polymer addition. After introducing a mean extensional rate, they put forward a formula for t_t, which is identical with equation (4). They chose the relaxation time derived by Bird et al. [12]

$$t_p = \frac{[\eta]\ \eta_s\ M}{A\ k\ t} \tag{5}$$

as the characteristic polymer time scale ($[\eta]$ = intrinsic viscosity, η_s = shear viscosity of the solvent, A = Avogadro's constant, k = Boltzmann's constant) and then plotted DR' (=DR+1) as a function of the Deborah number. They were able to show that results obtained in their experiments with pipes of different diameters and at different temperatures produced a single curve when the above graphical respresentation was used.

Extensional fields are present when turbulent kinetic energy is produced by means of the burst-process. Donohue et al. [13] have shown that the number and intensities of bursts in a turbulent flow are reduced by the addition of polymers. A turbulent time scale, which was derived from measurements of Donohue et al. [13] and which is identical with equation (2), has been presented in [14], [15]. The relaxation time derived by Bird et al. [12] was used for the polymer time scale, as in [10], [11] (refer equation (5)), and an equal Deborah number De as in [10], [11] was obtained.

$$De = \frac{[\eta] \cdot \eta_s \cdot M}{A\,k\,T} \frac{\tau_w}{\eta} \qquad (6)$$

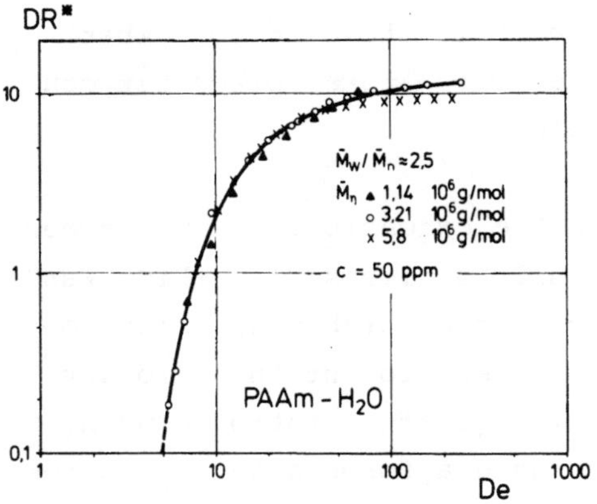

Fig. 10. Drag reduction DR* as function of the Déborah number De for PAAm samples with different average molecular weights

Fig. 10 shows $DR^* = DR/(c[\eta])$ as a function of the Deborah number (η = viscosity of the solution) for different molecular weights and for a constant concentration c = 50 ppm. It can be seen that our experimental results lie approximately on a single curve; the influence of the polymer molecular weight is well described by the relationship $DR^* = f(De)$. The onset point occurs at a higher Deborah number, as the constant in equation (4) has been arbitrarily given the unit value. This relationship shows the influence of τ_w (or Re) as well as M correctly and supports the conclusions made earlier in this paper regarding the experimental results of our work.

References

1. Patterson, R.W.; Abernathy, F.H.: Turbulent Flow Drag Reduction and Degradation with Dilute Polymer Solutions. J. Fluid. Mech. 43, 4 (1970) 689-710.

2. Berman, N.S.: Drag Reduction of the Highest Molecular Weight Fractions of Polyethylene Oxide. Phys. of Fluids 20, 5 (1977) 715-718.

3. Kulicke, W.-M.; Kniewske, R.; Klein, J.: Preparation, Characterization, Solution Properties and Rheological Behaviour of Polyacrylamide. Prog. Polym. Sci. 8 (1982) 373-468.

4. Kulicke, W.-M.; Böse, N.: [η] - M-Relationship for Polyacrylamide in Aqueous 0.1 M $Na_2\,SO_4$ Solution. Polym. Bulletin 7 (1982) 205-210.

5. Lumley, J.L.: Drag reduction by additives. Ann. Reviews of Fluid Mech. 1 (1969) 367-383.

6. Elata, C.; Lehrer, J.; Kahanovitz, A.: Turbulent Shear Flow of Polymer Solutions. Israel Journal of Technology 4, 1 (1966) 87-95.

7. Rouse, P.E. Jr.: A Theory of the Linear Viscoelastic Properties of Dilute Solutions of Coiling Polymers. J. Chem. Phys. 21 (1953) 1272-1820.

8. Fuller, G.G.; Leal, L.G.: Flow birefringence of dilute polymer solutions in two-dimensional flows. Rheol. Acta 19 (1980) 580-600.

9. Bueche, F.: Mechanical Degradation of High Polymers. J. Appl. Polymer Sci. IV, 10 (1960) 101-106.

10. Durst, F.; Haas, R.; Interthal, W.; Keck, T.: Polymerwirkung in Strömungen - Mechanismen und praktische Anwendungen. Chem.-Ing.-Tech. 54 (1982) Nr. 3, 213-221.

11. Durst, F.; Haas, R.; Interthal, W.: Laminar and turbulent flows of dilute polymer solutions: a physical model. Rheol. Acta 21 (1982) 572-577.

12. Bird, R.B.; Hassager, O.; Armstrong, R.C.; Curtiss, C.F.: Dynamics of Polymeric Liquids 2. New York: J. Wiley & Sons 1977.

13. Donohue, G.L.; Tiederman, W.G.; Reischman, M.M.: Flow visualization of the near-wall region in a drag-reducing channel flow. J. Fluid Mech. 53, 3 (1972) 559-575.

14. Gampert, B.; Wagner, P.: Experimentelle Untersuchung turbulenter Rohrströmungen hochreiner wäßriger Polyacrylamid-Lösungen, Rheol. Acta 21 (1982) 578-581.

15. Wagner, P.: Der Einfluß molekularer Parameter auf das Verhalten turbulent strömender Polymerlösungen. Dissertation, Universität-GH-Essen, 1984.

Session 4
Molecular Parameters II

Influence of Molecular Parameters on Laminar Non-Newtonian and on Turbulent Flows of Dilute Polymer Solutions

M.N. LAYEC-RAPHALEN and Y. LAYEC

Laboratoire d'Hydrodynamique Moléculaire
Faculté des Sciences, 6, Avenue Le Gorgeu
29283 - Brest - cedex
FRANCE

Summary

We show that the hydrodynamic radius of Poly(ethyleneoxide) macromolecules may be very different in water and water-isopropanol mixture (90%/10%). This dimension is a function of the age of the solution in pure water. The influence of the solvent and the age (i.e. of the hydrodynamic radius-) on the properties in laminar and turbulent flows is studied.

Introduction

The addition of long chain macromolecules to a fluid submitted to a turbulent flow can reduce the drag up to 80%. Polyethylene-oxide (PEO) is a polymer of particular practical interest in drag reduction experiments. PEO is known to form aggregates or association complexes in aqueous or organic solvents. Many experiments are concerned with low molecular weight samples. However, Cuniberti (1), Carpenter (2) and Güner and Güven (3) reported results on molecular weights in the 10^5-10^6 range.

On the other hand, PEO is subject to oxidative and bacterial attack, chain scission in a shear field, thermal degradation and aging (Bailey et al (4), Shin (5), Mc Gary (6), Bortel and Lamot (7)). The influence of degradation on the flow properties in laminar or turbulent condition has been studied by Barnard and Sellin (8), Ting and Little (9), Fisher and Rodriguez (10), Kenis (11), Gold et al (12)... . In some cases, it is not clear if the loss of the properties of PEO solutions is due to degradation, dissociation or even very slow dissolution of networks (Hinch and Elata (13), Dunlop et al (14), Laufer et al (15), Gadd (16)).

The Influence of Polymer Additives
on Velocity and Temperature Fields
IUTAM Symposium Essen 1984
Editor: B. Gampert

Bailey (4) suggested that the addition of isopropanol to the aqueous solutions of PEO prevents the degradation.

We have measured by quasi-elastic light scattering the hydrodynamic radius of PEO's samples in 2 solvents: water and water-isopropanol mixture (90%/10%) as a function of the age of the solutions. We study the influence of the solvent on the properties of the 2 higher molecular weight samples (WSR 301 and Coagulant) in laminar (non-newtonian viscosity) and turbulent (drag reduction) flow conditions.

Measurement techniques

1) Quasi-elastic light scattering (QELS)

QELS experiments are performed in the homodyne mode, using a photon correlation spectrometer (Layec (17)). The single clipped photocount autocorrelation function $C(\vec{K}, t)$ of the scattered light is provided by a 50ns ATNE correlator. The experimental points are fitted using the method of cumulants (Koppel (18)).

$$C(\vec{K}, t) = A \exp 2\left(\bar{\Gamma} t + \frac{1}{2}\mu_2 t^2 - \frac{1}{6}\mu_3 t^3 + \dots\right) + B \qquad (1)$$

$\vec{K}$ is the scattering wave vector, $\bar{\Gamma}$ is an average characteristic frequency related to the translation diffusion coefficient D or the hydrodynamic radius R_H. If $KR_G < 1$:

$$\bar{\Gamma} = DK^2 = \frac{k_B\ T}{6\ \pi \eta_o\ R_H}\ K^2 \qquad (2)$$

where k_B is the Boltzmann constant, η_o the viscosity of the solvent, T the absolute temperature.
The ratio $\mu_2/\bar{\Gamma}^2$ is a measure of the polydispersity of the sample.

2) Non-newtonian viscosity

The viscosity of the solutions is measured in a Ostwald type capillary viscometer connected to an automatic time recorder and to a pressure regulator, allowing measurements to be performed at shear rates between $500s^{-1}$ and $4000s^{-1}$ (Layec-Raphalen and Wolff 19)).

3) Drag-Reduction

The experimental set-up is a pipe of diameter d = 2 mm and length L = 1.5 m. The driving pressure allowing the fluid to flow from a reservoir through the pipe into a second reservoir is adjustable between 0 and 4 Bar. The maximum Reynolds number is about 14000. For the degradation studies, the solutions flow back through the pipe into the initial reservoir after each pass. The return flow is laminar (R_E < 500) and is postulated not having effect on the degradation.

PEO samples - Preparation of the solutions

Table I : Characteristics of the PEO samples

Denomination	Provider	Molecular weight	Polydispersity	Intrinsic viscosity
SE 70	Toyo Soda	$6.6\ 10^5$ (a)	1.1 (b)	
SE 150	Toyo Soda	$1.2\ 10^6$ (a)	1.1 (b)	
WSR 301	Union Carbide	$2.6\ 10^6$ (c) $2.9\ 10^6$ (d)		1040
Coagulant	Union Carbide	$4.9\ 10^6$ (c) $6.1\ 10^6$ (d)		1700

(a) from low angle light scattering ($\theta = 5^o$) in water at 25^oc

(b) from GPC in water at 25^oc

(c) $[\eta] = 1.03\ \ 10^{-2}\ M^{0.78}$ (Shin (5))

(d) $[\eta] = 5.94\ \ 10^{-2}\ M^{0.657}$ (Kato et al (20))

The solvents are distilled water and a mixture of water (90%) and isopropanol (10%). In order to obtain reproducible results, each solution is mixed directly to its final concentration by sprinkling the required weight of powder at the surface of the solvent and the dissolution is acheived without agitation or heating.

Experimental results

1) QELS

a) in distilled water: The hydrodynamic radius R_H of the molecule is a function of the age of the solution. For WSP 301 and Coagulant, the dimensions just after dissolution are too large to be measured in our experimental set-up (R_H > 2000 A). Experimental results on the evolution of SE 70 dilute solutions have been already published (Layec and Layec-Raphalen (21)). Figure 1 represents the variation of the hydrodynamic radius P_H of SE 70 and SE 150 over a fifty days period. The evolution of R_H with time t may be fitted by a law such as

$$R_H = R_{H\infty} + \beta \exp(-\gamma t) \qquad (3)$$

$R_{H\infty}$ represents the hydrodynamic radius at the final state of the solution. β and γ are constants depending on the concentration. In the table II, we have reported the values of R_H and for SE 70 and SE 150 the values of $(R_H)2$ obtained after 2 days.

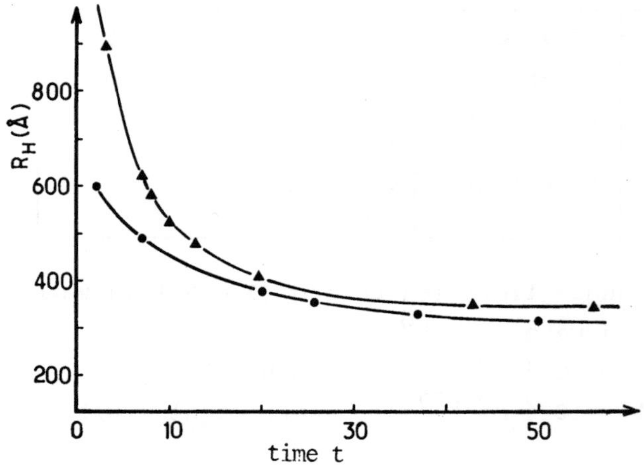

Fig.1 : Hydrodynamic radius of SE 70 (c = 540 ppm) ● and SE 150 (c = 280 ppm) ▲ as a function of the age of the solution. Solvent is water.

Using $R_G/R_H \sim 1.5\text{-}1.6$ (Yamakawa (22)), $[\eta] = 1.03\ 10^{-2}\ M^{0.78}$ (Shin (5)) and the Flory-Fox relation (23) we obtain the molecular weights associated with R_H. The final molecular weights are in good agreement with those given by the providers.

The evolution is compatible with a dissociation of initial entanglements. From the value of $\mu_2/\bar{\Gamma}^2$, we can deduce a high polydispersity. We can notice, in addition, that filtration through 0.8 Millipore filters or centrifugation of fresh solutions of WSR 301 lead to a rapid disentanglement up to the final dimension given in table II.

b) in water isopropanol mixture (90%/10%)

The hydrodynamic radius is constant over a forty days period and very similar to the limiting value in water (Table II). This would tend to prove that the action of isopropanol is a stabilization of individual macromolecules, associations being hindered. Such a behaviour is also observed in methanol, where the polydispersity is lower than in distilled water.

Table II : Hydrodynamic radius of SE 70 and SE 150 as a function of the age of the solutions and related molecular weights in water and water-isopropanol mixture (90%/10%). Concentrations as on Fig.1.

Solvent	water					water-isopropanol mixture	
Polymer	$(R_H)2$	M_2	$(R_H)_\infty$	M_∞	$\mu_2/\bar{\Gamma}^2$	R_H	$\mu_2/\bar{\Gamma}\ 2$
SE 70	600	$2.6\ 10^6$	280	$7\ 10^5$	0.5 to 0.8	280	0.3
SE 150	1100	$7\ \ 10^6$	340	10^6	"	350	"
WSR 301			700	$3.4\ 10^6$	> 0.6		

2) Non-newtonian viscosity

The non-newtonian viscosity of high molecular weight PEO samples is very different in water and water-isopropanol mixture.

In water, we observe a shear thickening of the solutions. This effect is in some cases due to end effects in the capillary (Moan et al (24), Ouibrahim (25), Ambari (26). However, we cannot detect any end effect and we correlate the shear thickening to an association of the macromolecules due to the shear rate. A dimerization process may explain the variations of the reduced specific viscosity with the concentration (Layec-Raphalen and Wolff (27)) and with the shear rate (Fig.2.).

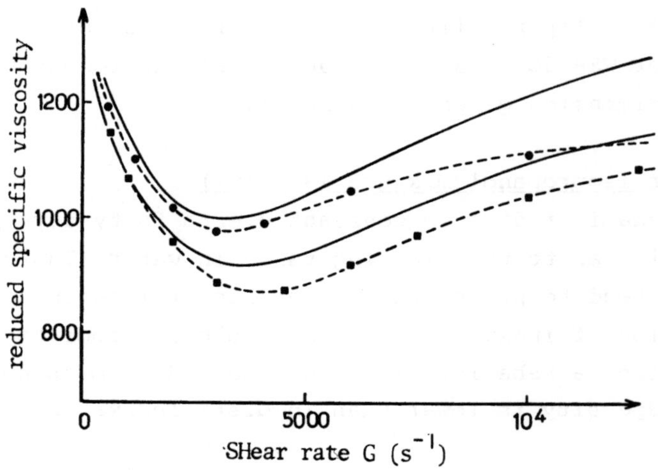

Fig.2 : Reduced specific viscosities of PEO solutions in distilled water versus the shear rate: —— calculated from a dimerization process. --- Experimental ■ c = 60 ppm; ● c = 100 ppm.

On the contrary, in water-isopropanol mixture, the PEO is shear thinning (Fig.3). The variations of the reduced specific viscosity with the concentration (Huggins constant k_H) and with the shear rate (intrinsic viscosity limit $[\eta]_\infty$) are those usually observed for flexible polymers in bad solvents (Wolff (28)).

The values of the limiting viscosity numbers $[\eta]_O$ are similar in the 2 solvents; $[\eta]_O$ = 1040 cm^3/g in water, $[\eta]_O$ = 1100 cm^3/g in water-isopropanol mixture.

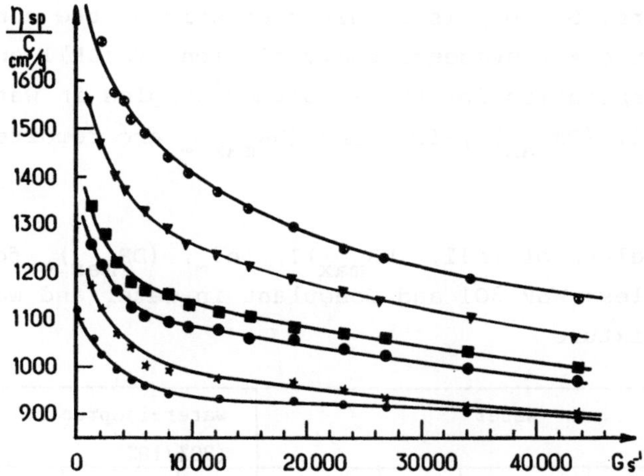

Fig.3 : Reduced specific viscosities of PEO solutions in a Water-isopropanol mixture (90%/10%) versus the shear rate.

★c = 119 ppm; ● c = 299 ppm; ■c = 395 ppm; ▼c = 640 ppm; ● c = 991 ppm; ⊙limiting viscosity number

3) Drag-Reduction effectiveness and resistance to the shear degradation

We measure no DR effectiveness of the 2lower molecular weight samples (SE 70 and SE 150) in our experimental conditions (R_e < 14000).

In order to compare the results obtained with WSR 301 and Coagulant as a function of the solvent and the age of the solutions, we use the Virk's relation:

$$c/DR = (c + [c])/DR_{max} \tag{3}$$

where [c] is the intrinsic concentration which gives $DR_{max}/2$. The values of [c]1 and $(DR_{max})1$ are given in the table III.

The study of the degradation as a function of the applied pressure shows that the asymptotic value DR_∞ of DR when the number of passes n through the pipe increases is independent of the pressure. So, DR_∞ is a characteristic of the concentration. The Figure 4 represents the variation of (DR) 1 and $(DR)_\infty$ with the concentration for the 2 polymer samples in water. The values of [c]1, $(DR_{max})1$, $[c]_\infty$ and $(DR_{max})_\infty$ are reported in table III.

Table III : Values of [c]1, $(DR_{max})1$, $[c]_\infty$, $(DR_{max})_\infty$ for the 2polymer samples: WSR 301 and Coagulant in water and water-isopropanol mixture

Solvent	Water				water-isopropanol mixture (90%/10%)			
Polymer	[c]1	$(DR_{max})1$	$[c]_\infty$	$(DR_{max})_\infty$	$[c]_1$	$(DR_{max})1$	$[c]_\infty$	$(DR_{max})_\infty$
WSR 301	2.3	55	25	31	2	54	35	31
Coagulant	1.7	55	17	39	1.6	54	25	38

As reported in table III, the values of [c] and DR_{max} at the first pass (n = 1) and when $n \to \infty$ are very similar in water and water-isopropanol mixture. Nevertheless, it can be noticed that the rate of degradation is very different in the 2 solvents (fig. 5).

In addition, drag reduction experiments performed on WSR 301 solutions as a function of the age of the solutions show that the initial and limiting values of the effectiveness at n = 1 and $n \to \infty$ are constant. The rate of degradation is a function of the storage (fig. 6).

Non newtonian viscosity measurements performed on the WSR 301 solutions after degradation lead to an intrinsic viscosity $[\eta]_O = 700\ cm^3/g$, i.e. to a molecular weight of about $M = 1.5\ 10^6$.

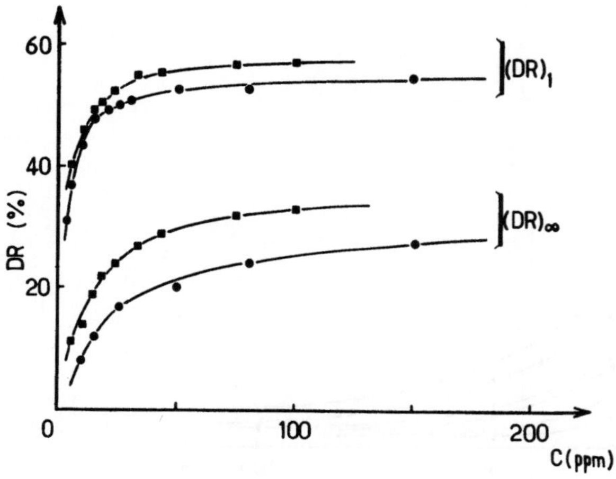

Fig.4 : Drag reduction at the first pass (n = 1) and at n → ∞ versus concentration c. ● WSR 301; ■ Coagulant. Solvent is water.

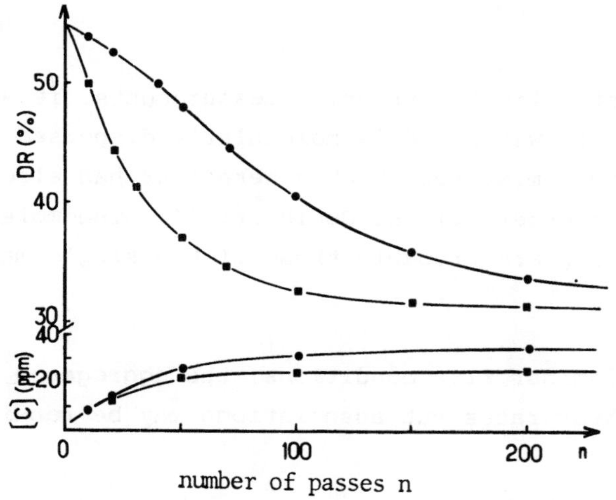

Fig.5 : Maximum drag reduction DR_{max} and intrinsic concentration [c] as a function of the number of passes n. WSR 301. ■ water; ● water-isopropanol mixture.

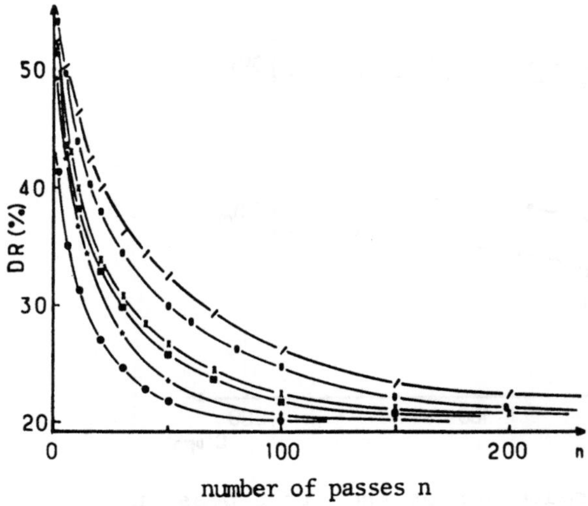

Fig.6 : Influence of the age of the solution on the resistance to the shear degradation. WSR 301. c = 50 ppm. p = 0.8 Bar. x 2 days; ▮ 3 days; / 7 days; ■ 21 days; ★ 30 days; ● 33 days. Solvent is water.

Discussion

From the Quasi-elastic light scattering measurements, it appears that PEO aggregates in water and is molecularly dispersed in alcool or water-alcool mixtures. Such a behaviour has also been observed by Carpenter (2) and Cuniberti (1). The molecular dimensions after aging are probably those of the single macromolecule.

When submitted to laminar flow conditions, the aggregates are destroyed at low shear rates but associations may be rebuilt at higher shear rates.

The initial drag reduction is the same in water and water-isopropanol mixture and the loss of effectiveness is due to the degradation of the macromolecules. The final properties are similar in the 2 solvents, but, the rate of degradation is greater in distilled water.

Bibliography

1. Cuniberti, C.: Evidence for aggregation of Poly(ethylene-oxide). Polymer 16 (1975) 306-307

2. Carpenter, D.K.; Santiago, G.; Hunt, A.H.: Aggregation of polyexyethylene in dilute solution. J. Polym. Sci., Polym. Symp. 44 (1974) 75-92

3. Güner, A.; Güven, O.: Molecular association in aqueous solutions of high molecular weight poly(ethyleneoxide). Makromol.Chem. 179 (1978) 2789-2791

4. Bailey, F.E., Jr.; Koleske, J.V.: Poly(ethylene oxide), Acad.press N.Y. (1976)

5. Shin, H.: PhD Thesis. Reduction of drag in turbulence by dilute polymer solutions. MIT (1965)

6. Mc Gary, C.W.jr.: Degradation of Poly(ethylene oxide)/J. Polym.Sci. XLVI (1960) 51-57

7. Bortel, E.; Lamot, R.: Untersuchung des Abbaus hochmolecularer Polyäthylenoxide. Makromol. Chem. 178 (1977) 2617-2628

8. Barnard, B.J.S.; Sellin, R.H.J.: Degradation of dilute solutions of drag reducing polymers. Nature Phys. Sci. 236 (1972) 12-14

9. Ting, R.Y.; Little, R.C.: Characterization of drag reduction and degradation effects in the turbulent pipe flow of dilute polymer solutions. J. Applied Polym. Sci. 17 (1973) 3345-3356

10. Fisher, D.H.; Rodriguez, F.: Degradation of drag reducing polymers. J. Appl. Polym. Sci 15 (1971) 2975-2985

11. Kenis, P.R.: Turbulent flow friction reduction and hydrodynamic degradation of polysaccharides and synthetic polymers. J. Appl. Polym. Sci. 15 (1971) 607

12. Gold, P.I.; Amar, P.K.; Swaidan, B.E.: Friction reduction degradation in dilute PEO solutions. J. Appl. Polym. Sci. 17 (1973) 333

13. Hinch, E.J.; Elata, C.;: Heterogeneity of dilute polymer solutions. J. N. N. Fl. Mech. 5 (1979) 411-425

14. Dunlop, E.H.; Cox, L.R.: Role of molecular aggregates in liquid drag reduction by polymers. Phys. of Fluids 20-10 (II) (1977) S203

15. Laufer, Z.; Jalink, H.L.; Staverman, A.J.: J. Polym. Sci.; Polym. Chem. Ed. 11 (1973) 3005

16. Gadd, G.E.: Effect of drag reducing additives on vortex

stretching. Nature 217 (1968) 1040-1042

17. Layec Y.: to be published

18. Koppel, D.E.: Analysis of Macromolecular polydispersity in intensity correlation spectroscopy: the method of cumulants. J. Chem. Phys. 57 (1972) 4814-4820

19. Layec-Raphalen, M.N.; Wolff, C.: On the shear thickening behaviour of dilute solutions of chain macromolecules. J. N. N. Fl. Mech. 1 (1976) 159-176

20. Kato, T.; Nakamura, K.; Kawaguchi, M.; Takahashi, A.: Polym. J. 13-11 (1981) 1037

21. Layec, Y.; Layec-Raphalen, M.N.: Instability of dilute poly(ethyleneoxide) solutions. J. Phys. Lett. 44 (1983) L121-L128

22. Yamakawa, H.: Modern theory of polymer solutions. New-York, Harper et Row 1971

23. Flory, P.J.: Principles in polymer chemistry. New York, Cornell Univ. press 1963

24. Chauveteau, G.; Ghoniem, S.; Moan, M.: Effet del'entrée sur les écoulements laminaires de solutions de polymère. C. R. Acad. Sci. (Paris)

25. Ouibrahim, A.; Fruman, D.H.: Characteristics of HPAM dilute polymer solutions in three elongational flow situations. J. N. N. Fl. Mech. 7-4 (1980) 315-332

26. Ambari, A.: Ecoulement des solutions aqueuses diluées de polyoxyethylene dans une fente rectangulaire très allongée en mince paroi. C.R. Acad. Sci. (Paris) 289 (1979) 1-4

27. Layec-Raphalen, M.N.; Wolff, C.: Shear rate dependence of the association of high molecular weight macromolecules in dilute solution. Rheology, Vol. 2, 303-, Ed. G. Astarita et al., Plenum Press, 1980

28. Wolff, C.: Comparaison des courbes de viscosité intrinsèque non-newtonienne obtenues avec différents types de polymères. J. Phys. 32 (1971) 263

The Peculiarities of the Structure of Wall Turbulence and the Mechanism of Drag Reduction by Polymer Additives

L.I. SEDOV, V.A. IOSELEVICH, V.N. PILIPENKO

Institute of Mechanics
Moscow University, Moscow, USSR

The steady attention of investigators to the problem of turbulent drag reduction by polymer additives is connected both with the search of reasonable answers to the question, why insignificant quantities of a polymer (at concentrations about 10^{-6} - 10^{-4} g/sm^3) are capable of changing the structure of wall turbulence efficiently, and also with practical supplements, the number of which is permanently increased.

Numerous experimental data show that the influence of polymer molecules on mean and pulsating characteristics of turbulent flow is localized in a thin turbulent layer. The dilute polymer solution flow may be accompanied by macromolecule adsorption on the wall [1]. In connection with this fact, the following question arose: what molecules are namely responsible for the drag reduction: the molecules which are on the wall, or the molecules moving in the fluid volume. It is known from experimental data that the substitution of the wall material (it means also the change of absorption characteristics) is not resulted in the change of the value of turbulent drag reduction. The detailed analysis of the influence of macromolecular adsorption on Toms effect was made in [2]. The results of experiments always showed, that the turbulent friction is not dependend on the adsorption thickness on the wall.

It is clear, that the flow drag is the result of the interaction between the molecules in the fluid volume with the turbulent disturbances.

The Influence of Polymer Additives
on Velocity and Temperature Fields
IUTAM Symposium Essen 1984
Editor: B. Gampert

In the very dilute polymer solutions the macromolecules practically do not interact with one another, submolecular structures are absent. In the solution at rest long linear molecules are rolled up into slightly asymmetric balls saturated by the solvent. Their sizes are small as compared with characteristic scales of the wall turbulence.

In the steady laminar shearing flow the molecular balls are rotated with periodically changeable angle velocity, then they are deformed by the consequent subjection to extension and compression and achieve the preferred orientation. The more the speed of the shift, the nearer (according to the average time) the direction of the axis of stretching of molecular balls towards the streamlines. Nevertheless even at highly large sliding velocity characteristics of the developed turbulent flow, the macromolecules stretch (extend) insignificantly. The maximum size of a ball in a shifting flow only twice or three times outnumbers its corresponding size in the steady solution and in accordance with the order of magnitude it equals approximately 10^{-5} sm. In this case the presence of polymer molecules resulted in the increase (though it is slight enough at consentrations about 10^{-6} - 10^{-4} g/sm^3) of shear viscosity.

The macromolecules behave themselves in another way in the laminar flow with stretching. In this case even moderate stretching velocity seems to be enough to unfold the molecular ball, extracting the molecule up to its maximum permissible length (limited by the number, geometry and sizes of chemical bond). As is followed from experimental facts and theoretical estimates, the unfolding and flow orientation take place practically immediately when achieving the critical value of the sliding velocity, connected with maximum relaxation time of the macromolecule. The stretches oriented along the direction, and molecules, stretched up to their maximum length, form a system of "threads" in the flow, which intensely depressed the flow with stretching. The fluid, flowing past such molecules, spends the greater part of its power to get over their viscous resistance, the longer their length and the more their number, i.e. the higher the solution concentration. The resistance to the stretching is resulted in the sharp increase of elongated viscosity as compared with Truton's viscosity of Newton's solvent.

The wall turbulent flow is a kind of "mixture", containing both elements of shearing flows, and the elements of flows with stretching. The ejection of the fluid, restrained at the wall, to the outer region of the boundary layer, and which periodically brake the steady flow, and the invasion of the accelerated fluid to the wall region are realized in a form of flooding jets, i.e. flows with longitudinal velocity gradient. With the growth of dynamical speed (or with the increase of Reynold's number) the characteristic stretching velocity in such cases increases and, finally, the reciprocal quantity becomes commensurate with maximum relaxation time of polymer molecules. The last molecules, being turned, prevent from the development of jet flows, i.e. they descend the intensity of ejections, lower their frequency, and, finally, suppress the turbulent transverse transfer, conditioned by jet flow elements; therefore, they descend the friction resistance. Hence, a movable lattice from oriented molecules arises in each of flow elements with stretching in the wall turbulent flow, which prevents from the further development of jet flows, i.e. blow-out.

The given presentation of the mechanism of the interaction of polymer molecules with turbulent flow admits a peculiar theoretical examination. The presence of polymer addition besides the increase of longitudinal viscosity is resulted in the appearance of such rheological solution properties as elastic plasticity, pseudo-plasticity, anisotropy. In [3] the influence of different rheological fluid characteristics on the wall turbulence is theoretically analyzed within the limits of monoharmonic approximation, which affords to take into account turbulent blows-out. Different variants of rheological behaviour were considered. For all that we succeded to show, that the decrease of turbulent friction arose only in mediums, possessing

From the theory of dimension and given notions of the mechanism of friction descent by polymer additives, it is followed, that

$$F = F\left(\eta_\delta\ ,\ \frac{\nu_e}{3\nu}\right)$$

where ν_e - longitudinal kinematic solution viscosity.

Using the known solution [6] of the problem of two-dimensional laminar flow of the model polymer solution with stretching for the description of wall turbulent flows, the following approximate expression may be received

$$\frac{\nu_e}{3\nu} \simeq 1 + 6c[\mu]N \quad \frac{F\cdot\left(\frac{v_*}{v_{*o}}\right)^2 - 1}{F\cdot\left(\frac{v_*}{v_{*o}}\right)^2 + \frac{1}{2}}$$

where c - mass concentration of a polymer, [μ] - characteristic viscosity, N - the number of Kunov's segments in the molecular chain; v_{*o} - critical (onset) value of dynamic velocity, corresponding to the turning of molecular balls, connected with maximum relaxation time of macromolecules θ ,

$$v_{*o} \simeq \frac{\nu}{2\theta}$$

The treatment of experimental data afforded to receive the following system of formulae for the parameters

$$F = 1 - \left[1-F_{min}(\eta_\delta)\right] f_1\left(\frac{c}{c_{opt}}\right) f_2\left(\frac{v_*}{v_{*o}}\right) ,$$

$$c_{opt} \simeq 6.8\cdot 10^2/[\mu]N ,$$

$$f_1 \simeq 1.03\left\{1-\exp\left[-10.4\left(\frac{c}{c_{opt}}\right)/\left(1+2\left(\frac{c}{c_{opt}}\right)^{3/2}\right)\right]\right\} ,$$

$$f_2 \simeq 1 - 1/\left[1+3.5\ \ln\left(\frac{v_*}{v_{*o}}\right)\right] ,\quad F_{min} \simeq 1/(4.08\ \ln \eta_\delta - 15.16)$$

the possibility of increasing the elongated viscosity in flows with stretching. The rest rheological factors practically do not influence on the depression of wall turbulence. The calculation results, carried out for reasonably changed elastic plastic solution model, are in agreement with the known experimental results with dilute polymer solutions.

The given information affords to define more precisely the developed semiempirical turbulent theory of polymer solutions [4,5]. In this case it is necessary to connect the length of the mixing path or of turbulence viscosity with the increased longitudinal solution viscosity in the presence of macromolecules, and the last - with molecular characteristics of a polymer and with its solution concentration.

The existence of the diagram of the maximum drag reduction, practically independent of the form of the used polymer additives and situated between laminar and turbulent curves of solvent resistance, shows, that macromolecules afford to descend only a part (though considerable) of turbulent resistance of the friction. It may be assumed that at maximum drag reduction the polymer additives almost depress the "jet" part of turbulent disturbances, conditioned by the flow elements with longitudinal stretching. The estimates show, that the part of turbulent resistance, conditioned by these disturbances, is about 80%. At maximum value of Toms effect, the part of disturbances, connected with shearing turbulence, is obviously preserved. The maximum value of drag reduction (or minimal value of the parameter in damping multiplier of the formulae for the length of mixing path F

$$\ell = \kappa y[1-\exp(-AF\eta)] ,$$

where y - the distance from the wall, $\eta = yv_*/\nu$, $\kappa \simeq 0.4$, $A \simeq 0.0385$) is connected only with dimensionless conditioned parameter $\eta_\delta = \delta v_*/\nu$, in which δ - the radius of the tube, the semiwidth of the channel or the thickness of the boundary layer.

Here c_{opt} - optimal concentration, corresponding to maximum drag reduction.

The difined semiempirical theory of the turbulence of the diluted polymer solutions gives quite satisfactory results when calculating the flows in tubes and channels.

References

1. Pilipenko, V.N.: The influence of additives on wall turbulent flows. The results in science and technology. VINITI AN USSR, Series Mech. of fluids and gas, 1980, 15, 156-257.

2. Grygoryan, S.S.; Damaskin, B.B.; Ioselevich, V.A., etc: About the mechanism of Toms effect. DAN USSR, 248, N 5 (1979) 1074-1076.

3. Pilipenko, V.N.; Kireyko, G.V.: The influence of viscoelastic properties of fluids on wall turbulence. DAN USSR, 267, N 3 (1982) 578-583.

4. Sedov, L.I.; Vasetskaya, N.G.; Ioselevich, V.A.: About the calculations of turbulent boundary layers with small polymer additives. In collection of articles: Turbulent flows. Science (1974) 205-220.

5. Sedov, L.I.; Ioselevich, V.A.; Pilipenko, V.N.; Vasetskaya, N.G.: Turbulent diffusion and degradation of polymer molecules in a pipe and boundary layer. J. Fluid Mech., 94, N 3 (1979) 561-576.

6. Peterlin, A.: Hydrodynamics of linear macromolecules. Pure and Appl. Chem., N 12 (1966) 563-586.

The Influence of Molecular Mass Distribution, Size and Elasticity of Macromolecules on Friction Reduction Effect

A.P. MATJUKHOV, B.P. MIRONOV, I.A. ANISIMOV

Institute of Thermophysics of the Siberian Branch of the USSR Academy of Sciences, 630090, Novosibirsk 90, Ulitsa Lavrentyeva 1, USSR

The value of hydrodynamic friction drag reduction by means of small additives of water-soluble high-molecular polymers is known to depend on many factors [1,2]. These factors are, first of all, the molecular mass of a polymer, the linearity of a molecular structure with a small number of lateral branchings which determine elasticity; the concentration of a polymer; the range of tangential stresses; the degree of mechanical destruction etc.

The present article considers the size of macromolecules as one of the basic parameters. In so doing, the article mainly uses the materials obtained, in due time, by the authors while studying various characteristics of polymer solutions [3-5]. Some other publications are being analysed, too.

As has already been mentioned, the size of asymmetric polymer molecules in a solution largely determines the value of turbulent friction reduction. It is illustrated in fig. 1. This diagram represents the data for neutral macromolecules of polyethyleneoxide (PEO) within the range of molecular masses (1-5) 10^6 and those of polyacrylic acid, their molecular mass being $M = 1.2\ 10^6$, whose water solutions are polyelectrolytes. Hydrodynamic tests were carried out in an installation with coaxial cylinders described in ref. [4]. The diameter of the external rotating cylinder is 0.11 m, its height being 0.070 m. The speed of rotation of the external cylinder is 17 m/sec.

The efficiency was calculated according to the formula

The Influence of Polymer Additives on Velocity and Temperature Fields
IUTAM Symposium Essen 1984
Editor: B. Gampert

$$f = 100 \cdot (\tau_o - \tau_n)/\tau_o \text{ , \%} \quad (1)$$

where τ_o is the shear stress in a pure liquid, τ_n is the shear stress in a polymer solution.

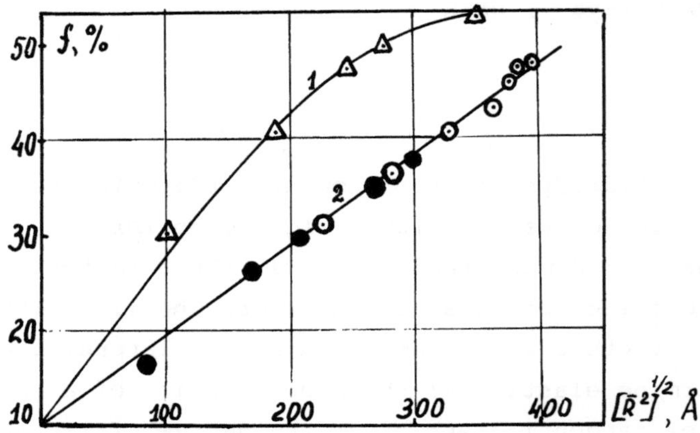

Fig.1. Value of Toms effect vs. size of macromolecules 1-PEO, 2-PAA

The size of PEO macromolecules was determined on the basis of light diffusion, by a "Fica" photogoniodiffusionmeter, according to a standard procedure (fig.1, curve 1). In this case the Ph of the solution was changed and the hydrodynamic activity was measured in the installation with coaxial cylinders. The size of PAA macromolecules R has been found out to vary within the range of 110-300 A. In this case the hydrodynamic activity of the solutions with their flowing between the coaxial cylinders increased from 15 up to 37 %. Ref. [6] reports the similar results with the same PAA and in the same hydrodynamic installation. By using Ph within a wide range, the authors succeeded in obtaining the efficiency of PAA turbulent friction as being reduced to 48%. To determine the RMS radius of rotation $[\bar{R}^2]^{1/2}$ for macromolecules, we used a procedure to calculate it from internal dynamic characteristics. The rotational-relaxation time t_c for a macromolecule segment, measured by the electron paramagnetic resonance (EPR) method was taken as this dynamic characteristic. t_c was measured by a "Varian" E-3 spectrometer (made in the USA). t_c being known, one can determine the size of the segments (according to [7]) considering it as an ellipsoid of rotation whose long axis is equal to the length of the segment and using the Perry form for the friction coefficient of an axisymmetric ellipsoid in a flow. If the geometry of the segment is spherical, then the value of the RMS radius of rotation $[\overline{R}^2]^{1/2}$ can be calculated, by using the friction coefficient, according to the Stokes formula. The size of the RMS radius of inertia of a macromolecule depends on the segment size [4]:

$$[\overline{R}^2]^{1/2} = Z \cdot \alpha \cdot \ell^2/6 \quad . \quad . \quad . \qquad (2)$$

where α is the swelling coefficient, Z is the number of segments and ℓ is the length of a segment. The present method does not allow us to obtain $[\overline{R}^2]^{1/2}$, since α is unknown. Thus we obtain the product $[\overline{R}^2]^{1/2}/\alpha$. However, if one uses the data obtained for the same polymer under similar conditions by the light diffusion method, one can find α and complete the calculation of $[\overline{R}^2]^{1/2}$ for PAA (see ref. [6]). The light points constituting the upper part of curve 2 in fig.1 have been measured by the EPR method. The lower dark points have been measured, accordingly, by the light-diffusion method.

As is seen, the change in the size of macromolecules at the cost of electrostatic tension affects the efficiency of turbulent friction reduction in a linear way, whereas the variation at the cost of molecular-mass increasing for neutral PEO molecules affects the efficiency in a more complicated way.

The data on $[\overline{R}^2]^{1/2}/\alpha = f(Ph)$ obtained in ref. [6] for similar PAA and polymetacrylic acid (PMA) of molecular mass 2.5 10^5 are given in fig.2. Curve 1 in fig.2 corresponds to the PAA data, curve 2 does to the PMA data. Both PMA and PAA solutions were used in the above-described installation with coaxial cylinders. The point of the minimum value of $[\overline{R}^2]_o^{1/2}$ with respect to Ph for these polymers characterizes their isoelectric points which correspond to the case when the elasticity of macromolecules is ideal. The ratio of the experimentally obtained values for $[\overline{R}^2]^{1/2}$ to $[\overline{R}^2]_o^{1/2}$ (the value when the elasticity of macromolecules is ideal) yields a certain measure of rigidity (elasticity) of PAA and PMA macromolecules for various Ph of the medium, on the assumption that the swelling coefficient α = const within the whole range of Ph variation. In this case one may use the data given in fig.2 for both polymers (PAA and PMA).

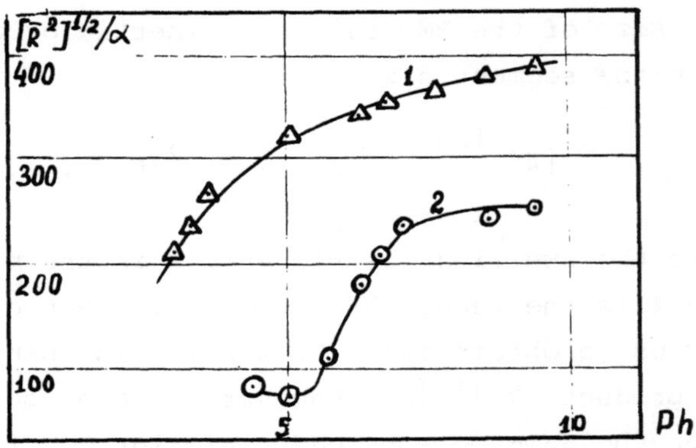

Fig.2. Size of macromolecules vs. Ph.
1-PAA, 2-PMA

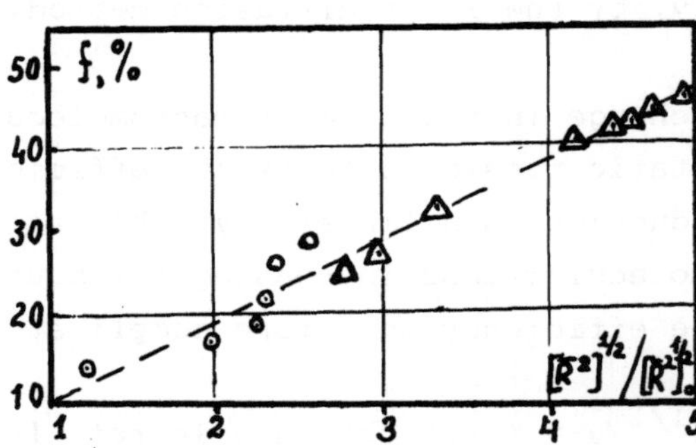

Fig.3. Value of Toms effect vs. elasticity of macromolecular chains

In fig.3 curve 1 represents the dependence for turbulent friction reduction with increasing the rigidity (elasticity) of macromolecules. It has been found out that these data can be generalized by one and the same straight line for two different polymers (PAA and PMA) within the range of the sizes and elasticities considered. It is well-known [9] that the RMS size of molecules can be calculated from the molecular weight and the characteristic viscosity of a polymer, i.e. according to the Flory-Fox formula

$$[\bar{R}^2]^{3/2} = [\eta]_o M/\Phi \quad . \quad . \quad . \qquad (3)$$

where $[\eta]_o$ is the zero-shift viscosity, M is the molecular mass and Φ is the Flory constant.

On this basis one can obtain additional information about the dependence of the friction drag reduction effect on the size of macromolecules. Let us, to this effect, use the results of ref. [4].

The authors of ref. [4] studied the efficiency of a polymer solution of 4 PEO samples with the similar statistical-average molecular mass $M_w = 4.6 \cdot 10^6$, but different molecular-weight distributions. Hydrodynamic tests were performed in the above-described installation with coaxial cylinders.

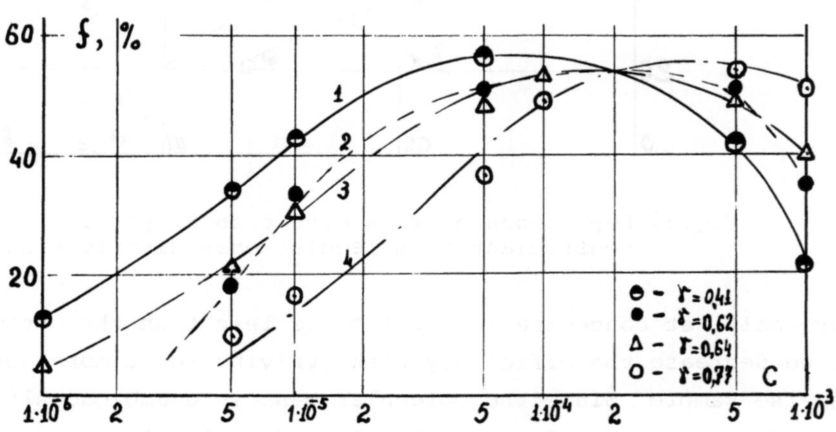

Fig.4. Drag reduction in a flow of PEO solution vs. concentration (the samples have the same molecular mass, but different dispersion).

In fig.4 the obtained data concerning the efficiency of polymer solutions on the concentration of a polymer in a solution are given. The diagram shows that though the efficiency is the same at a concentration optimal for each polymer sample, however, the behaviour of the curves at concentrations different from the optimal one considerably differs. The same polymer samples were tested in an ultracentrifuge. The dispersion coefficients of the molecular-weight distribution $\gamma = \sqrt{Mn/Mw^{-1}}$ were calculated from the molecular-mass distribution. Fig.5 shows the dependences obtained for the efficiency of polymer solutions with various concentrations on the value of the dispersion coefficient. One can see that the hydrodynamic efficiency at a concentration optimal for each polymer sample in a solution does not depend on the dispersion coefficient of the molecular-mass distribution. At concentrations which are less than the optimal one, one obtains an ever growing efficiency with striving for a more monodisperse sample.

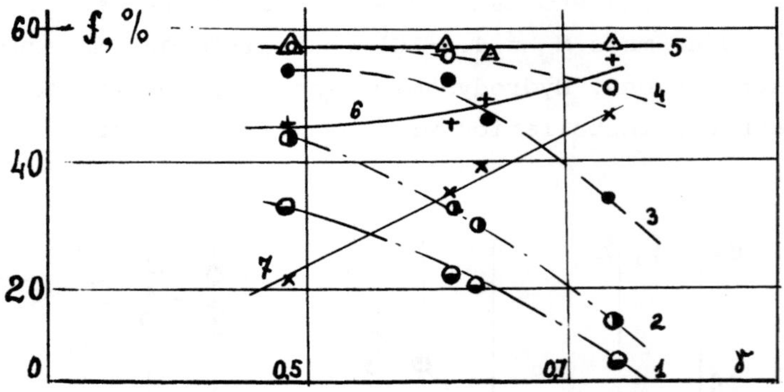

Fig.5. Dependence of Toms effect on dispersion coefficient of molecular-mass distribution

Conversely, at concentrations which are less than the optimal one we decrease the efficiency with striving for a more monodisperse sample. Since the molecular mass is unambiguously related to the size of macromolecules (formula 3), one can assume that the dependences for the dispersion coefficient plotted according to the size of macromolecules will be similar. In ref. [5] the mechanism for destruction of polymer solutions was studied, where one of the basic aims was an attempt to find out what happens to macromolecules during the destruction process. The direct answer to this question can be obtained by following free radicals in a flow, which are bound to appear in case when the destruction is accompanied by the breaking-up of chemical bonds in macromolecules. Such a registration allows us to determine the speed of breaking-ups in polymer chains under various hydrodynamic regimes and, consequently, the kinetics of variation in the molecular mass and the size during the destruction. Since the expected concentration of radicals generated during the mechanical destruction of low-concentration polymer solutions is bound to be very low, the use of the commonly applied methods to follow free radicals (the EPR method, in particular) does not seem to be possible in this case.

Quite recently a new method to investigate processes with the participation of free radicals has been proposed, which is noted for a very high sensitivity [10-11]. The ability of luminol (hydrozide of triaminophtalic acid) to react with free radicals underlies this method. This reaction in the alkaline medium in the presence of oxigene is accompanied by chemiluminescence. In ref. [10] the above method was used to follow free radicals appearing with ionizing irradiation of solid carbohydrates. During the subsequent dissolution of the carbohydrates in the alkaline solution in the presence of oxigene luminescence could be observed. According to the data of refs. [10-11], the intensity of the glow has been established to be proportional to the concentration of radicals existing in the solution within a wide range of irradiation. The high sensitivity of this method has been used in the present work to follow free radicals which may be formed in the course of destruction of polymer solutions with their flowing. With this end in view, an installation was assembled, which will be described below.

The polymer solutions under study were poured between two coaxial cylinders made of organic glass. The diameter of the external rotating cylinder amounted to 40 mm, the diameter of the internal one did to 36 mm. The height of the internal cylinder was 30 mm. By twisting the spring of a dynamometer connected with the internal cylinder, one could determine the tangential stresses on the wall and their variation in the course of destruction. The angular velocity of rotation was measured by way of determining the frequency of the chopping of a light beam, issuing from a LED, with the help of a shutter mounted on the motor axis. The measurements were taken at the temperature 21 $\pm$ 0.5^{o}C. The temperature was controlled by a thermocouple whose junction was on the wall of the internal cylinder. The cylinders were enclosed in a light-tight housing connected, by means of a plastic 10 mm dia. light guide, with a photomultiplier PEM-79 recording the chemiluminescence of the solution. The luminescence was measured under the photon-counting regime. The speed of the photons counting, the temperature, the shear stresses and the speed of rotation were automatically recorded by a digital printing device, in time intervals specified by a programming unit. Air was supplied to the working space through the hollow axis of the internal cylinder to provide the solution with oxigene (while blowing the solution with argon, no chemiluminescence could be observed).

Aqueous WSR polyethyleneoxide solutions of molecular mass $M = 1.7 \cdot 10^{6}$ and various concentrations ($C = 5 \cdot 10^{-4}$, 10^{-3}, $2 \cdot 10^{-3}$ of mass fractions (m.f.)) have been investigated. Additives of KOH ($C = 6 \cdot 10^{-2}$) and luminol ($C = 10^{-6}$) were introduced into the solutions. The efficiency of drag reduction was determined as $(\tau_o - \tau_p)/\tau_o$ (where τ_p are the shear stresses for a polymer solution, τ_o those for water) and did not practically differ in polymer solutions which either contained or did not contain the above-mentioned additives. The alkali and luminol additives were introduced into the solution just before the experiment, by way of injecting 0.5 ml of the solution containing the alkali and the luminol in appropriate concentrations into the volume of the aqueous polymer solution (9.5 mm). In this case the change in the concentration of the polymer was taken into account. After having introduced the solution under study into the space between the cylinders, the external cylinder was set in rotation. The rotation was accompanied by the appearance of the solution glow which ceased when the cylinder stopped. The glow rapidly intensified, as the velocity of the cylinder increased, and reached its maximum when the rotation became constant and then slowly decreased, as the polymer was broken down. Since the glow could be observed during the rotation of the cylinder only, it is natural to attribute it to the appearance of free radicals generated as a result of breaking-ups of polymer molecules in the flow. The destruction of the solution can be characterized by the amount of the specific work A_{sp} spent on the transfer of one unit of the liquid volume:

$$A_{sp} = \frac{1}{h} \int_{0}^{T} \tau(t) \cdot V(t) dt \quad . \quad . \quad . \qquad (4)$$

where h is the width of the gap, T is the time interval, V(t) is the velocity, $\tau(t)$ is the shear stress. The destruction and the measurement of A_{sp} were carried out according to the procedure described in ref. [12], in a thermostatted installation with coaxial cylinders, their peripheral velocity being V = 23 m/s and 17 m/s; the size of the gap between the cylinders h = 2 mm; the diameter of the external rotating cylinder 110 mm; the height of the internal one 53 mm.

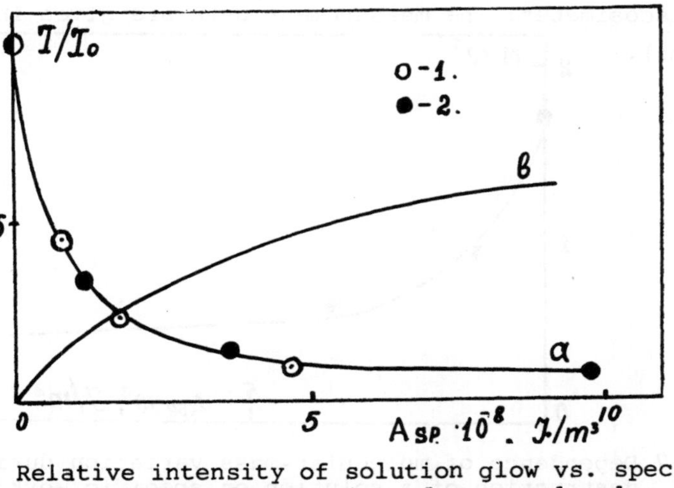

Fig.6. Relative intensity of solution glow vs. specific work spent on destruction of macromolecules

Fig.6 shows the diagram of variation in the ratio J/J_o for broken-down solutions ($C = 10^{-3}$ m.f.) depending on A_{sp}, where J_o was assumed to be the intensity of the glow of the quick-prepared solution (see curve a).

By integrating the considered curve, one can obtain the dependence of the light sum J_Σ on A_{sp}, which is proportional to the number of breaking-ups of molecular bonds per unit volume of the solution (see curve b). Assuming that the mean molecular mass varies proportionally to the number of breaking-ups which a polymer molecule has undergone from the beginning of the destruction, one can write

$$M_i(A_{sp}) = M_o/[J_\Sigma(A_{sp})/N\cdot\alpha_1)+1] \quad . \quad . \qquad (5)$$

where M_i is the mean molecular mass of a broken-down polymer, M_o is that of a quick-prepared one, N is the number of polymer molecules per unit volume of the solution, J_Σ is the light sum, α is the number of photons per radical (full quantum output).

It is interesting to compare the dependence of the molecular-mass variation during the destruction on A_{sp} (see formula (5)) with the experimental data for M obtained by the viscosimetric method. To achieve this, the molecular mass of a broken-down polymer was measured at various A_{sp} by using an Engler capillary viscosimeter. The measurement data are given in fig. 7 (circles).

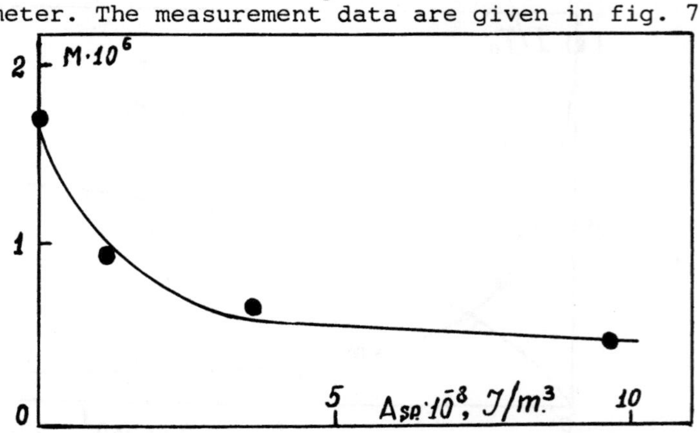

Fig.7 Dependence of molecular-mass variation during destruction of a solution on specific work

In the same figure the solid curve represents the dependence $M = F(A_{sp})$ calculated according to formula (5) and the data in fig. 6 at $\alpha_1 = 0.8\ 10^{-7}$. The figure shows that the experimental and the calculated data are in good agreement and can be described by a single dependence.

Thus the authors of ref. [11] unambiguously established that the process of destruction may be attributed to the breaking-up of macromolecules, which, naturally, leads to a decrease of the average-statistical size: the more considerable the destruction, the less is the size of macromolecules.

References

1. Little, R.C.: Flow properties of polyox solutions. J. and E.C., Fundamentals, vol. 8, 1969, p. 557.

2. Hoyt, J.W.: The effect of additives on fluid friction. Trans. ASME, J. Basic Eng., 1972, series D., p. 258-285.

3. Kobets, G.F.; Matjukhov, A.P.: The influence of physical parameters of polymer solutions on drag reduction in a turbulent flow. IFZh, vol. 25, N 6, 1973.

4. Matjukhov, A.P.: The influence of molecular-weight distribution of a polymer on hydrodynamic drag reduction in a turbulent flow of a solution. IFZh, vol. 33, N 1, 1977.

5. Anisimov, I.A.; Anisimov, O.A.; Mironov, B.P.; Molin, Yu.N.: Chemiluminescent registration of free radicals generated during destruction of aqueous PEO solutions. DAN SSSR, vol. 267, N 5, 1982, p. 1136-1139.

6. Anisimov, I.A.; Mironov, B.P.: Dependence of Toms effect on size of polymer macromolecules. In: "Vliyanie polimernykh dobavok i uprugosti poverkhnosti na pristennuyu turbulentnost", ITF, Novosibirsk, 1978, p. 7-15.

7. Perrin, F., J. Phys. Radium, 1934, vol. 7, N 5, p. 497.

8. Tager, A.A.: Physics and chemistry of polymers. M., "Nauka", 1968, 536p.

9. Tsvetkov, V.N.; Eskin, V.E.; Frenkel, S.D.: Structure of macromolecules in solutions. M., "Nauka", 1964, 719 p.

10. Klimentov, A.S.; Mikhalchenko, G.A.: Chemistry of high energies. 1973, vol. 7, N 6, p. 513-518.

11. Atary, N.A.; Ettiger, K.V., Nature, 1974, Vol. 247, p.193-198.

12. Anisimov, I.A.; Mironov, B.P.: Dependence of destruction of aqueous polyethyleneoxide solution on work of friction forces. In: "Turbulentnye sdvigivye techeniya nenyutonovskikh zhidkostei", Novosibirsk, ITF, 1981, p. 14-38.

Characterization of Dilute Polyacrylamide and Polystyrene Solutions by Means of Porous Media Flow

R. Haas* and W.-M. Kulicke+

* Hoechst AG, 6230 Frankfurt 80, FRG

+ Institut für Technische Chemie
Technische Universität, 3300 Braunschweig, FRG

Summary

Viscoelastic flow behaviour of dilute polymer solutions in porous media is described as a method for characterization of polymer-solvent-temperature systems. Porous media flow tests provide information on the solution state of polymer solutions and the molecular weight of the polymers used. Furthermore, flow-induced and thermally induced degradation effects - frequently observed in polymer solutions - can be characterized by the measurement of viscoelastic effects in flow through porous media. Decrease of molecular weight and changes of the conformation of macromolecules in solution are important parameters in these processes.

1. Introduction

With increasing velocity gradients, dilute solutions of high molecular weight polymers exhibit interesting rheological properties. Depending on the type of flow, shear - or elongational flow, the viscosity may be lowered or raised. Whilst in conventional viscometry in the dilute solution state ($[\eta]c \ll 0.2$) a slight decrease in viscosity is observed, a sudden increase in viscosity, in some cases by more than a power of ten, is noted in elongational flow with the same polymer solutions. The observed change of viscosity can be attributed in both types of flow to drastic deformation of individual molecules. Flow in a porous medium represents a particulary suitable form of elongational flow for highly diluted polymer solution [1, 2], with the aid of which it is possible to determine the viscoelastic polymer effect by simple measurement of the pressure drop $\Delta p/\Delta L$ and superficial velocity $\overline{v}$.

The Influence of Polymer Additives
on Velocity and Temperature Fields
IUTAM Symposium Essen 1984
Editor: B. Gampert

Serveral authors [1 - 3] describe the non-Newtonian flow behaviour of dilute polymer solutions in porous media as an increase of extensional viscosity based on finitely extensible, nonlinear-elastic dumbbells (FENE-dumbbell-theory [4]). Conformity between theory and experiment is partially so good that the relationships found can be used for characterization of polymer solutions. A few significant examples are demonstrated in the following with the aid of selected polymer-solvent-temperature systems.

2. Flow effects of dilute polyacrylamide solutions in porous media

For the present fluid-mechanical tests a homologous series of polyacrylamide (PAAm) samples was used whose mean weights, M_w, of the molar masses were determined by means of a low-angle laser-light-scattering photometer. The PAAm samples exhibit virtually the same molecular weight distribution $M_w/M_n = 2.5$; the intrinsic viscosity $[\eta]$ was determinded in a Zimm-Crothers rotational viscometer, since the polymer solutions are subjected to a very low shear rate in this instrument. The porous media flow tests were carried out with the aid of an instrument such as described in Refs. [1, 2]. Reference is also made to these studies as regards the test procedure and evaluation of the measured data.

Fig. 1 shows the characteristic porous flow behaviour of dilute PAAm solutions with different molar masses M_w. For calculating the Reynolds number, $Re = \bar{v} d \rho / (\eta (1-n))$ with n= porosity and d= sphere diameter of parcticles in Fig. 1a, the measured zero-shear-rate viscosity of the individual PAAm solutions was used which corresponds quite accurately with $\eta_p = \eta_s \cdot (1+[\eta]c)$ for highly diluted solutions.

Fig. 1a shows three typical classification ranges for the individual PAAm solutions:

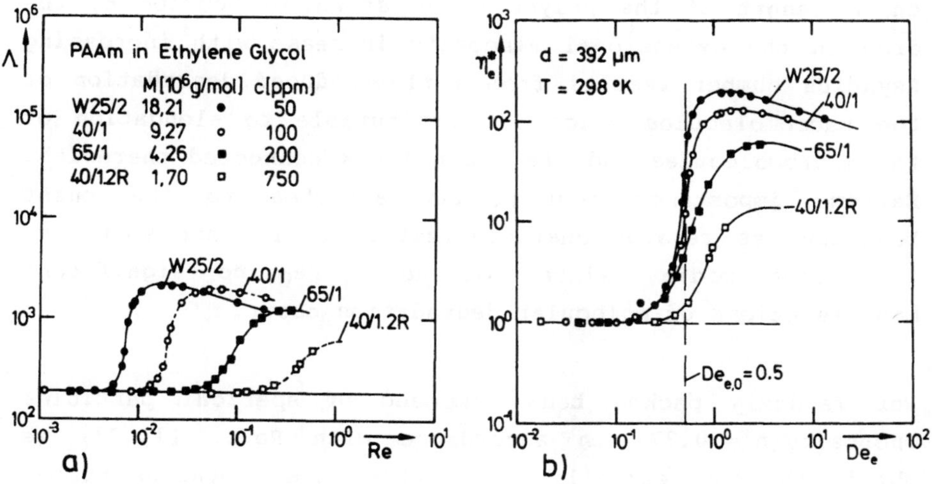

Fig. 1 Porous-media flow behaviour of dilute PAAm solutions with different molecular weights

1. quasi-Newtonian flow in the range of small Reynolds numbers characterized by zero-shear-rate viscosity η_p;

2. onset behaviour with incipient viscoelasticity characterized by the onset Reynolds number, Re_0, of the individual PAAm solutions;

3. the viscoelastic flow range characterized by the maximum value of the increased resistance coefficient, Λ , with a gradual decrease for $Re >> Re_0$.

If modified coefficients, η_e^* and De_e, are used, as suggested in Refs. [1, 2] on the basis of the FENE dumbbell model, Fig. 1b is obtained. The onset behaviour is described by the onset Deborah number, $De_{e,0} = \dot{\varepsilon}_{e,0} \cdot \tau = 0.5$ with $\dot{\varepsilon}_{e,0}$ = critical elongation rate of the porous media flow and τ = relaxation time of the polymer solution, whilst the maximum value of the attainable increase of the extensional viscosity in normalized form only depends on the chain length of the polymers. As shown in Section 6, the drop in the extensional viscosity increase with increasing Reynolds number results from a flow-induced degradation of the macromolecules which is attributable to elongation of the macromolecules and the chain break connected therewith. Extreme importance consequently attaches to the onset behaviour as regards characterization of polymer solutions in porous media, since for $Re \leq Re_0$ no significant manifestations of molecular degradation occur.

For randomly packed beds composed of sperical particles (porosity n = 0.37) in accordance with Refs. [1, 2], we obtain the theoretically predictable onset Reynolds number as

$$Re_{0,th} = \frac{\rho_s \cdot A \cdot k \cdot T \cdot d^2}{2\, k_1 \cdot (\eta_s)^2 \cdot (1-n)(1+[\eta]c) \cdot M_w \cdot [\eta]} \qquad (1)$$

where A = Avogadro's constant, k = Boltzmann's constant and T = absolute temperature. Equation (1) can be obtained by substituting

$$\dot{\varepsilon}_e = k_1 \cdot (1-n) \cdot \eta \cdot Re/(\rho\, d^2) = \text{effective elongation rate} \qquad (2)$$

$$\tau = [\eta] \cdot \eta_s \cdot M_w/(AkT) = \text{relaxation time of the polymer solution} \qquad (3)$$

in the well-known equation

$$De_e = \dot{\varepsilon}_e \cdot \tau = \text{effective Deborah number} \qquad (4)$$

at onset condition (De = 0.5). The elongational flow factor k_1 for randomly packed beds using PAAm samples with $M_w/M_n = 2.5$ is $k_1 = 8$.

3. Characterization of the solution state of PAAm solutions

In Fig. 2 is shown the dependence on concentration of the increase of elongational viscosity for a PAAm sample of $M_w = 18.2 \cdot 10^6$ g/mol in the solvent 0.5 m NaCl ($[\eta] = 2800$ cm^3/g). Surprisingly, a destinct increase of the resistance coeficient, Λ, can still be measured even at c = 3 ppm. Cleary seen here is the sensitivity of flow through porous media to viscoelastic effects which, in contrast to conventional viscometric measurements, still provides evaluable results at extremely small polymer concentrations. In Fig. 2 are shown in addition to the c-values also the values $[\eta]c$ which can be considered as a measure of the solution state of the PAAm solutions. From light-scattering measurements [3] the critical concentration, c*, was calculated for the selected polymer-solvent-temperature system above which overlapping of macromolecules and thus dominant macromolecule interaction must be counted on. In this case, c* = 85 ppm, which corresponds to a value $[\eta]c = 0.24$.

In Fig. 2 it is clearly shown that already at c = 25 ppm or $[\eta]c \geq 0.07$ there is a shift of Re_0 measured in the direction of smaller values which can no longer be correlated with the aid of eq. (1), $Re_{0,th} \sim (1+[\eta]c)^{-1}$. By simple means porous media flow thus furnishes measured data which clearly classify the dilute and semi-dilute solution state. The actions of individual molecules can thus be determined rheologically.

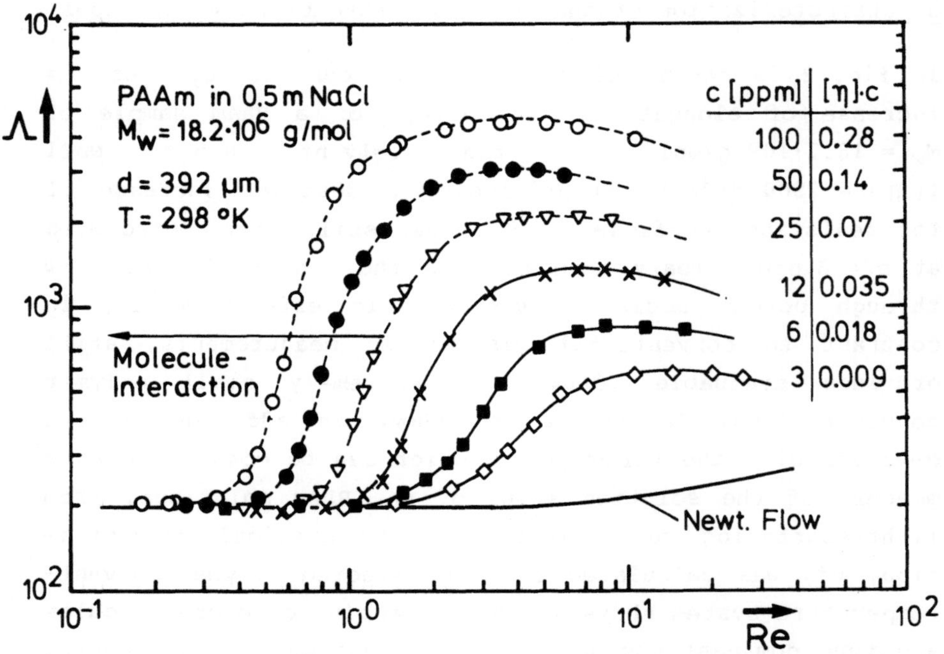

Fig. 2 Characterization of the solution state of PAAm solutions by means of porous media flow.

4. Determination of the relaxation time of polymer solutions

By appropriate rearrangement of eq. (1) to obtain an expression according to eq. (3) it is possible by means of a porous media flow test to determine the relaxation time of a dilute polymer solution. For this purpose Re_0 is determined for the polymer solution to be tested. If $[\eta]$ and the fixed values n, d and k_1 are known, it is possible to determine T without knowing M_w.

As an example of this, Fig. 3 shows the porous flow behaviour of a PAAm sample ($M_w = 7 \cdot 10^6$ g/mol) under solvent conditions near the Θ-condition. Ethylene glycol (η_s = 16.6 mPas, $[\eta]$ = 676 cm^3/g, a = 0.54) and a salt/water/methanol mixture (η_s = 1.64 mPas, $[\eta]$ = 580 cm^3/g, a = 0.5) were used as solvents. The $Re_{0,th}$-values calculated by eq. (1) are marked for both PAAm solutions by vertical arrows. Conformity between theory and experiment is satisfactory. The relaxation times determined by the actual tests are also given in Fig. 3 in comparison with the calculated values with known molecular weight (eq. (3)).

As regards the thermodynamic quality of the solvent, it can be seen from Fig. 3 that under equal porous media flow conditions (n, d, k_1 = const), transition to the non-Newtonian flow state is the steeper the more the polymer-solvent-temperature system approaches the Θ-condition.

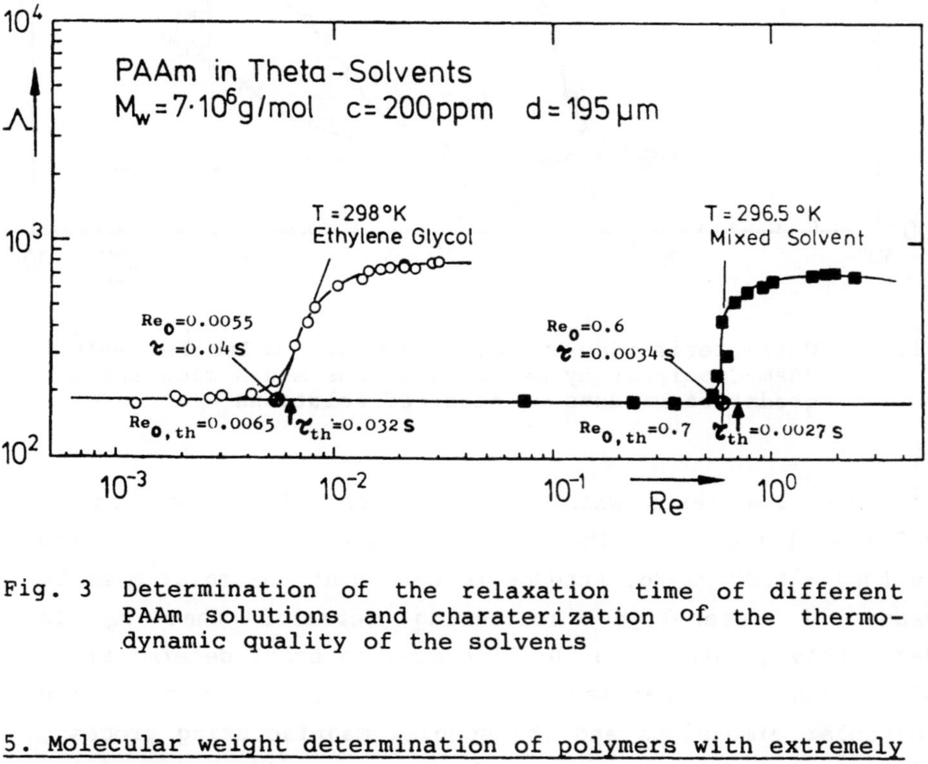

Fig. 3 Determination of the relaxation time of different PAAm solutions and charaterization of the thermodynamic quality of the solvents

5. Molecular weight determination of polymers with extremely high molar masses

As already described by Haas [2] and also by Kulicke and Haas [3] the viscoelastic flow behaviour of dilute polymer solutions in porous media can be used for the molecular weight determination of polymer samples with high molar masses ($M > 10^6$ g/mol). As an example of this an extremely high molecular weight PAAm copolymerisation product is used

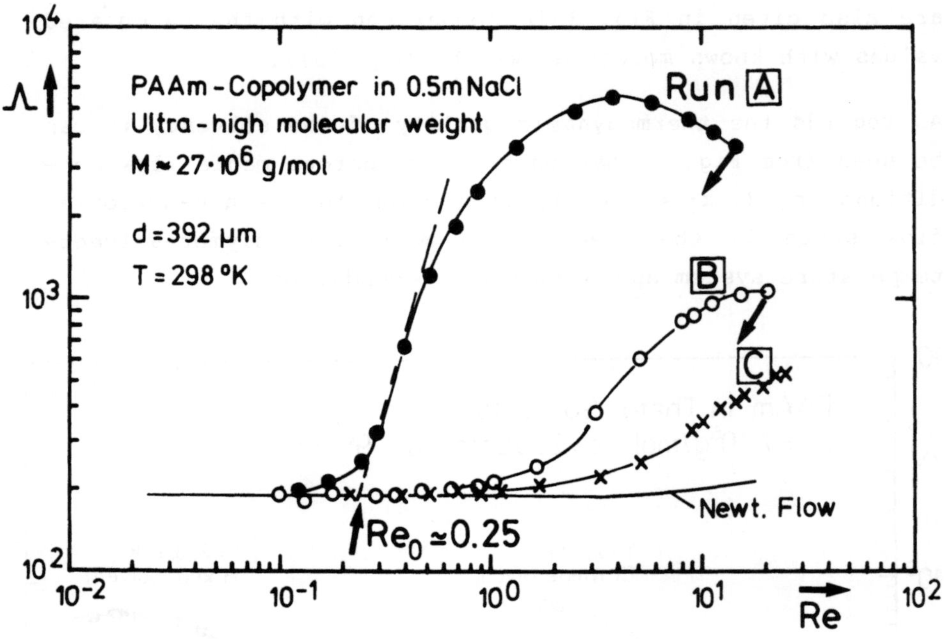

Fig. 4 Characterization of an ultra-high molecular weight PAAm-Copolymer by means of porous media flow and degradation effects of strained solutions

for the flow tests which exhibits nonionic properties in 0.5 m NaCl solution. The intrinsic viscosity of the system is $[\eta] = 5200$ cm^3/g. Attemps to determine the molar mass by means of laser-light scattering revealed that M_w is definitely greater than $20 \cdot 10^6$ g/mol; an exact determination of M_w cannot be carried out at present. Due to the large molecular dimensions and the special manufacturing process, difficulties were encountered during the purification of the polymer solutions which must always precede a light-scattering test (see Ref. [5]). In Fig. 4 the viscoelastic flow behaviour of a 50 ppm solution of such a sample is shown. The rheological behaviour of this polymer solution (curve A) is similar to the results of the PAAm solutions in Figs. 1 - 3. The onset Reynolds number is $Re_O = 0.25$. For calculating M_w, eq. (1) is solved for M_w. If the corresponding measured data are inserted, a molar mass of $M_w = 27 \cdot 10^6$ g/mol is calculated.

6. Characterization of the shear- and elongational flow-induced degradation of macromolecules

The question is frequently discussed as to what flow-induced mechanisms cause a break of the chain of macromolecules in solution and the decrease of the molar mass or decrease of viscosity this engenders. The viscoelastic flow behaviour of dilute polymer solutions reacts very sensitively to changes in molecular weight degradation effects of all kinds. This applies both to mechanically and thermally induced influences and also to aging effects [3] which are frequently observed in polymer solutions.

Flow-induced degradation effects are to be expected in particular if the flow gradients in shear- or elongational flow are so strong that drastic decoiling of the macromolecules (onset effect) is enforced. Fig. 4 also shows how non-Newtonian flow behaviour of a polymer solution with high molecular weight changes in the viscoelastic Re-range if it is subject to systematic flow strain. Curve A corresponds to the unstrained polymer solution. A part of the volume of this solution was forced through the porous medium at Re = 15 in the highly effective elongational flow range and then tested again for viscoelasticity. Curve B shows the result. Repetition of the test with the degraded polymer solution at Re = 20 reinforces this effect (Curve C). With increasing onset Reynolds number and decreasing elongational viscosity maximum a pronounced decrease of the mean molar mass of the macromolecules and thus also a decrease of the solution viscosity results from the viscoelastic flow behaviour of the strained solutions.

This result can be demonstrated quantitatively by the example of polystyrene of high molecular weight (M_w = 23.6·10^6 g/mol) in toluene whose viscoelastic flow behaviour is shown in Fig. 5, curve 1. The polymer concentration, c = 300 ppm, corresponded to the semi-dilute solution state ($[\eta]c$ = 0.674) so that by means of sample taking at certain Reynolds numbers (see curve 1) it was possible to determine the molar mass M_w of the polymer sample with light-scattering technique. Curve 1 shows clearly that the mean mass M_w of the polymer sample decreases with increasing Re and drops in the elongational viscosity maximum from 23.6·10^6 g/mol to 14.3·10^6 g/mol.

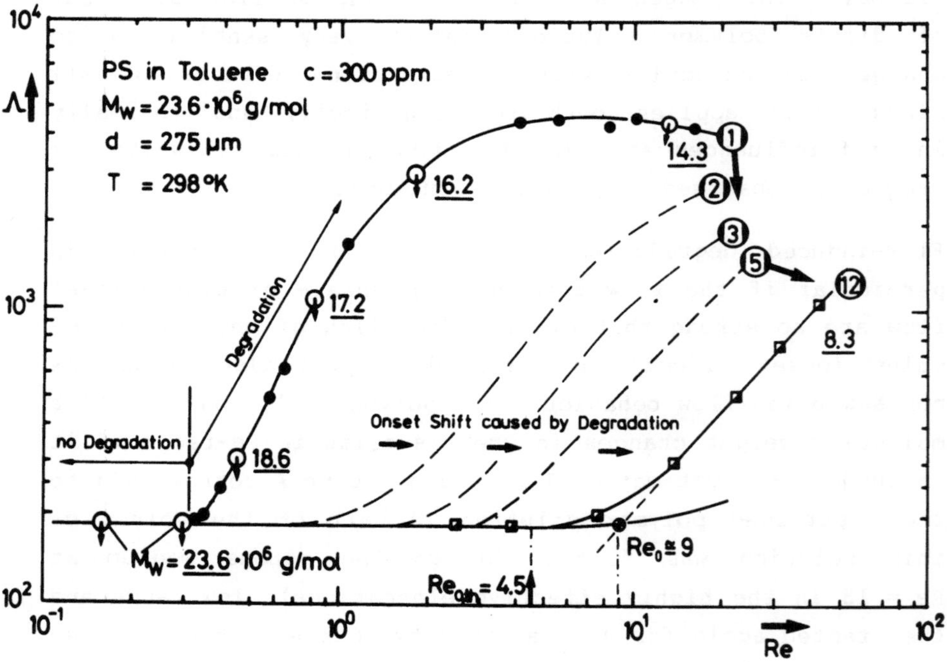

Fig. 5 Degradation effects of a high-molecular-weight polystyrene solution during laminar flow through a porous medium

If the polymer solution, which has been separated in the elongational viscosity maximum, is remeasured in the test section, a further decrease of M_w takes place until finally, after 12 test runs, a molecular weight of $M_w = 8.3 \cdot 10^6$ g/mol is reached. If it is premised, for example, that a wider molecular weight distribution is obtained due to the degradation process and that consequently eq. (1) can be used for calculation of the onset Reynolds number, we have: $Re_{O,th} = 4.5$ for curve 12. The measured value, $Re_O = 9$ indicates that M_w/M_n for the degraded sample from curve 12 is smaller than 2.5.

Typically, however, the molecular weight degradation is not restricted to very high molar masses; degradation occurs in all molecular weight ranges which fulfil the condition, De = 0.5, which is necessary for the case of elongational flow. This may become important, for example, for gel permeation chromatography [6].

References

1. Durst, F.; Haas, R.; Dehnströmungen mit verdünnten Polymerlösungen: Ein theoretisches Modell und seine experimentelle Verifikation. Rheol. Acta 20 (1981) 179 - 192.

2. Haas, R.; Einfluß von Hochpolymeren auf Sickerströmungen. Dissertation (1982) Universität Karlsruhe, FRG.

3. Kulicke, W.-M.; Haas, R.; Flow behaviour of dilute polyacrylamide solutions through porous media. I. Influence of chain length, concentration and thermodynamic quality of the solvent. II. Indirect determination of extremely high molecular weights and some aspects of viscosity decrease over long time intervals. Ind. & Eng. Chemistry, Fundam. (in press) (1984).

4. Bird, R.B.; Hassager, O.; Armstrong, R.C.; Curtiss, C.F.; Dynamics of Polymeric Liquids, Vol. 2, Kinetic Theory. New York: J. Wiley & Sons 1977.

5. Kulicke, W.-M.; Kniewske, R.; Purification of Aqueous and Non-aqueous Polymer Solutions for Light Scattering Measurements and a Comparison of Molecular Weights Determined with a Low-angle and a Wide-angle Photometer. Makromol. Chem., Rapid Commun. 1 (1980) 719 -727.

6. Kulicke, W.-M; Böse, N.; Bestimmung der Molmassenverteilung sowie der Stabilitätsgrenze von Polyacrylamiden unter Benutzung einer kombinierten Ausschlußchromatographie/Kleinwinkel-Laser-Streulichtphotometer Anlage. Coll. & Pol. Sci. 262 (1984) 197 - 207.

The Effect of Grafted and Ungrafted Guargum on Turbulent Flow of Water and on Hydraulic Conductivity of Soil

R. P. SINGH,* J. SINGH,** S. R. DESHMUKH,* and A. KUMAR**

Materials Science Centre* and Department of Agricultural Engineering,** Indian Institute of Technology, Kharagpur 721302, India.

Summary

A detailed investigation has been undertaken to study the turbulent drag reduction caused by commercial guargum, purified guargum and newly synthesized graft copolymers of guargum, in recirculatory turbulent flow of water. The shear stability of these systems has been studied in the same recirculatory turbulent flow system. The relative performance of these systems for the turbulent drag reducing effectiveness and the shear stability has been explained on the basis of molecular behaviour of drag reducing polymers.

A detailed study has been performed on the percolation of water added with guargum and guargum and carboxymethyl cellulose mixture through a soil column. The prediction equation of polymer added water through soil column have been developed. A molecular interpretation has been put forward for both the phenomena discussed in this paper.

Introduction

Guargum (GG) is a very prominent drag reducing polymer for water. However it is biodegradable. It has been observed recently that branched polymers enhance drag reduction effectiveness and shear stability [1]. Hence in the present investigation, polyacrylamide chains have been grafted onto purified guargum in order to enhance the drag reduction effectiveness, shear stability and to decrease the biodegradation. The comparative studies on these properties of purified and grafted guargums have been made in the present investigation.

The percolation is a major loss of water in paddy fields. Recently guargum has been applied in agriculture in fields of drag reduction in sprinkler irrigation and evaporation control

The Influence of Polymer Additives
on Velocity and Temperature Fields
IUTAM Symposium Essen 1984
Editor: B. Gampert

[2]. Some preliminary conducted studies have indicated the scope of examining the possibility of application of guargum for controlling the percolation losses in paddy fields [2]. In the light of above specified facts, the study of the effect of guargum on hydraulic conductivity of soil under treatments of (i) puddling (ii) puddling in the presence of guargum solution (iii) puddling in the presence of guargum and carboxymethyl cellulose (CMC) and guargum and sodiumtetraborate solutions has been undertaken. A mathematical equation relating cumulative volume of percolated water and time has been developed for various treatments controlling percolation losses.

The molecular parameters and solvation of guargum have been correlated with above two aspects of the applications of guargum.

Experimental

A commercial guargum supplied by Hindustan Chemicals and Gums, Bhiwani, India, has been used. To remove the protein, fat and fibrous impurities, a method given by Whitcomb et al. [3] has been adopted to purify the guargum. The various amounts of polyacrylamide chains have been grafted onto guargum main chain using redox initiators by solution polymerisation technique. The details of grafting, optimisation of grafting reaction, separation of graft copolymers and characterisation and analysis of graft copolymers have been given elsewhere [4] .

The drag reduction (DR) percentage has been measured by Turbulent Flow Rheometer supplied by Dr. J.W.Hoyt at Reynolds number of 14000.

The relative viscosity of the solution as a function of time is taken as the criterion for the study of biodegradation.

Percolation studies have been conducted by collecting the percolated water through the standard columns of two types of soils S_1, loamy sand (9.5% clay, 19.0% silt and 71.5% sand) and S_2, sandy loam (12.75% clay, 28.75% silt and 58.5% sand).

The soil columns were prepared to conduct the study with soil passed through 20 mesh sieve. The extreme care was taken while filling the soil so that uniform compaction and natural settlement could be achieved. An electric motor driven stirrer was used to puddle the soil. The cumulative volumes of water percolated at definite time intervals were recorded [5] .

Results and Discussion

Turbulent Drag Reduction by Commercial,Purified and Grafted Guargums; their Shear Stability and Biodegradability: Guargum has been the subject of study by numerous workers in the field of drag reduction [6-9]. The DR data reported in the literature are by the commercial guargum only and have the same features in causing the DR as in the present case (Fig.1). The purified guargum causes higher DR than the commercial one and also the DR percentage vs. composition curve levels off at about 400 ppm which is understandable because in the purified guargum the fibrous and extreneous materials are extracted out leaving pure drag reducing guaran in the residue.

The main factors to enhance the drag reduction are the molecular weight and the extension of polymer chain [9] . However such polymers are prone to molecular chain scission. According to Beuche's midpoint break theory [10], the polymer chain scission does not occur at random but occurs predominantly in the centre of the chain. In the case of linear polymer, this kind of chain scission will reduce the molecular weight to approximately one half and hence drag reduction ability is greatly reduced. In case of branched polymers the reduction of molecular weight is not drastic probably because the shear forces acting on the polymer are distributed among the branches. The occurrence of the reduction in molecular weight will probably correspond to that of the individual branch rather than the whole molecule. It has been shown in literature[1] as well as by Singh et al. [11] that branching definitely enhances the shear stability of the polymers and also increases the drag reduction effectiveness. On the above basis three grades of polyacrylamide grafted guargum, i.e. Gm_1-Gm_3 have been synthesized(Table 1). As is evident from Fig.1 the DR percentage is higher in the case of Gm_3 where the degree of grafting is maximum among the systems studied so far. In case of Gm_2, the DR is lower than the commercial and purified guargum. It is welknown fact that during grafting, due to continuous stirring, the polymer main chain scission occurs [12] and even there is sufficient grafting, the molecular weight of the graft copolymer may be lower than the original backbone molecule. These results indicate that the drag reduction effectiveness increases with the increase in the polyacrylamide content of the graft copolymers.

The shear degradation studies have been performed at 400 ppm concentration in case of commercial guargum, purified guargum and grafted guargums Gm_2 and Gm_3 by recirculation of same solution through the test capillary of the turbulent flow rheometer. One recirculation is depicted as one unit pass number. In all these cases (Fig.2), it has been found that there is no change in DR percentage till 40 pass number. It is apparent that the shear resistance of polymers is not lowered by the grafting.

Recently the various aspects of biodegradability have been discussed by Kopećk and Ulbrich [13]. The biodegradation occurs by microbial and enzymatic degradations. In case of guargum, the biodegradation can occur by both the processes. The Fig.3 depicts the variation of relative viscosity with respect to time for all the studied samples. As commercial guargum is prone to both the type of biodegradation, degrades first followed by purified guargum where the possibility of enzyme being extracted out during the purification is obvious. Hence it is degraded by microbial attack only. However due to the grafting of the stable polyacrylamide chains on the guargum, it has been found that Gm_1-Gm_3 are not at all affected by the biodegradation within the period of observation i.e. 10 days. One point has to be noted that since the chain scission takes place during grafting reaction, the relative viscosity is lower in case of grafted guargum samples. However relative viscosity increases with degree of grafting.

It may be concluded that by grafting polyacrylamide chains onto guargum backbone, highly drag reduction effective, shear stable and biodegradation resistant drag reducing polymers can be developed.

The Effect of Guargum and Guargum and Carboxymethyl Cellulose on Variation of Percolation Volume with Time: The four treatments have been given to soils for the present studies,(i)puddling in presence of water, (ii) puddling in presence of guargum solutions of 0.025%, 0.05%, 0.075% and 0.10% concentrations by weight, (iii) puddling in presence of guargum and sodium tetraborate and (iv) puddling in presence of guargum and carboxymethyl cellulose.

The relation between cumulative volume percolated and time is curvilinear in all the cases which tends to approach to a linearity after 800 minutes. Fig.4 shows a typical result. It is observed that puddling itself has considerable effect in controlling the percolation. This effect is further intensified by addition of guargum. It was also observed that addition of higher doses of guargum donot produce proportional benefit. The slope of the curve represents the volume flow rate of water which tends to a constant value after 800 minutes. Sodium tetraborate is crosslinking agent for guargum hence the amount of guargum required to bring the same reduction in percolation has been found to be less. The addition of sodium tetraborate (STB) increases the effectiveness of guargum in reducing percolation loss. The effect of CMC on the variation of percolation volume with time was studied. The nature of the curves was found to be similar to what was found in the case of GG. The synergism [14] has been found in DR caused by GG and CMC mixture hence an attempt was made to study the effect of mixtures of GG and CMC mixed in proportion of 1:1 and 3:2. The trend was the same as found earlier. The mixture containing higher percentage of GG was found to be more effective in both the soils even though the concentration of the mixture was kept the same. The relationship between the cumulative volume of percolation (V) and time (t) was found to be of the form

$$V = a + bt + c\ e^{dt} \qquad (1)$$

The values of constants were determined by least square method. These equations can be differentiated which will yield the rate of cumulative volume of flow with time. For values of t greater than 800 minutes coefficient 'b' represents the slope of the linear portion.

Guargum in aqueous solution is found to be highly solvated [15]. Hence the flow of water in presence of guargum becomes slow. At the same time due to the obstructed flow of hydrated gum molecules through soil pores, the flow should further become slower. The above mechanism of retarded flow is further vindicated by enhanced reduction in percolation by crosslinked guargum when due to chemical crosslinking a number of guargum molecules make a giant aggregated molecular entity. Naturally due to bigger size, the flow of its aqueous solution as well as its passage through pores become highly obstructed, consequently there is further reduction in hydraulic conductivity and percolation losses.

Acknowledgement

Authors thank the Department of Science and Technology (Govt. of India) for the financial support.

References

1. Little, R.C.; Hansen, R.J.; Hunston, D.L.; Oh-Kil Kim; Patterson, R.L.; Tang, R.Y.: The drag reduction phenomenon. Observed characteristics, improved agents and proposed mechanism. I and EC Fundamentals 14 (1975) 283-296.

2. Sanker, G.: Application of drag reducing polymers in sprinkler irrigation and some other related fields. M.Tech. thesis (1980) Indian Institute of Technology, Kharagpur, India.

3. Whitcomb, P.J.; Gutowski, J.; Howland, W.W.: Rheology of guar solutions. J. Appl. Polymer Sci. 25 (1980) 2815-2827.

4. Singh, R.P.; Purohit, N.K.; Malhotra, J.P.; Chaturvedi, P.N.; Deshmukh, S.R.; Rao, G.R.: DST Project on Development

and application of polymeric drag reducing agents for sprinkler irrigation, slurry transport and aerofoil systems, HCS/DST/855/80 (1st April 1983 - 31st January 1984).

5. Akhilesh Kumar: Application of polymers for controlling percolation losses under ponded conditions. M.Tech.thesis (1982) Indian Institute of Technology,Kharagpur,India.

6. Hoyt, J.W.: The effect of additives on fluid friction. Trans ASME J. Basic Engg. (1972) 258-285.

7. Hoyt, J.W.: Turbulent friction measurements. U.S.Patent 3,327,522 (1967).

8. Elata, C.; Lehren, J. ; Kahanovitz, A.: Turbulent shear flow of polymer solutions. Israel J.Technol. 4 (1966) 87-95.

9. Virk, P.S.: Drag reduction fundamentals. A.I.ChE Journal 21 (1975) 625-656.

10. Beuche, F.: Mechanical degradation of high polymer. J. Appl. Polymer Sci. 4 (1960) 101-106.

11. Singh, R.P.; Majumdar, S.; Reddy, G.V.: Turbulent drag reduction in recirculatory flow of water by grafted polymers. Proc. IUPAC (1982) 726.

12. Singh, Y.P.: Studies on multiphase polymeric systems. Ph.D. thesis (1983) Indian Institute of Technology, Kharagpur, India.

13. Kopeck, J.; Ulbrich, K.: Biodegradation of biomedical polymers. Prog. Polym.Sci. 9 (1983) 1-38.

14. Reddy, G.V.: The studies on turbulent drag reduction and drag reducing polymers. Ph.D. thesis (1983) Indian Institute of Technology, Kharagpur, India.

15. Majumdar, S.; Holay, S.H.; Singh, R.P.: Adiabatic compressibility and solvation of aqueous solutions of drag reducing polymers. Eur.Polym.J. 16 (1980) 1201-1206.

Table 1. The details of grafting reaction*

Sample No.	Acrylamide (Mol/L)	HNO_3 (mMol/L)	Ce(IV) (mMol/L)	Guargum (g/L)	Monomer conversion	$[\eta]$** (mL/g)
Gm_1	0.1408	50	2.5	10	0.52	260.0
Gm_2	0.3052	50	2.5	10	0.64	272.8
Gm_3	0.7004	50	2.5	10	0.74	408.7

* i) Ceric ammonium nitrate / HNO_3 redox initiator system is used in the grafting reaction.

ii) All reactions were carried out under nitrogen atmosphere at 29 ± 1°C. Reaction time was kept three hours

** For viscosity measurements, solutions were prepared in 1 m $NaNO_3$ solution. Measurements were taken at 30°C.

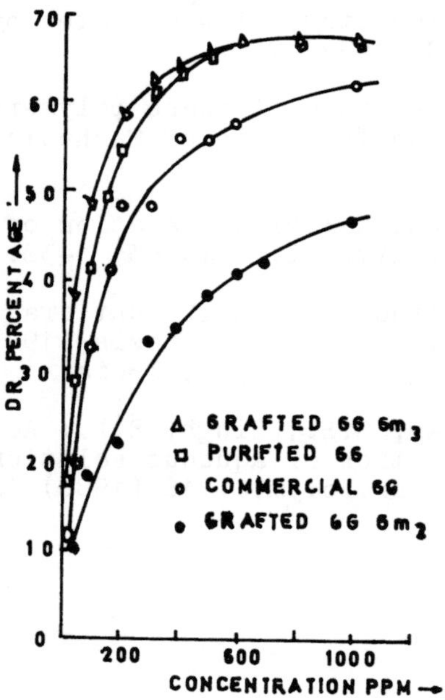

Fig.1. DR percentage vs. concentration in case of commercial guargum, purified guargum, and grafted guargums at Reynolds number 14000.

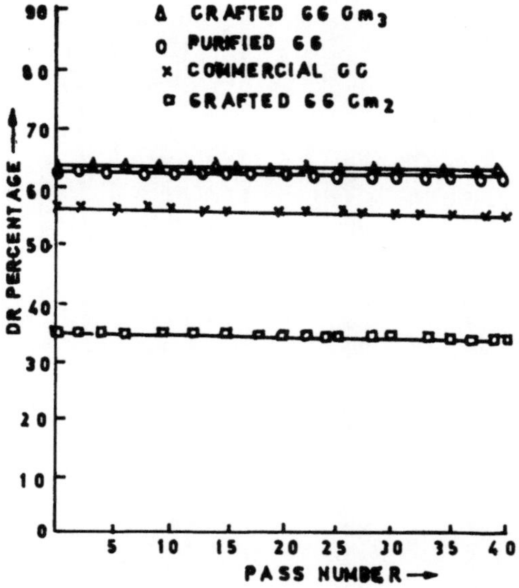

Fig.2. DR percentage vs. pass number to study shear resistance

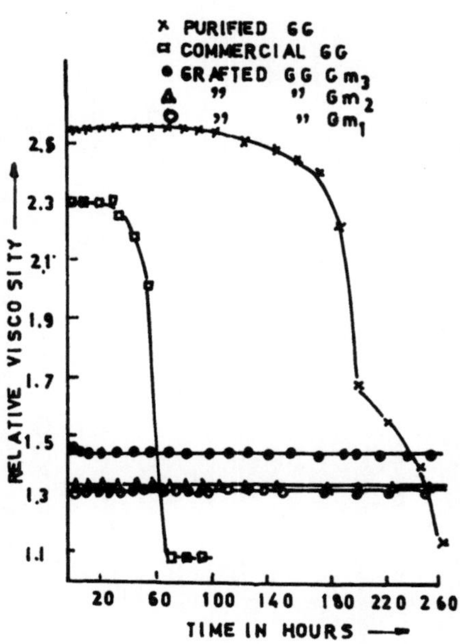

Fig.3. Relative viscosity vs. time to study biodegradation

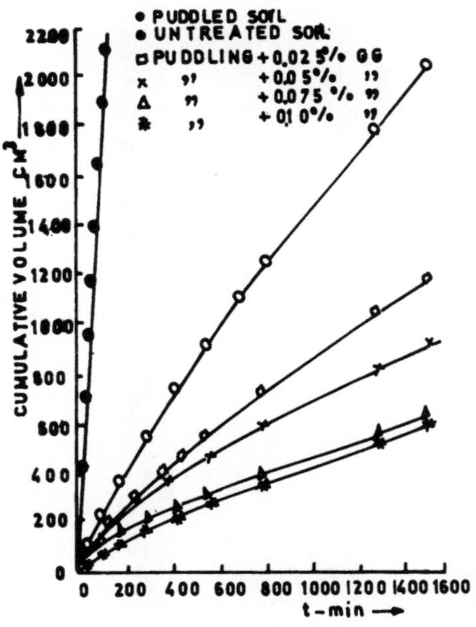

Fig.4. Effect of guargum on variation of percolation volume with time for soil-S_1

Session 5
Molecular Parameters III

Some Fringe Effects of the Flow on the Macromolecules in Dilute Polymer and on Their Non-Newtonian Behaviour

M.N. LAYEC RAPHALEN* and C. WOLFF**

* Laboratoire d'Hydrodynamique Moléculaire
Faculté des Sciences, F 29283 Brest Cédex

** Laboratoire de Mécanique et Rhéologie
Ecole Nationale Supérieure des Industries Textiles
11 rue Alfred Werner, F 68093 Mulhouse-Cédex

Introduction

Diluted solutions of high molecular weight polymers display generally in a laminar simple shear flow, a shearthinning effect, which is a peculiar case of non Newtonian behaviour [Lohmander (1), Wolff (2), Kotaka et al. (3), Yang (4)]. Shearthinning has been explained for rigid macromolecules, by their orientation in the flow direction partly counterbalanced by Brownian motion [Peterlin (5), Saito (6)], while several origins of this behaviour have been proposed for flexible macromolecules: Internal viscosity [Cerf (7), Bazua et al. (8), de Gennes (9)], shear rate dependence of the excluded volume [Fixman (10)], non Hookian restoring forces between the chain segments [Rheinhold et al. (11), Fong et al. (12)], orientation effects [Chikahisa (13)]; an unified theory taking into account some of these parameters has been presented by Bird et al. (14).

The shearthinning behaviour has many applications in the field of lubrication [Wolff et al. (15)], sizing [Schutz (16)], oiling [Schick (17)] etc... But flexible macromolecules diluted solutions may give rise to amazing other hydrodynamic behaviours. In a laminar simple shear flow, under some conditions, they give rise to shearthickening effects [Burow et al. (18), Quadrat et al. (19), Wolff (20), Layec-Raphalen et al. (21), Quibrahim (22)], which may also exist or persist in the semi-dilute regime [Moan et al. (23), Haas et al. (24)]. The origin of this effect remains disputed and will be discussed later.

The Influence of Polymer Additives
on Velocity and Temperature Fields
IUTAM Symposium Essen 1984
Editor: B. Gampert

The applications of shearthickening may be important in enhanced oil recovery and in lubrication.

In turbulent flow conditions, both rigid and flexible macromolecules diluted solutions exhibit drag reduction, which was described for the first time by Toms (25) and often reviewed since [Hoyt (26), Berman (27), Virk (28)]. Many practical applications have been proposed for this effect: increase of the range of water-hose nozzles, better efficiency of fluidized beds and central heating, increase of the flow rate in pipe lines, increase of the speed of torpedos etc... But the fundamental origin(s) of drag reduction is still debated.

Besides theses hydrodynamical behaviours, which may be explained by the resistance to the flow or to conformational changes of the individual single macromolecules, some other effects are a consequence of the change of the number and the size of the particles present in the flow. These fringe effects are degradation, and destruction or formation of aggregates, microgels and associations. These changes of molecular weight may be due to the flow or préexist at rest: as it will be shown in the following sections, they will have a feedback effect on the hydrodynamical and rheological behaviour.

I. Degradation of macromolecules in diluted solutions

The degradation of a macromolecule, i.e. the scission of the chain in two or more parts needs a certain amount of energy to break the bonding between two successive atoms of the skeleton. This energy may come from flow and/or temperature elevation and/or irradiation. The two latter origins have been widely studied and reviewed [Jellinek (30)]. The theorical value of the breakage force per volume unit and per bond has been calculated by Levinthal and Davison (31) and is $F_c \simeq 8.5.\ 10^{-4}$ dyne. Harrington and Zimm (32) have estimated the average hydrodynamic force F acting on a bond: $\langle F \rangle = \dot{\gamma}\ (\eta-\eta_s)/nz$; $\dot{\gamma}$ is the velocity gradient, η and η_s the viscosities of the solution and of the solvent, n the number of macromolecules per unit volume and z the average of the projections of the end to end distance on the direction of the flow. Thus, the degradation of flexible macromolecules should occur at a critical shear rate or shear stress so that $\langle F \rangle = F_c$.

But a quantitive estimation of the forces acting really on a given point of the macromolecule in a flow is difficult because it is necessary to know simultaneously the conformation of the molecule and the flow field in its vicinity. Unfortunately, concerning the conformation, very few results are available [Cottrell et al. (33), Smith et al. (34)]. For rigid molecules, supplementary H bonds lead to a much higher bonding energy so that rigid macromolecules are more difficult to degrade; the case of semi rigid or wormlike polymers as, for example, polyelectrolytes, should be intermediate.

We will concentrate here on mechanical, i.e. flow degradation in laminar flows. Two types of experimental arrangements have been employed. North et al. (35) and Yu et al. (36) have shown that mechanical degradation can occur at high shear rate in concentric cylinders viscometers, i.e. in well defined simple shear flows and in weak flows [Tanner (37)] conditions. However the situation is complicated by the effect of the duration of the shearing which looks as an important parameter and suggests that the degradation could be a fatigue process. North et al. have also proved that the scission takes place at the mid-point of the chain. A multiple pass viscometric arrangement has been employed by Culter et al. (38, 39) who have measured the intrinsic viscosity after a fixed number of passes under a given external pressure and by Ghoniem (40) who has measured continuously the flow rate under a given external pressure. They have found that the degradation takes place in the entry flow region of the tube or/and in the stretching region outside the capillary entrance, and is thus due to extensional flow. These results are apparently in contradiction with the rotating cylinder viscometer's measurements. Ghoniem et al. (41) have obtained a good correlation between the values of the end pressure losses and the appearance of secondary flows in the upstream reservoir; moreover, under a given external pressure the size of the vortex in the secondary flows decreases sharply when the number of passes increases (figure 1) indicating a drastic loss of the elastic properties of the solution. The loss of viscosity measured by: $\Delta\eta = (\eta_i - \eta_F)/\eta_i$ where η_i and η_F are the initial and final (i.e. the plateau value after an infinite number of passes) viscosities of the solution, increases linearly with the shear rate $\dot{\gamma}_i$ corresponding to the first pass (figure 2). This indicates the existence, at each shear rate above the critical shear rate, of a limiting molecular weight [Ram (42)]. In one of the experiments, Harrington et al. (32) have also employed a capillary tube, but of very small diameter (50 m) in order to achieve very high shear rates (10^6 s^{-1}). The capillary entrance was protected by a sintered glass of which porosity was in order of 10 μm. Thus the part of extensional flow of this arrangement is not well defined. This may explain the discrepencies between the theorical value (8.510^{-4} dyne) and the value of the breakage forces calculated by these authors on the basis of the experimental critical shear rate within the capillary, i.e. $4\ 10^{-6}$ dyne in a good solvent and $4\ 10^{-7}$ in a poor solvent. However, the qualitative aspect of these results is very interesting: it shows that degradation is easier in a poor solvent because of the weak interactions and also that a semi rigid macromolecule is more difficult to break than a flexible one.

More quantitative degradation experiments seem necessary in order to understand the role of the polydispersity and of the concentration of the sample, of the fatigue process and of the quality of the solvent. Attention should be also given to ensure that the solutions submitted to degradation are constituted only by individual macromolecules without microgels or aggregates. From the results obtained in laminar well defined flows, it is possible to predict the degradation behaviour of diluted polymer solutions in turbulent flows. As the energy developped in such flows is much larger than in laminar flows, the degradation should be enhanced, and the efficiency of the turbulent drag reduction will decrease with the increase of the number of passes [Gampert et al. (43)]. A very large number of degradation experiments have been made in turbulent flow, specially

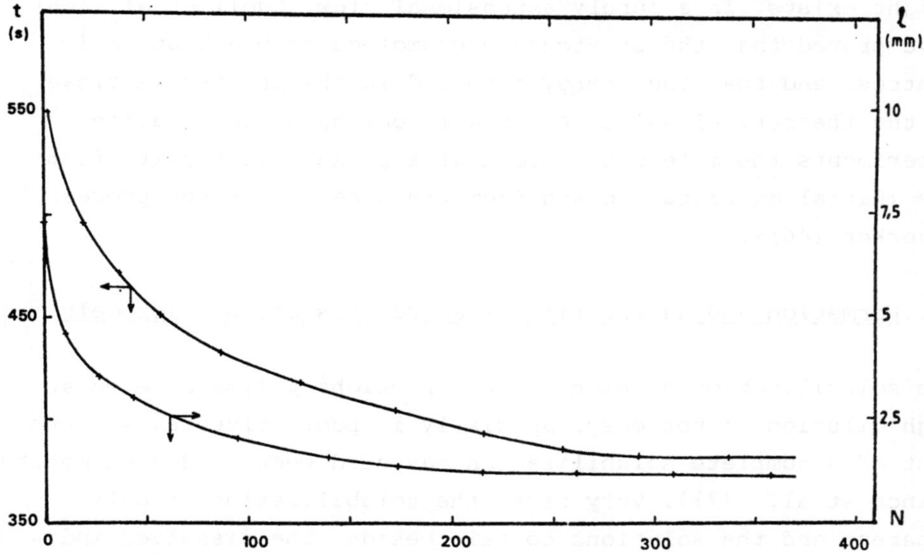

Fig. 1. Flowtime t, and size l of the vortex in the upstream reservoir as a function of the number of passes N. Non ionic polyacrylamide; M_w = 4.5 10^6; concentration: 2000 ppm in water; diameter of the capillary: 0.3 mm; total pressure loss: 9 10^4 Pa.

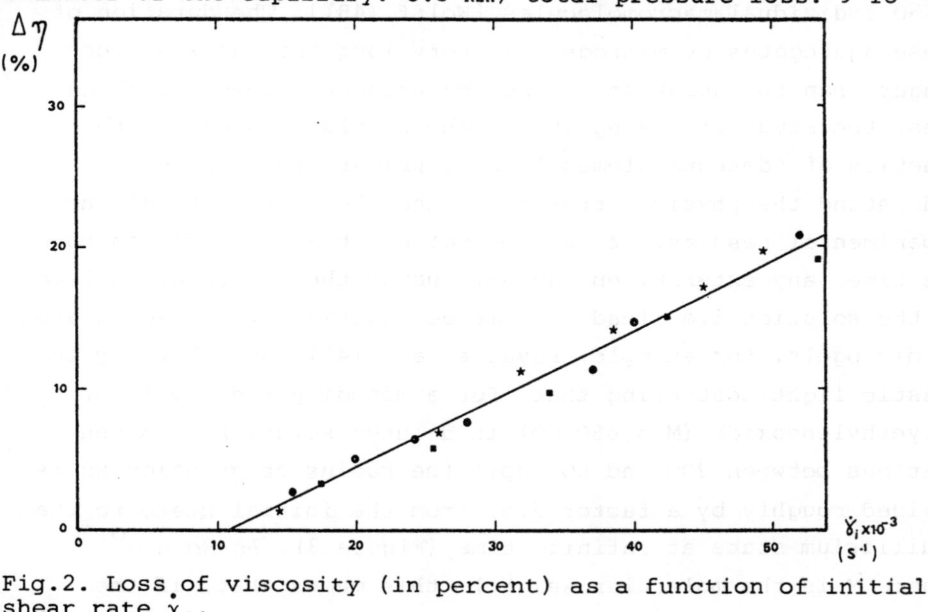

Fig.2. Loss of viscosity (in percent) as a function of initial shear rate $\dot{\gamma}_i$.

with stirring devices [see for example Nakano et al.(44)]; they show clearly that the high molecular weight fraction is primary affected by the degradation, and that a limiting molecular weight exists. In a purely extensional flow, Odell et al. (45) have proved that the stretched macromolecules break at their centres, and that the energy involved in the process is close to the theoretical value. As in all continuous degradation experiments the molecular weight at a given time results from the initial distribution and from the kinetics of the process [Maerker (46)].

II. Formation and destruction of aggregates and/or microgels

The solubilization of high molecular weight polymers, even at high dilution is not easy, specially in poor solvents. Achievement of a complete solubilization may need weeks and even months [Hinch et al. (47)]. Very often the solubilization is only apparent and the solutions contain besides the dissolved individual macromolecules, non visible aggregates with up to several tens of macromolecules, and even optically visible microgels; calculation has shown that freshly diluted solutions of Polyox WSR 301 contain aggregates which are constituted, in average by 50 individual macromolecules [Wolff (48)]. The duration of these aggregates or microgels is very long and, anyway, much longer than the duration of the measurement process involved. Thus, the study of the ageing of these solutions and of the kinetics of "desentanglement" is of primary importance to understand the physical properties and the scattering of the experimental results. It must be noticed that in addition to the time, any external energy may change the "aggregation level" of the solution i.e. lead to some degradation of the aggregates or microgels. For example, Layec et al. (49) have shown by inelastic light scattering that, for a monodisperse sample of polyethyleneoxide (M = 660000) in diluted solutions (concentrations between 200 and 500 ppm) the radius of gyration Rg is divided roughly by a factor 2.5, from the initial state to the equilibrium state at infinite time (Figure 3). As $Rg \propto M^{0.5}$ (where M is the molecular weight), this means that, if the conformation remains the same, the average molecular weight of the particle has been reduced by a factor 6; this result is

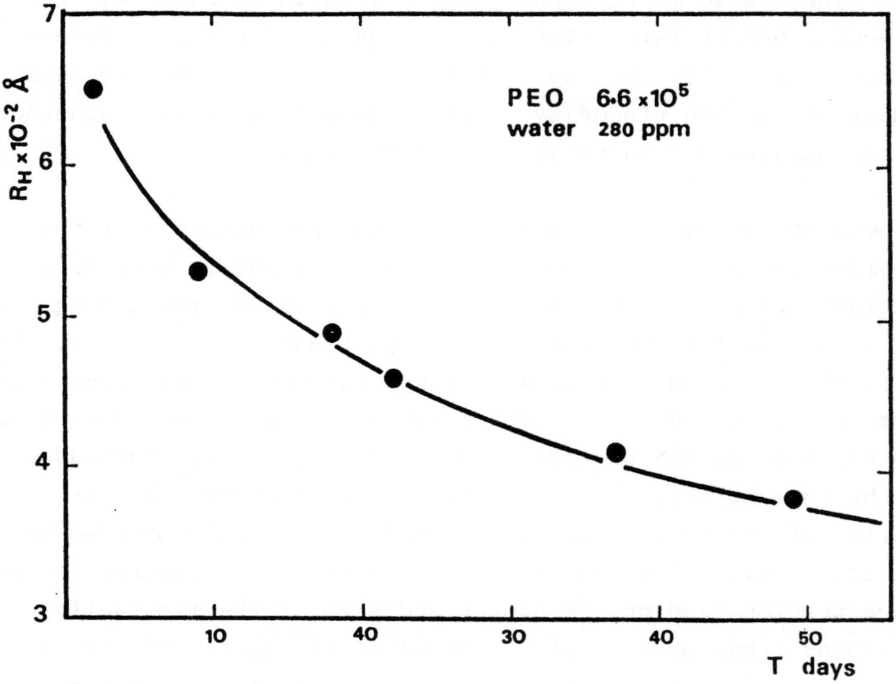

Fig.3. Variation of the radius R_H with ageing of solution. Polyethyleneoxide M_w = 660000; concentration c = 280 ppm.

confirmed by the reduction of the scattered intensity. If this change of radius of gyration would come from a conformational change, for example from an ellipsoid of revolution into a sphere, at constant volume, the aspect ratio of the ellipsoid of revolution should be 9: such an anisotropy is unlikely at rest and should be very easy to detect by light scattering or by non Newtonian viscosity measurements. The same authors have also shown that centrifugation or filtration of the solution accelerates the disentanglement process. On the other hand, Kulicke et al. (50, 51) have studied non ionic polyacrylamide copolymers by viscosity and light scattering. They have observed only a 18 % reduction of the radius of gyration within the first 35 days of the experiment; they do not attribute this reduction to any degradation or disentanglement but to a conformational change. These results will be presented and discussed in detail by their authors in this symposium. In contradiction with these results, Shyluk et al. (52) concluded from viscometric measurements that a drastic decrease of the molecular weight takes place during ageing of polyacrylamide; Narkis et al. (53), on the same type of polymer, observe also a decrease of the viscosity, accompagnied by a larger reduction of the Huggins hydrodynamic interaction constant.

As pointed out by Hinch et al. (47), the consequences of the progressive disappearance of the aggregates and/or microgels initially present in the solution are multiple: with ageing, the solutions lose their elastic properties [Brennan et al. (54), James (55)]. In the presence of large particles, the transition from laminar to turbulent regime is delayed to higher values of the Reynolds number and the drag reduction becomes independent of the concentration; on the contrary, with individual (or nearly so) macromolecules, the transition is well observed at the usual critical value, and the drag reduction depends on the concentration (and on the number and size of the eventually remaining large particles) [Stenberg et al. (56)]. In the same way, ageing diminishes the drag in flow around spheres [White (57), Kit et al. (58)] or cylinders and through porous media [Naudascher et al. (59)] at low Reynolds numbers. In laminar simple shear viscometric experiments, it is unlikely that the energy of the flow will be sufficient to disentangle the large particles; if it does, the corresponding flow rate or pressure loss measurements would become erratic, or, at least, time dependent [Peterlin et al. (60)]. If it does not and if the measurement time is sufficiently short in comparison to the characteristic time of the disentanglement process, the concentration dependence of viscosity should be linear and the solution should behave like any polydisperse polymer solution, displaying, because of the average high molecular weight and the polydispersity [Wolff (61)], an important shearthinning at relatively low shear rate.

III. Formation of associations in flow

Experiments in dilute suspensions demonstrated that, in the process of collision in the flow of large particles, they form temporary doublets. The kinetics of the formation and release of these doublets have been established [Goldsmith and Mason (62)]. Higher degrees of associations have also been observed and studied with methacrylate polymers or copolymers [Wagstaff et al. (63), Chaffey et al. (64)]. But these results cannot be transposed easily to polymer solutions. Indeed, the movement and the conformation of the macromolecules depend on the Brownian motion, on the draining of the solvent and on the thermodynamical interactions between the segments of the chain and the solvent. However, at least in poor and theta solvents where the chains are "self-knoten" and where the disentanglement times are relatively large [Brochard et al. (65)] the existence of temporary associations is likely. Because of their short duration with respect to the macroscopic time of many experiments, the consequences of these associations on the flow properties should be different from those observed with aggregates or microgels. In peculiar the number of "dipolymers" or "multipolymers" present in the solution will not depend on the time but on the concentration and, eventually on the flow conditions. The effect of the associations on the viscosity at a given shear rate has been calculated [Wolff et al. (66)] in the case of a doublet formation and also of higher degrees of association, assuming a mass action law for the balance between single and multiparticles. The most striking result is that the reduced viscosity $\eta_{sp}/c = (\eta-\eta_s)/\eta_s c$ where η_s and η are the viscosities of the solvent and of the solution of concentration c do nomore obey to the Huggins law:

$$\eta_{sp}/c = [\eta] + k_H[\eta]^2 c + O(c^2)$$

where k_H is the Huggins constant. In the presence of associations, the concentration dependence of η_{sp}/c becomes non linear (Figure 4). This theoretical result agrees with several experiments [Quadrat et al. (19), Layec-Raphalen.et al. (21)] in which the absence of artefacts, like absorption and draining has been carefully checked. By superimposing the theoretical and experimental curves, it is possible to obtain the intrinsic viscosity $[\eta]_1$ of the single particle of molecular weight M_1, the ratio of the association constant K to M_1 and the ratio r of the intrinsic viscosity of the doublet to that of the single macromolecule (Figure 5). It must be noticed that the curvature in the η_{sp}/c vs c plot appears clearly only at low concentration (i.e. typically at viscosity ratio $\eta/\eta_s \leqslant 1.2$) so that the effect may be hidden, leading then to the erroneous conclusion of an "individual" particle behaviour.

By this method, the shear rate dependence of K/M_1 and r have been obtained [Layec-Raphalen et al. (67)]; the value of K increases drastically with the shear rate, more than it would be expected from a simple collision theory. By introducing this experimental law together with the shear rate dependence of the intrinsic viscosity $[\eta]_1$ of the individual particle, in the relation giving the previous non linear concentration dependence of η_{sp}/c , it has been shown [Wolff (68)] that the shear rate dependence of the reduced viscosity of the solution containing associations is rather complex: when the shear rate increases, the reduced viscosity decreases first, reaches then a minimum followed by a shearthickening behaviour, and, finally a plateau value or even a second shearthinning at high shear rates (figure 6). The abcissa of the minimum depends on the concentration, as observed experimentally [Layec-Raphalen et al. (21)]. This type of behaviour has been observed with several types of high molecular weight flexible polymer dilute solutions in poor solvents. In the case of polyethyleneoxide of high molecular weight (for example Union Carbide Polyox WSR 301), by application of the previous method, the shear rate dependence of the intrinsic viscosity $[\eta]_1$ of the individual particle has been obtained (figure 7); the behaviour is purely shearthinning. However, its characteristics, i.e. the initial decrease and the asymptostic behaviour are not typical of the molecular weight given by the producer ($M_1 \simeq 4.10^6$) but of a much larger one, corresponding to an aggregated particle which preexists in the

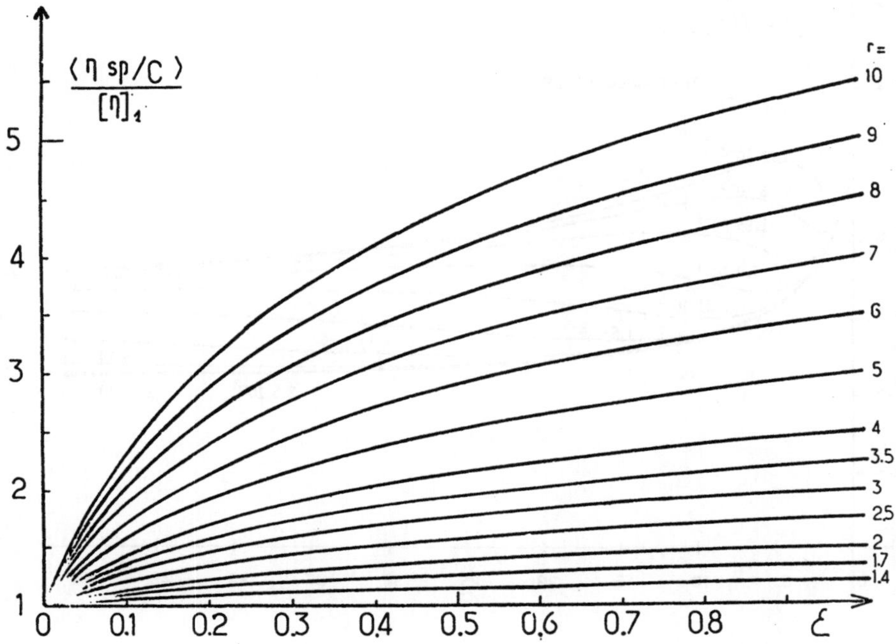

Fig. 4. Reduced viscosity $\langle\eta_{sp}/c\rangle$ of the solution of single and dipolymers to the intrinsic viscosity $[\eta]_1$ of the single polymer versus $\varepsilon = Kc/M_1$ for different values of r.

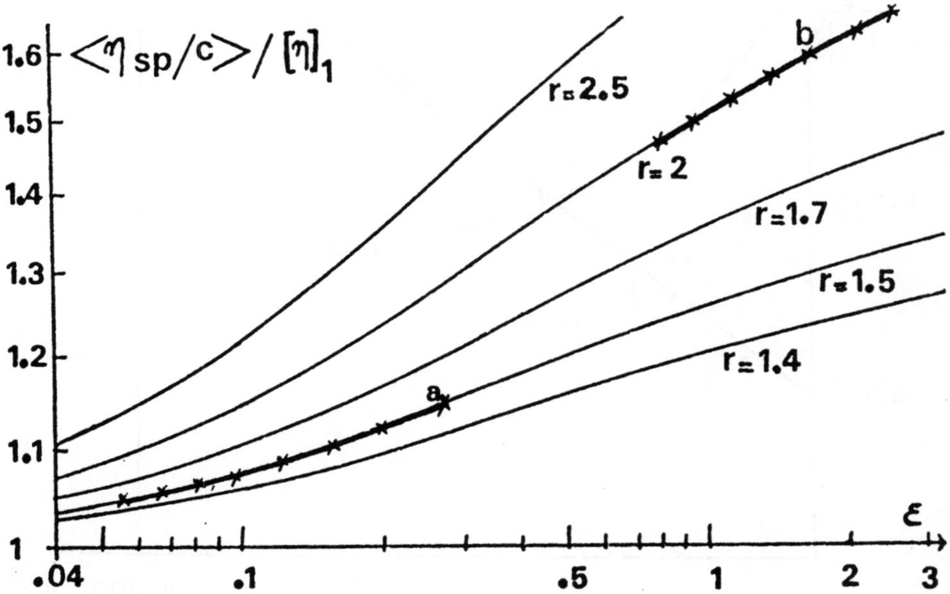

Fig.5. Determination of $[\eta]_1$, K/M_1 and r by superimposition of experimental curves log $\eta_{sp}/c = f(\log c)$ over the theoritical network. Polystyrenes in decalin.

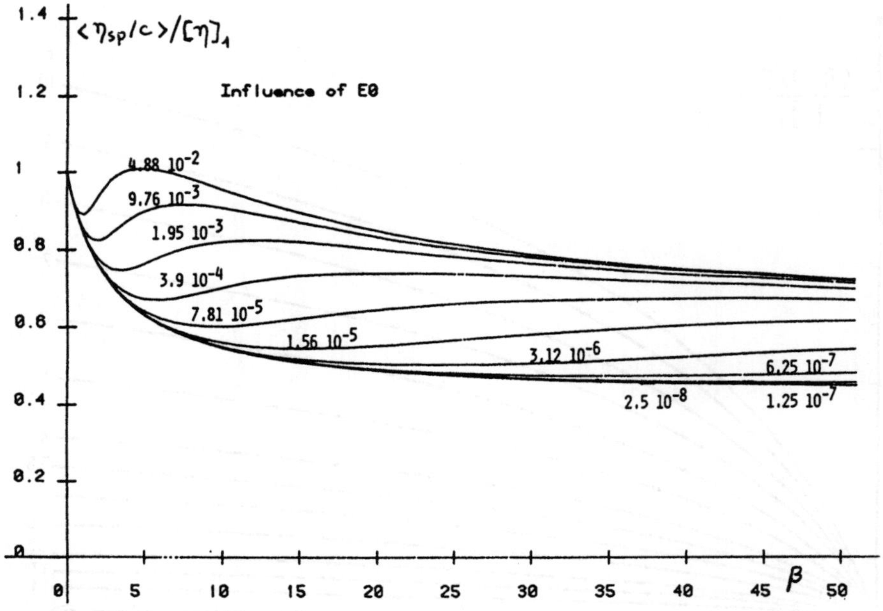

Fig.6. Example of variation of $<\eta_{sp}/c>/[\eta]_1$ with the Deborah number $\beta = M_1[\eta]_1\dot{\gamma}\eta_s/RT$ for different values of the reduced concentration $\varepsilon_o = K_o c/M_1$

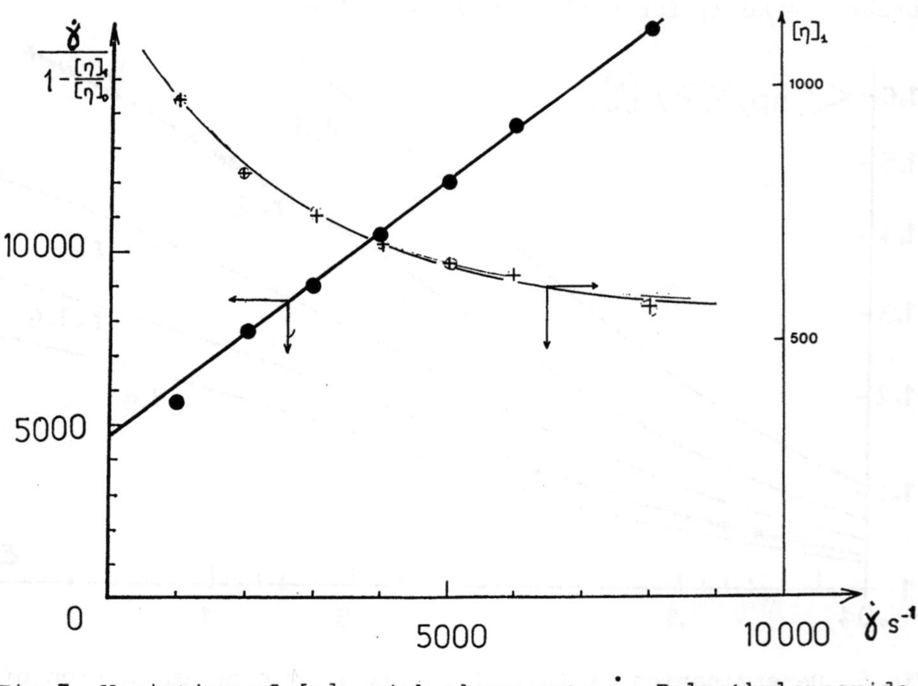

Fig.7. Variation of $[\eta]_1$ with shear rate $\dot{\gamma}$. Polyethyleneoxide WSR 301 in water

solution as shown and discussed above. Thus the flow properties of Polyox WSR 301 in laminar und turbulent flow may be explained nearly quantitatively by the existence, at rest, of the hugh aggregateds (which break up slowly with time, or rapidly in strong flows) able to form themselves doublets associations of short time of life. Nevertheless, more experimental and theoretical work is necessary for a general proof of the previous scheme and a reduction of the number of parameters of the theory.

IV. Coil stretch transition

The conformational transition of the flexible macromolecules from a gaussian (or almost gaussian) coil to a stretched thread has often been employed as a convincing explanation of behaviours which were not predictable by more classical theories. But according to de Gennes (69) such a transition is unlikely in a weak flow; its experimental evidence in strong flows has been claimed by several authors from pressure losses measurements in short capillaries or in slits [Quibrahim (22), Ambari (70)] and from flow birefringence measurements in extensional flows [Pope et al. (71), Cressely et al. (72)]. It must be noticed that the experimental conditions involved are very closed to those of degradation. Assuming that, even in weak flow conditions, a limited transition can occur from the gaussian coil into a deformed coil described by a prolate ellipsoid of revolution [Wolff (73)] at a critical Deborah number $\beta^{*} = M[\eta]_{o}\eta_{s}/RT \simeq 8$; ($[\eta]_{o}$: zero shear intrinsic viscosity), the calculation [Dupuis et al. (74)] show that diluted solutions of flexible single macromolecules may display a shear-thickening behaviour in simple shear flows. This effect depends on the polydispersity, on the deformability of the macromolecule, and indeed, on the shear rate (figure 8). However as it is a diluted solution calculation where only single macromolecules are considered, the concentration dependence obeys to the Huggins law. It is not excluded that aggregation, associations and coil deformed coil transition play together and lead to the complex behaviours observed experimentally.

Conclusion

The diluted solutions of high molecular weight flexible macromolecules display, specially in poor solvents, some fringe effects like shear degradation, existence of aggregates or microgels because of the slowness of the solubilization process, formation of labile associations in flow: these effects must be carefully analysed and taken into account for the explanation of their hydrodynamic behaviour both in weak and strong flows.

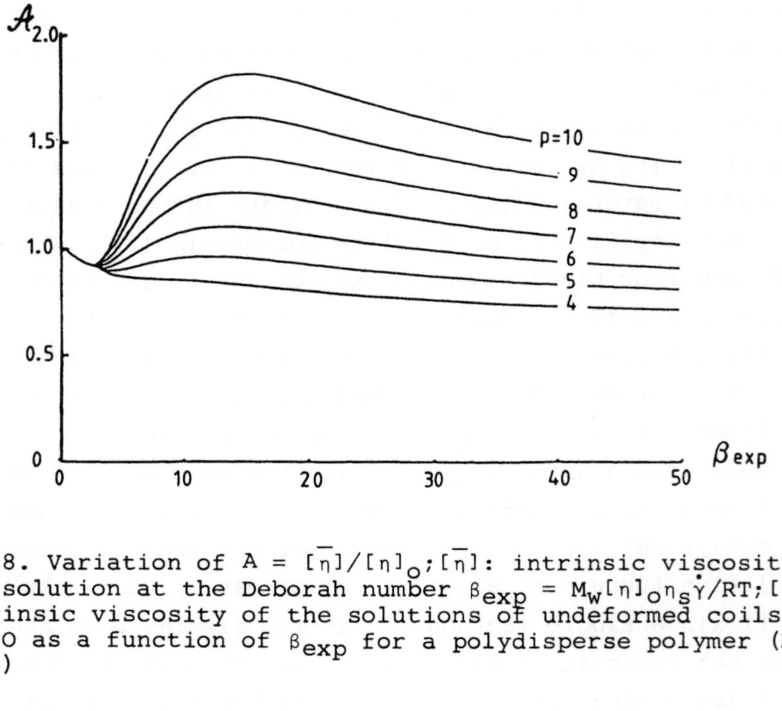

Fig.8. Variation of $A = [\bar{\eta}]/[\eta]_o$; $[\bar{\eta}]$: intrinsic viscosity of the solution at the Deborah number $\beta_{exp} = M_w[\eta]_o\eta_s\dot{\gamma}/RT$; $[\eta]_o$: intrinsic viscosity of the solutions of undeformed coils at $\dot{\gamma} = 0$ as a function of β_{exp} for a polydisperse polymer (M_w/M_n = 1.15)

References

1. Lohmander, U.: Non-newtonian flow of dilute macromolecular solutions studied by capillary viscometry. Makromol. Chem. 72 (1964) 159-173

2. Wolff, C.: Recherches expérimentales sur la viscosité non newtonienne des solutions très diluées de macromolécules en chaînes souples. J. Chim. Phys. 65,9 (1968) 1569-1579

3. Kotoka, T.; Suzuki, H.; Inagaki, H.: Shear rate dependence of the intrinsic viscosity of flexible linear macromolecules. J. Chem. Phys. 45,8 (1966) 2770-2773

4. Yang, J.T.: Non newtonian viscosity of PBLG solutions. J.Am. Chem. Soc. 80 (1958) 1783-1788

5. Peterlin, A.: Über die Viskosität von verdünnten Lösungen und Suspensionen in Abhängigkeit von der Teilchenform. Zeitschrift für Phys. 111 (1938) 232-263

6. Saito, N.: The effect of the brownian motion on the viscosity of solutions of macromolecules. J. Phys. Soc. Jap. 6,5 (1951) 297-301

7. Cerf, R.: Sur la théorie des propriétés hydrodynamiques des solutions de macromolécules en chaînes. J. Chim. Phys. 66 (1969) 479-488 / Cerf, R.: Dynamics of linear polymeric chains. Chem. Phys. letters 24,3 (1974) 317-322

8. Bazua, E.R.; Williams, M.C.: Rheological properties of internal viscosity models with stress symmetry. J. Polym. Sci. 12 (1974) 825-848

9. De Gennes, P.G.: Origin of internal viscosity in dilute polymer solutions. J. Chem. Phys. 66 (1977) 5825-5826

10. Fixman, M.: Polymer dynamics: Non-newtonian intrinsic viscosity. J. Chem. Phys. 45 (1966) 793-803

11. Rheinhold, C.; Peterlin, A.: Hydrodynamics of linear macromolecules. I. Finite segment length. J. Chem. Phys. 44 (1966) 4333-4341

12. Fong, J.T.; Peterlin, A.: Matrix algebra and eigen values for the bead/spring model of polymer solutions. J. Res. Nat. Bur. of Standards B 8013,2 (1976) 273-284

13. Chikahisa, Y.: Non-newtonian intrinsic viscosity of coiled chain polymers. J. of the Phys. Soc. of Japan 21, 11 (1966) 2324-2331

14. Bird, R.B.; Dodson, P.J.; Johnson, J.L.: Non-newtonian polymer solution rheology based on a finitely extensible bead-spring chain model. J. Non Newt. Fl. Mech. 7 (1980) 213

15. Wolff, C. (Ed) Polymères et lubrification, colloque international du CNRS n° 233, Brest (1974), CNRS, Paris (1975)

16. Schutz, R.A.: Theoretical and practical aspects of sizing today and tomorrow. Proc. 3d Intern. Sizing Symp., Shirley Institute (Ed.) Manchester 1977, Paper n° 1 pp 1-9

17. Shick, M.J.: Friction and lubrication of synthetic fibers in surface characteristics of fibers and textiles, Schick M.J. (Ed), Marcel Dekker, 1975, pp 1-65

18. Burow, S.; Peterlin, A.; Turner D.T.: Increasing viscosity of polymer solutions with increasing shear stress. Polymer letters 2 (1964) 67-70

19. Quadrat, O.; Bohdanecky, M; Munk, P.: Influence of thermodynamic quality of a solvent upon non-newtonian viscosity of PMMA solutions. J. Polym. Sci. C 16 (1967) 95-102

20. Wolff, C.: Sur l'augmentation de la viscosité intrinsèque de certaines solutions macromoleculaires avec le gradient de vitesse. C.R. Acad. Sci. Paris C 264 (1967) 1364-1367

21. Layec-Raphalen, M.N.; Wolff, C.: On the shear thickening behaviour of dilute solutions of chain macromolecules. J. Non Newt. Fl. Mech. 1 (1976) 159-173

22. Ouibrahim, A.: Rheology of polymer solutions in laminar capillary tube flow. Phys. Fluids 21,1 (1978) 4-8

23. Moan, M.; Chauveteau, G.; Ghoniem, S.: Entrance effect in capillary flow of dilute and semi dilute solutions. J. Non Newt. Fl. Mech. 5 (1979) 463-474

24. Haas, R.; Durst, F.: Die Charakterisierung viskoelastischer Fluide mit Hilfe ihrer Strömungseigenschaften in Kugelschüttungen. Rheologica Acta 21 (1982) 150-166

25. Toms, B.A.: Some observations on the flow of linear polymer solutions through straight tubes at large Reynolds numbers. Proc. Int. Congr. Rheol. North Holland (1948), vol. II, pp 135-141

26. Hoyt, J.W.: The effect of additives on fluid friction. J. Basic Engng, Trans ASME 94 (1972) 258-285

27. Berman, N.S.: Drag reduction by polymers. Ann. Rev. Fl. Mech. 10 (1978) 47-64

28. Virk, P.S.: Drag reduction fundamentals. AICHE J. 21,4 (1975) 625-655

29. Conference on Drag Reduction BHRA Cambridge 1977

30. Jellinek, H.H.G. (Ed): Degradation and stabilization of polymers, Elsevier Sci. Publ. 1983

31. Levinthal, C.; Davison, P.F.: Degradation of DNA under hydrodynamic shearing forces. J. Mol. Biol. 3 (1961) 674-683 / Levinthal C.; Davison, P.F.: Shear degradation and the molecular weight of DNA. J. Chim. Phys. 58 (1961) 887-890

32. Harrington, R.E.; Zimm, B.H.: Degradation of polymers by controlled hydrodynamic shear. J. Phys. Chem. 69 (1965) 161-175

33. Cottrell, F.R.; Merrill, E.W.; Smith, K.A.: Conformation of PIB in dilute solutions subjected to a hydrodynamic shear field. J. Polym. Sci. A2,7 (1969) 1415-1434 / Cottrell, F.R.; Merrill, E.W.; Smith, K.A.: Intrinsic viscosity and axial extension ratio of PIB in dilute solutions J. Polym. Sci. A2,8 (1970) 289-294

34. Smith, K.A.; Merrill, E.W.; Banijamali, H.: Elongation of drag reducing macromolecules by a pure straining motion. Polymères et lubrification, colloque CNRS n° 233, Brest (1974), C. Wolff (Ed); CNRS, Paris 1975, pp 341-344

35. North, P.F.; Champion, J.V.: Hydrodynamic degradation of DNA J. Chim. Phys. 71,5 (1974) 1282-1284

36. Yu, J.F.S.; Zakin, J.L.; Patterson, G.K.: Mechanical degradation of high molecular weight polymer in dilute solutions. J. Appl. Polym. Sci. 23 (1979) 2493-2512

37. Tanner, R.T.; Huilgol, R.R.: On a classification scheme of flow fields. Rheol. Acta 14 (1975) 959-962

38. Culter, J.D.; Zakin, J.L.; Patterson, G.K.: Mechanical degradation of dilute solutions of high polymers in capillary tube flow. J. Appl. Polym. Sci. 19 (1975) 3235-3240

39. Culter, J.D.; Mayhan, K.G.; Patterson, G.K.; Sarmasti, A.A.; Zakin, J.L.: Entrance effects on capillary degradation of dilute polystyrene solutions. J. Appl. Polym. Sci. 16 (1972) 3381-3385

40. Ghoniem, S.: Rheological behaviour and degradation of polymeric solutions in modelized porous media. Dr. Ing. Thesis, University of Brest (1979)

41. Ghoniem, S.; Chauveteau G.; Moan, M.; Wolff, C.: Mechanical degradation of semi-dilute polymer solutions in laminar flows. Can. J. of Chem. Engng. 59 (1981) 450-454

42. Ram, A.: Polymers and their effectiveness in drag reduction. Colloque CNRS n° 233, Brest (1974); Wolff, C. (Ed); CNRS, Paris, 1975, pp 271-282

43. Gampert, B.; Wagner, P.: Turbulent flow with polymer additives. Arch. Mech. 34,4 (1982) 493-502

44. Nakano, A.; Minoura, Y.: Effect of solvent and concentration on scission of polymers with high-speed stirring. J. Appl. Polym. Sci. 19 (1975) 211-2130

45. Odelle, J.A.; Keller, A.; Miles, M.J.: A method for studying flow-induced degradation: verification of chain halving. Polym. Comm. 24 (1983) 7-10

46. Maerker, J.M.: Shear degradation of polyacrylamide solutions. Proc. 49th Annual Fall Meet. Soc. Petrol. Eng. of AIME, Houston (1974)

47. Hinch, E.J.; Elata, C.: Heterogeneity of dilute polymer solutions. J.N. New. Fl. Mech. 5 (1979) 411-425

48. Wolff, C.: On the real molecular weight of polyethylene oxide of high molecular weight in water. Can. J. Chem. Eng. 58 (1980) 634-636

49. Layec, Y.; Layec-Raphalen, M.N.: Instability of dilute PEO solutions. J. Phys. Letters 44 (1983) 121-128 / Layec, Y.; Layec-Raphalen, M.N.: Private communication

50. Kulicke, W.M.; Klein J.: Zur Frage der Instabilität von Polyacrylamidlösungen. Angew. Makromol. Chem. 69 (1977) 189-210 / Kulicke, W.M.; Kniewske, R.; Klein, J.: Preparation, Characterization, Solution Properties and Rheological Behaviour of Polyacrylamide Prog. Polym. Sci. 8 (1982) 373-468

51. Kulicke, W.M.; Kniewske, R.: Long term change in conformation of macromolecules in solution. Poly(acrylamid-co-sodium acrylate)s. Makromol. Chem. 182 (1981) 2277-2287 / Kulicke, W.M.; Hörl, H.H.: Zur Characterisierung und dem ungewöhnlichen Viskositätszeit-Verhalten von mäßigen Polystyrolsulfonatlösungen. Angw. Makromol. Chem. 116 (1983) 149-164

52. Shyluk, W.P.; Stow, F.S. Jr.: Aging and loss of flocculation activity of aqueous polycrylamide solutions. J. Appl. Polym. Sci. 13 (1969) 1023-1036

53. Narkis, N.; Rebhun, M.: Aging effects in measurements of polyacrylamide solution viscosities. Polymer 1 (1966) 507-512

54. Brennan, C.; Gadd, G.E.: Aging and degradation in dilute polymer solutions. Nature 215 (1967) 1368-1370

55. James, D.F.: Method for measuring normal stresses in dilute polymer solutions. Trans. Soc. Rheol. 19 (1975) 67-72

56. Stenberg, L.G.; Lagerstedt, T.; Lindgren, E.R.: Mechanical mixing of polymer additives in turbulent drag reduction. Phys. Fluids 20 (1977) 5276-5280

57. White, A.: Some observations on the flow characteristics of certain dilute macromolecular solutions in viscous drag

reduction. CS Wells ed, Plenum Press 1969, 297-311

58. Kit, E.; Poreh, M.: Drag of circular cylinders in dilute polymer solutions. Second Intern. Conf. on drag reduction, BHRA. Cambridge 1977, paper F2, 15-24

59. Naudascher, A.; Killen, J.M.: Onset and saturation limit of polymer effects in porous media flows. Phys. Fl. 20 (1977) 5280-5283

60. Peterlin, A.; Quan, C.; Turner, D.T.: Role of molecular aggregates in the flow properties of PMMA in Arochlor. Polym. Lett 3 (1965) 521-524

61. Wolff C.: On non Newtonian intrinsic viscosity of flexible coil macromolecules. 23e Intern. Congress IUPAC, Boston 1971. Macromolecular preprints, Vol. 1, p. 23-27

62. Goldsmith, H.L.; Mason, S.G.: The microrheology of dispersions. Rheology; theory and applications. Eirich, F. (Ed), Academic Press 1967, vol. 4 p. 85-250

63. Wagstaff, I.; Chaffey, C.E.: "Shearthinning and thickening rheology". I. Concentrated acrylic dispersions. J. Colloid Interface Sci. 59 (1977) 53-62

64. Chaffey, C.E.; Wagstaff, I. "Shear thinning and thickening rheology". II. Volume fraction and size of disperses particules. J. Colloid Interface Sci. 59 (1977) 63-75 / Chaffey C.E. "Mechanismus and equations for shearthinning and thickening in dispersions". Colloid and Polymer Sci., 255 (1977) 691-698

65. Brochard F.; de Gennes, P.G.: "Dynamical scaling for polymer in theta solvents". Macromolecules 10 (1977) 1157-1161

66. Wolff, C.; Silberberg, A.; Priel, Z.; Layec-Raphalen, M.N. "Influence of the association of macromolecules in dilute solution on their reduced viscosity". Polymer, 20 (1979) 281-287

67. Layec-Raphalen, N.M.; Wolff, C.: "Shear rate dependance of the association of high molecular weight macromolecules in dilute solutions. " Rheology. Astarita, G.; Marrucci, G.; Nicolais, L. (Ed), Plenum Press 1980, vol. 2, pp. 303-308

68. Wolff, C. "Non newtonian behaviour of associations of macromolecules in dilute solutions". Adv. in colloid and Interface Sci., 17 (1982) 263-274

69. de Gennes, P.G.: "Coil stretch transition of dilute flexible polymers under ultrahigh velocity gradients". J. Chem. Phys. 60 (1974) 5030-5042

70. Ambari, A.: "Ecoulement des solutions aqueuses diluées de polyoxyethylene dans une fente rectangulaire très allongée en mince paroi." C.R. Acad. Sc. Paris, 289 B (1979) 1-4

71. Pope, D.P.; Keller, A.: "A study of the chain extending effect of elongational flow in polymer solutions." Colloid and Polymer Sc. 256 (1978) 751-756

72. Cressely, R.; Hocquart, R.: "Dynamic of flexible and large macromolecules in elongational flow using flow birefringence. Rheology. Astarita, G., Marrucci, G., Nicolais, L. (Ed), Plenum Press (1980), Vol. 2 pp. 377-383

73. Wolff, C.: "Molecular origin of the shear thickening of dilute polymer solutions." 2nd World Cong. of Chem. Eng., Montreal (1981), Proceed. Vol. 6, pp. 394-398

74. Dupuis, D.; Wolff, C.: "Consequences of a conformational change of flexible macromolecules on their solutions." Chem. Eng. Comm. (under press)

Aging of Aqueous Polymer Solutions

W.-M. Kulicke

Institut für Technische Chemie
Technische Universität
3300 Braunschweig

Summary

Aging of polymers in solution under the influence of light, microorganism, temperature and mechanical energy are often to be observed and in such cases the viscosity decreases with storage time, because chain scission of polymers occur.
By way of contrast, in the case of some water soluble polymers, one can observe a drastical decrease (or increase in the case of polyvinylalcohol) of solution viscosity with time, whereas the solution is stored at room temperature in the dark and without mechanical influence (e. g. shaking). Chain scission as a reason for this aging effect could be excluded on the basis of light scattering measurements and supports some recent investigations which showed that polyacrylamide at least is resistant to microbial degradation. However, the loss of the efficiency of technically important properties (for example flocculation activity) occurs simultaneously. Therefore this aging-effect is of scientific (mechanism) and of practical (handling) interest. This paper presents an overall view of polymers which show this unusual effect and reports a possible mechanism.

Introduction

It has been frequently observed that aqueous polyacrylamide (PAAm) solutions exhibit a time dependent viscosity. The effect expresses itself in a long term decrease of solution and intrinsic viscosity in the course of several weeks, even when the solutions are kept at room temperature in the dark and without shaking (mechanical influence). This behaviour has often been described however, the interpretation has been given in different ways and reviewed recently[1]. Possible mechanisms discussed were:

a) disentanglement of polymer chains ref.: 126 in 1)
b) weak link scission and disentanglement ref.: 127 in 1)
c) residual initiator ref.: 129 and 130 in 1)
d) attack by microorganism ref.: 133 in 1)
e) entanglements and aggregation ref.: 128 in 1)

The Influence of Polymer Additives
on Velocity and Temperature Fields
IUTAM Symposium Essen 1984
Editor: B. Gampert

Already in 1978[2], the author of this report together with J. Klein was able to show, that it is not likely that the aging-effect of aqueous PAAm solutions is caused by one of the above mentioned mechanisms. There is some evidence that a conformational change of the polymer molecules occurs[1,2]. However, in the meantime it was found that other aqueous polymer solutions exhibit similar behavior and in addition a viscosity increase with storage time could be observed, see fig. 1.

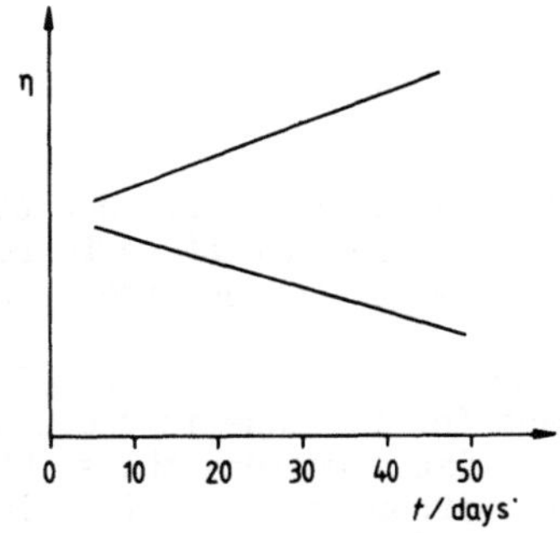

Fig. 1: Schematic representation of the unusual time dependent viscosity behavior of aqueous polymer solutions, where chain scission as a possible mechanism could be excluded.
viscosity increase: polyvinylalcohol[3]
viscosity decrease: PAAm[1,2], PAAm-co-acrylates[4], derivatives of PAAm[5,6], amylose[7], sulfonated polystyrene[8], poly(L-glutamic acid)[9] and graft-copolymers[11].

In all cases a conformational change of the polymer molecules with time seems to be the best explanation.

Results and Discussion

Polyacrylamide (PAAm):
The aging effect of high purity, non-branched PAAm-samples was examined for a number of different molecular weights in pure H_2O, see tab. 1. PAAm-samples with $M_w < 1 \cdot 10^6$ g/mol show no aging effect. The amount of the viscosity decrease with time increases with increasing M_w. However, the quantity of the viscosity loss can be influenced by repeated precipitation. In addition, the hygroscopy of solid PAAm becomes important when storing it for a prolonged period of time, see fig. 2. Storing the samples in air and possibly attackted by the presence of moisture, the intrinsic

M_w [g/mol]	time dependent viscosity
$6{,}9 \cdot 10^6$	yes
$5{,}3 \cdot 10^6$	yes
$2{,}2 \cdot 10^6$	yes
$1{,}6 \cdot 10^6$	yes
$1 \cdot 10^6$	no
794 000	no
500 000	no
165 000	no
60 000	no
38 000	no

Tab. 1:
Time dependent viscosity behavior under the influence of molecular weight.

viscosity, even after intensive drying, was significantly smaller (~ 25%) than directly after polymerisation. Elemental analysis, however, shows that the water content was only increased by 5% and therefore errors caused by incorrect concentrations were negligible, see fig. 2. On the other hand, the sample which was kept in an exsicator over a drying agent shows almost the same [η] as on the first day. Perhaps, because of the influence of moisture, water molecules are bonded at "active sides" of the macromolecule, which cannot be removed even by extensive drying. This in an indication of H-bonds. The existence of such H-bonds could be supported by viscosity measurements with H-bridge influencing

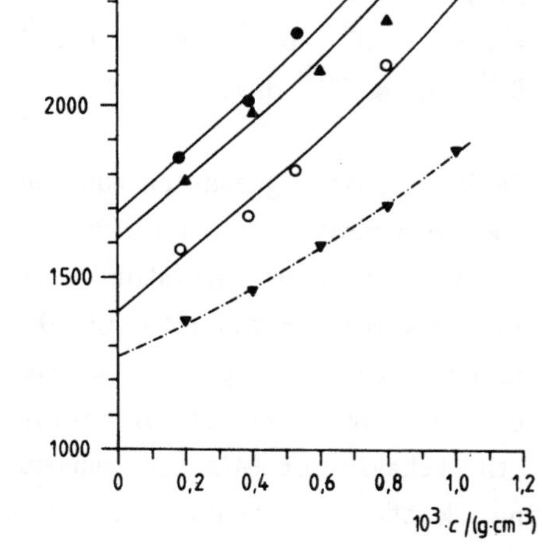

Fig. 2:
Reduced specific viscosity dependent on the concentration of PAAm in water at 25°C. ● 1st day in solution immediately after preparation. o 35th day in solution. ▲ Solid stored (exsicator) over 180 days (1st day in solution). ▼ Solid stored (under atmospheric moisture) over 180 days (1st day in solution).

substances as well as by IR-studies of the time dependent deuteration of PAAm[ref.: 67 in 1)]. It was shown that the incomplete deuteration of the protons versus time of NH_2-groups and water molecules demonstrates the dominating role of the functional side groups. It should be mentioned here, that the amount of the aging effect was not reproducable even when the measurements were carefully repeated. By preparing the same PAAm-sample 10 times, in 8 cases aging effects were observed and in 2 cases no viscosity time dependency occurred. This may be one explanation for the contradictory results reported so far and again reflects the complexity of such systems. The aging effect was also observed in size-exclusion-chromatography measurements. The viscometrically found observation of decreasing hydrodynamic volume in the course of time is confirmed by the SEC investigations. For this purpose the chromatograms of a gently shaken PAAm-solution, prepared in 0,1 m Na_2SO_4 containing biocide Kathon WT, were recorded at regular intervals[ref.: 86 in 1]. Very recently it was found by microbiologists that PAAm is resistant against microbial degradation whereas polysaccharids are not[10]. Therefore degradation caused by an attack of microorganisms can be disregarded.

An example of the time dependence of the reduced specific viscosity is shown in fig. 3. An almost parallel decrease of these curves can be seen. So, the explanation of the viscosity decrease by disentanglement seems to be inappropriate, because otherwise the changes in η_{sp}/c would depend on the concentration. Moreover, the viscosity measurements were performed in a concentration range $c < c_1^*$. Therefore, such drastic decreases in solution viscosity are not likely to be caused by disentanglement. .

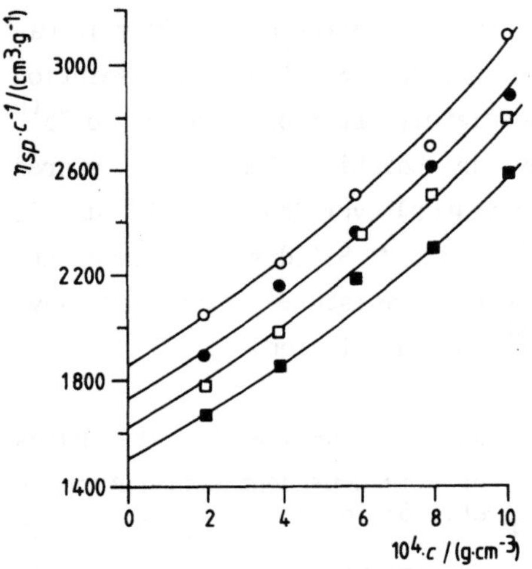

Fig. 3:
A typical example of reduced specific viscosity dependent on time at 25°C in H_2O.
$M_w = 5,05 \cdot 10^6$ o 1st day, ● 5th, ▫ 9th and ■ 27th day.

Oxidative chain degradation was seen to be responsible for the loss in viscosity caused by initiator residuals. However, instability was also detected when high purity samples were free a of residual initiator. Parallel to the viscometric studies, the behavior of PAAm in aqueous solution was evaluated by light scattering. Scattering intensity decreased with time which resulted in a decrease in the radii of gyration but the molecular weight remained unchanged. However the drastic change in dn/dc could not be reproduced in all PAAm-samples. Further investigations - com-

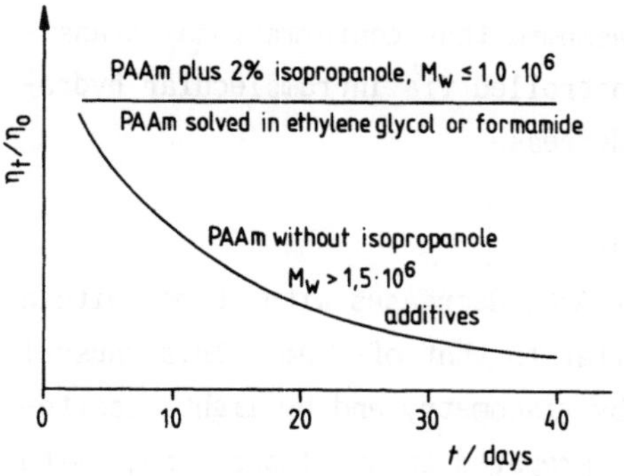

Fig. 4:
Schematic representation of the time dependent solution viscosity of PAAm.
η_o: viscosity at 1st day
η_t: viscosity at tth day

bined light scattering and viscosity measurements - have shown that although the viscosity decreases with time in the usual way, the scattering intensity and the dn/dc remain constant. It should be emphasized that even in this case the molecular weights remained unchanged. No degradation of the macromolecules occured. These results find support in an experiment [ref.: 94 in 1)] where constant molecular weight was observed by light scattering with a SOFICA photometer over a period of about 38 days.

The following results have been obtained so far and have been partly summarized in fig. 4:

a. The viscosity decrease did not occur in technical products only, but in high purity PAAm too, when the molecular weight exceeded $1{,}5 \cdot 10^6$.
b. Light scattering measurements yielded unchanged molecular weights; that means no chain degradation occurred.
c. The hydrodynamic volume became smaller with time as was confirmed by SEC measurements and flocculation tests.
d. The possible existence of hydrogen bonds in aqueous PAAm solutions was supported by the fact that structure forming or structure breaking additives influenced the solution instability. This was also supported by investigations.
e. The standardization of several solutions with a pH of 1 to 12 as well as the addition of hydroquinone, p-benzo-quinone, and Na_2SO_3 did not result in stable solutions.
f. No viscosity loss could be observed in formamide and ethylene glycol at a low temperature (239 K) in water, or when 2 vol % 2-propanol

to aqueous solutions.

g. Storing the solid samples under the possible influence of moisture, the intrinsic viscosity even after an intensive drying was significantly smaller than directly after polymerization.

From these experimental results it was assumed that conformational transitions within single polymer chains - controlled via intramolecular hydrogen bonds - were causing the viscosity decrease.

Poly(acrylamide-co-acrylate)(PAAm/AAcNa):

The solution viscosity of aqueous PAAm/AAcNa decreases with time within the span of weeks. The behavior is similar to that of PAAm. This unusual viscosity loss has been investigated by viscometry and by light scattering on high purity copolymer samples, prepared in our laboratory, with different copolymer compositions[4].

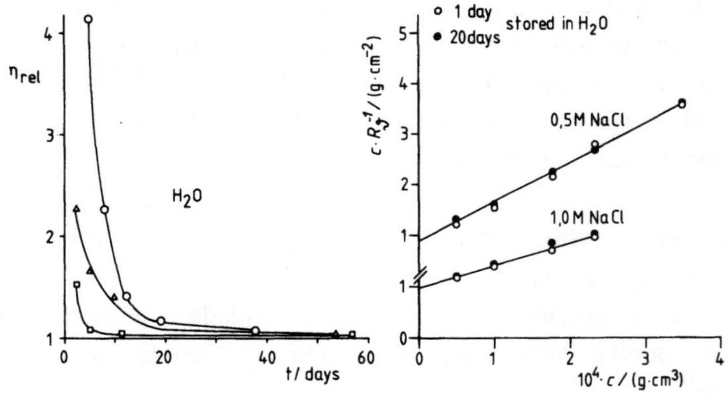

Fig. 5: Relative viscosity of aqueous PAAm/AAcNa-solutions (: 58,5 mol% AAm, : 82 mol% AAm, : 92 mol% AAm) dependent on storage time. $c = 3 \cdot 10^{-6}$ g/cm^3 and light scattering results from measurements on the 1st and 20th day in two different salt concentrations (λ_o = 633 nm; angle 6-7°, copolymer with 56 mol% AAm).

The level of reduction in the viscosity, dependent on the concentration of the solution, was measured at different times. High viscosities were observed on the first day and there was no evidence of a strong polyelectrolyter character. Measurements performed up to 60 days showed a drastic viscosity decrease. It would be interesting to find out whether the anomalous viscosity-time behavior is caused by chain scission or not. Light scattering measurements on polyelectrolytes are only possible in the presence of low molecular weight electrolytes. Therefore an aqueous copolymer solution was prepared and divided into two portions. Part 1 was measured in NaCl on the first day, part 2 was stored as a pure aqueous solution for 20 days and then measured in NaCl again. The results are shown in fig. 5. A sharp decrease in solution viscosity was observed, strongly dependent on copolymer ratio. The scattering intensities, however, did not change with time. In addition, light scattering measurements were performed on a wide angle photometer and from the obtained Zimm plots no change in M_w could be detected. Therefore, chain scission as a reason for the viscosity decrease with time can be omitted. It should be mentioned in this context, that no viscosity loss was observed in a NaCl-solution which had an adequately high salt concentration.

Poly(L-glutamic acid):
The phenomenon of the time viscosity decrease was also observed for poly-(L-glutamic acid) (PGA) by Cho et al[9]. Aqueous PGA-solutions exhibit a similar aging-effect at constant pH = 5.7 and 25°C as PAAm and PAAm/AAcNa. Time dependent estimations of the α-helix content from circular dichroism measurements show a decrease during the storage of the aqueous solutions from 52.3% (1st day) to 7.4% (26th day). The decrease in helicity was considerably lower at 10°C and this may mean that the partially helical conformation in the beginning is not the equilibrium conformation the authors reported.

Derivatives of PAAm:
Kationic and anionic derivatives of PAAm also exhibit the aging effect. The examples are poly(acrylamide-co-methacryloyloxyethyl-N,N,N,-trimethylammoniumchlorid)* and Poly(acrylamide-co-sodium-2-sulfoethylmethacrylathes) as well as poly(acrylamide-co-sodium-2-acrylamido-2-methylpropanesulfonates)[6].

Polystyrene sulfonate (PSsNa):
A homologous series of commercially available sodium-polystyrene sulfonate samples were characterized by viscometry and light scattering measurements. The molecular parameters were determined and the degree of sulfo-

* Special thanks to Dr. Wunderlich from the Röhm GmbH for providing the polymer sample.

nation was obtained from elemental analysis[8].

The viscosities were measured for dilute polymer solutions made from polymer samples having molecular weights ranging from 120 000 to 1 800 000 g/mol diluted by H_2O and aqueous 0.1 M Na_2SO_4. The viscosity of only the highest molecular weight sample was observed to decrease continnously with time, although. The solutions were kept in the dark and were nor shaken, see fig. 6. In concentrated salt solutions (0.5 M Na_2SO_4) and at low storage temperatures (-34°C) a viscosity decrease was not observed.

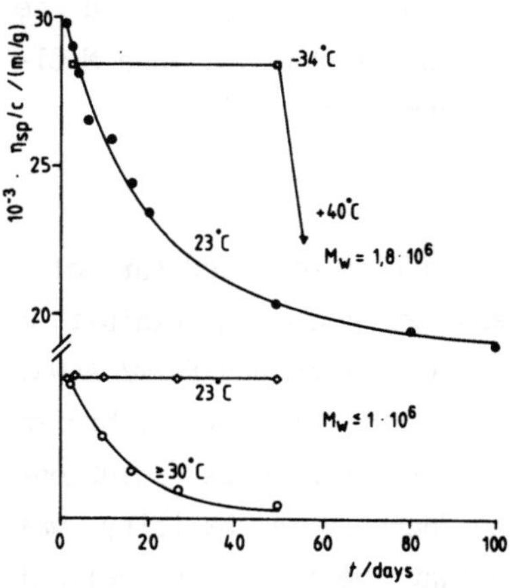

Fig. 6:
Viscosity-time-behavior of PSsNa-solutions of different molecular weights and stored at different temperatures, in the dark and without shaking.

From all the results one can conclude that a conformantional change of the polymer molecules probably caused this unusual time behavior.

Graft-copolymers:

Solutions of starch-acrylamide, graft copolymers, were prepared by J. J. Meister and the solution viscosity was researched with the goal to prepare highly viscous solutions[11]. He was able to prove, that graft copolymers also age. However, the data have another feature not found in the viscosities of the PAAm-solutions. That is, that graft-copolymer solution viscosity increases between 24 and 48 hours. After this a typical viscosity decrease can be observed as was mentioned earlier.

It may be of interest, that dextran-acrylamide, graft copolymers, prepared in a very similar way to starch-acrylamide, graft copolymers, donot show this aging effect, see page 28 in ref.: 3 in 6.

Amylose:

Uniform synthetic amylose samples were prepared and characterized by means of light scattering and viscosity measurements. A viscosity decrease very similar to that of PAAm was observed. The viscosity drop depends on solvent quality and at 0°C no viscosity decrease occurs[7].

Polyvinylalcohol (PVOH):
Viscosity measurements of aqueous and non-aqueous PVOH solutions exhibit an increase in viscosity with storage time only in aqueous solutions. This unusual effect was observed by different researchers. It has been suggested, that this phenomenon is caused by the association of resulting microgels (intermolecular) in the solution. This could also be supported by recent light scattering measurements carried out by Lechner[3].

Molecular model

The change of solution viscosity with time has been discussed by several researchers. The results discussed are very contraversial in the case of PAAm, only. This is mainly due to the fact, that a couple of researchers only used one or two samples for their entire investigations. But this is not enough for such a complex system. However, from all the systems researched so far, one can conclude that chain scission at least is not responsible for this.

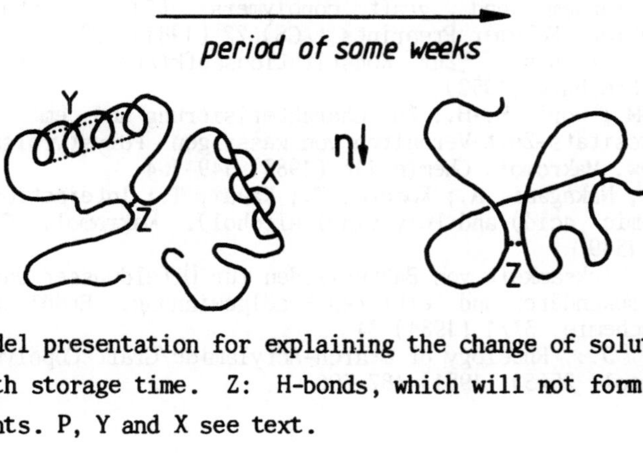

Fig. 7: Model presentation for explaining the change of solution viscosity with storage time. Z: H-bonds, which will not form extended segments. P, Y and X see text.

To explain the molecular origin of the change in solution viscosity with time (aging), a molecular model is presented see fig. 7. The viscosity increase (PVOH) with time may be explained by the intermolecular association process (H-bonds) forming partly cystalline (P) structures[3]. The viscosity decrease with time may be explained by an intra-molecular conformational transition of single polymer chains. Largely extended segments of type X and/or helical structures of type Y exist in the beginning. One may argue that H_2O molecules can attack the extended structures and destroy them by cleavages. However, new extended structures can arise from non-extended side groups by the formation of newly extended segments. Under the influence of an increasing entropy this to and fro leads to a decrease in viscosity over a period of several weeks and the conformation changes to a more compact flexible coil.

References

1. Kulicke, W.-M.; Klein, J.; Kniewske, R.: Preparation, characterization, solution properties and rheological behaviour of polyacrylamides. Progress in Polymer Science 8 (1982) 373-468.
2. Kulicke, W.-M., Klein, J.: Zur Frage der Instabilität von Polyacrylamidlösungen. Angew. Makromol. Chemie 69 (1978) 189-210
3. Lechner, M. D., Mattern, R.: Das Verhalten von Polyvinylalkohol in wässriger Lösung. Angew. Makromol. Chemie 123/124 (1984) 45-61
   Schurz, J.; Kashmola, T.; Falke, F.-J.: Rheologische Untersuchungen an Polyvinylalkohol-Lösungen. Angew. Makromol. Chemie 25 (1972) 51-67
   Braun, D.; Walter, E.: Zur Alterung von wässrigen Polyvinylalkohollösungen. Colloid & Polymer Sci. 258 (1980) 376-378
4. Kulicke, W.-M.; Kniewske, R.: Long-term change in conformation of macromolecules in solution, 2. poly(acrylamide-co-sodium acrylates). Makromol. Chemie 182 (1981) 2277-2287
5. Kulicke, W.-M.: (unpublished results)
6. Neidlinger, H. H.; Chen, G. S.; Park, L. S.; Mc Cormick, C. L.: Model acrylamide random and graft copolymers (I) - Synthesis and Characterization. Polymer Preprints (ACS) 22 (1981) 139
7. Bartl, H.: Thesis "Zur Konstitutionsaufklärung der Amylose" Universität Freiburg (1952)
8. Kulicke, W.-M.; Hörl, H.-H.: Zur Charakterisierung und dem ungewöhnlichen Viskositäts-Zeit-Verhalten von wässrigen Polystyrolsulfonatlösungen. Angew. Makromol. Chemie 116 (1983) 149-164
9. Cho, C. S.; Nakagami, A.; Komoto, T.; Kawai, T.: Interaction Between Poly(L-glutamic acid) and Poly(vinyl alcohol). Makromol. Chemie 179 (1978) 1345-1349
10. Näveke, R.: Wirksamkeit von Bakteriziden für Ölfeldwässer und Chemikalien in der sekundären und tertiären Erdölgewinnung. Erdöl und Kohle-Erdgas Petrochemie. 37/1 (1984) 34
11. Meister, J. J.: Rheology of Starch-Acrylamide/Graft Copolymer Solutions. J. Rheol. 25(5) (1981) 487-506

Model for Effects of Degradation on Polymer Drag Reduction

GARY K. PATTERSON

DEPARTMENT OF CHEMICAL ENGINEERING
UNIVERSITY OF ARIZONA
TUCSON, ARIZONA 85721

ABSTRACT

The effects of molecular weight decrease caused by degradation upon the drag reduction of a polymer in solution may be modeled by using either the drag reduction model of Virk or the correlation of Rodriguez, Zakin and Patterson. The results of the use of these two approaches is summarized in this paper. Their uses for engineering design considerations are compared.

INTRODUCTION

Basically, shear degradation of polymers in turbulent or laminar flow is caused by the stress imposed on the polymer molecules by the shearing motion of the liquid, whether in solution or a melt. When turbulent drag reduction is the objective, the degradation is primarily caused by the high shear stress region near the pipe wall or other surface where high shear stresses are produced.

Turbulent flow is not accurately characterized by averages of velocity and velocity gradient. The maximum shear stresses in the near wall region have even been observed to be much higher than the average wall shear stress, which is in a region least disturbed by the turbulent fluctuations. Much of the degradation of polymers must take place during short bursts of very high shear stress in the near wall region.

When degradation does occur, the molecular weight distribution is usually changed, because only the highest molecular weight components in a broadly polydisperse polymer degrade. This causes diminution of the high molecular weight "tail" and an approach to a final distribution with M_w/M_n less than 2, indicating a mechanism which does not cause randomly located breaks in the molecular chain (1). The mechanism does not, however, cause breaks strictly at the chain center (2). Since the final distribution of polymer molecular weights tends to be narrow, correlation of degraded average molecular weight and degree of drag reduction should be possible.

The rate of the degradation process is only now being intensively studied. The work is being done in a very high shear rate concentric cylinder viscometer which produces a couette flow of nearly uniform shear stress (2-4). Most previous work tended to concentrate on the extent of the degradation caused by long exposure to a given shearing condition and showed that final degradation level for a given polymer-solvent system is a function of shear stress and not of beginning molecular weight (2).

Results on degradation rate to date (3,4) indicate that polyiso-

The Influence of Polymer Additives
on Velocity and Temperature Fields
IUTAM Symposium Essen 1984
Editor: B. Gampert

butylene in oil degrades at a rate that can be corrected by the equation:

$$-\frac{1}{M}\left(\frac{dM}{dt}\right) = B\left(\frac{M-M_\infty}{M}\right) \tag{1}$$

where M is the weight average molecular weight and M_∞ is its final value at the current shear stress after a long time. B is a constant which is of the order of one $second^{-1}$ for polyisobutylene in oil. Definitive work has not yet been done in other polymer-solvent systems. Shear stress does not appear explicitly in the above equation but its effect is contained in the M_∞-level, since $M_\infty = f(\tau_w)$. Figure 1 shows typical rate data for a polyisobutylene solution.

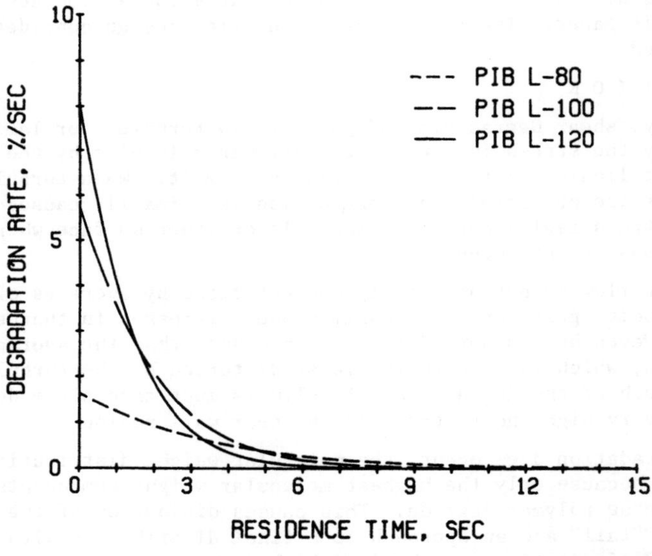

Figure 1 Degradation Rate Changes with Residence Time at $3.24 \times 10^4\ sec^{-1}$ Shear Rate for 1% by Weight of PIB L-80, L-100 and L-120 in Shell Vitrea 32 Oil at 25°C

EFFECT OF DEGRADATION ON DRAG REDUCTION

Two models are used to link drag reduction in pipe flows to molecular weight. One is the Virk model (5) given by:

$$\frac{1}{\sqrt{f}} = [4.0 + 70 \times 10^{-6} \left(\frac{c}{M}\right)^{\frac{1}{2}} \left(\frac{ML}{M_{seg}}\right)^{\frac{3}{2}}][\log_{10} N_{Re}\sqrt{f}$$

$$- (1/\sqrt{f^*} + 0.4)/4.0] + \frac{1}{\sqrt{f^*}} \quad (2)$$

where c is the polymer concentration in wppm
M is weight average molecular weight
M_{seg} is molecular weight of repeat unit
L is the number of chain links/repeat unit
f* is $8.8 \times 10^6/(R_G^3 \rho u^2)$
R_G is the rms radius of gyration for the polymer molecules

Figure 2 shows the predicted drag behavior for a particular polymer solution:

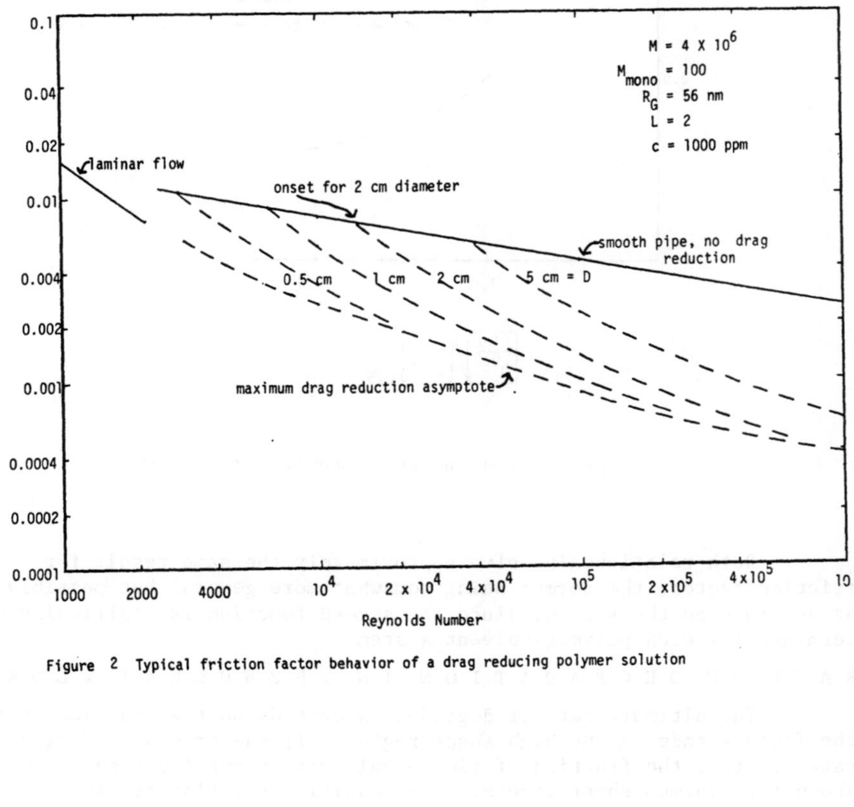

Figure 2 Typical friction factor behavior of a drag reducing polymer solution

The other model is the Rodriguez, Zakin, Patterson (RZP) correlation (6) which is f/f_{pv} = function $(\frac{\tau_1 \bar{V}}{D^{0.2}})$ where τ_1 is the Zimm first-order relaxation time given as $M(\eta-\eta_o)c/(0.586RT)$ for η and η_o as solution and solvent viscosity in poise and c as concentration in g/dl. The group $(\tau_1\bar{V}/D^{0.2})$ is a modified Deborah number N_{De} in which D (pipe diameter) to the 0.2 power is proportional to some important length term in the near wall momentum transport. The RZP correlation may be represented by $f/f_{pv} = 1+A(N_{De})+C(N_{De})^2+E(N_{De})^3$. Figure 3 shows the master curve shape that characterizes the RZP correlation (7).

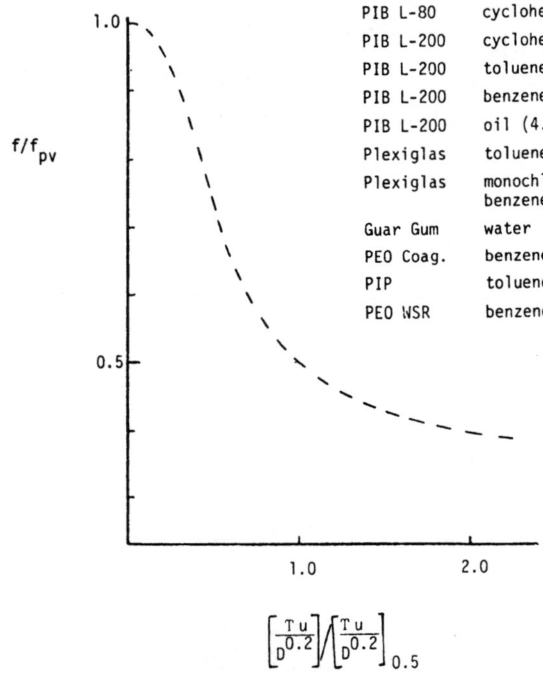

Solute	Solvent	$De_{0.5}(m^{0.8}) \times 10^3$
PIB L-80	cyclohexane	2.5
PIB L-200	cyclohexane	3.4
PIB L-200	toluene	3.4
PIB L-200	benzene	0.35
PIB L-200	oil (4.2 cp)	30.5
Plexiglas	toluene	1.1
Plexiglas	monochloro-benzene	1.1
Guar Gum	water	3.3
PEO Coag.	benzene	13.5
PIP	toluene	1.2
PEO WSR	benzene	1.4

Figure 3 Correlation master curve for drag reduction.

Both relationships give approximately the same result for friction factor, the former being somewhat more general but possibly not as accurate as the second, since the second function is empirically determined for each polymer-solvent system.

RATE OF DEGRADATION IN TURBULENT FLOW

The ultimate rate of degradation depends on the fraction of time the fluid spends in the high shear region. If the process is degradation rate limited, the fraction of time equals the volume fraction of the flow above the minimum shear stress. The resulting relationship is:

$$\frac{t_{deg}}{t_{tot}} \approx \frac{V_{deg}}{V_{tot}} = 2\ \tau_{min}/\tau_{wall} = \frac{4\tau_{min}}{f\rho\overline{V}^2} \tag{3}$$

where $\tau_{min} = f(M = M_\infty)$, the relationship between M_∞ and τ, the shear stress. Figure 4 shows typical final degradation (M_∞) results for a particular polymer solution.

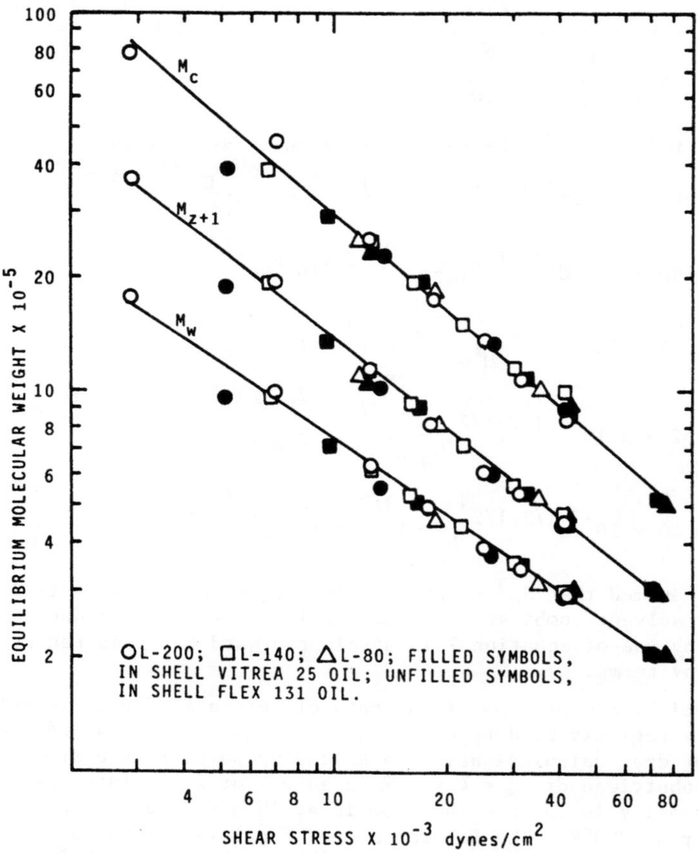

Figure 4 Dependence of Equilibrium MW on Shear Stress with Different Initial MWs and in Various Viscosities of Solvents for 1% by Weight of Polyisobutene.

The relationship of τ_{min} to M_∞ is conservative since degradation occurs only above M_∞. Ultimate degradation rate in a flow which is only partly degrading is then:

$$-\left(\frac{1}{M}\right)\frac{\partial M}{\partial t} \cong 2B\left(\frac{M-M_\infty}{M}\right)\left(\frac{\tau_{min}}{\tau_{wall}}\right) \qquad (4)$$

If the above ultimate degradation rate is combined with the relationship between drag reduction and molecular weight, a relationship for rate of drag reduction decrease by degradation results. For the RZP correlation that can be expressed as [where $d(f/f_{pv})/dt = (d(f/f_{pv})/dM)(dM/dt)_{ult}$]:

$$\frac{d(f/f_{pv})}{dt} = 4B\left(\frac{M-M_\infty}{M}\right)\left(\frac{\tau_{min}}{f\rho\overline{V}^2}\right)(A(N_{De}) + 2\,C(N_{De})^2 + 3\,E(N_{De})^3) \qquad (5)$$

where

$$N_{De} = \frac{M(\eta-\eta_o)\,\overline{V}}{0.586\,c\,D^{0.2}\,RT}\,. \qquad (6)$$

For the Virk relationship the rate expression is as follows:

$$\frac{df}{fdt} = 4B\,(M-M_\infty)\left[\frac{\tau_{min}}{f\rho\overline{V}^2}\right]\left[-4\cdot 2.303\left(\frac{1}{f}\right)^{3/2} - \left(\frac{\rho\overline{V}^2K_G}{8.8}\right)^{1/2}\left(\frac{f}{M}\right)^{1/2}\right.$$

$$+ 140\times 10^{-6}c^{1/2}f^{1/2}\left(\frac{L}{M_{seg}}\right)^{3/2}\log_{10}(N_{Re}f^{1/2})$$

$$- 2.303\cdot 70\times 10^{-6}\left(\frac{c}{M}\right)^{1/2}\left(\frac{ML}{M_{seg}}\right)^{3/2}\left(\frac{1}{f}\right)^{3/2}$$

$$- 52.5\times 10^{-6}c^{1/2}f^{1/2}\left(\frac{L}{M_{seg}}\right)^{3/2}\left(\frac{\rho\overline{V}^2K_GM}{8.8}\right)^{1/2}$$

$$\left.- 140\times 10^{-7}c^{1/2}f^{1/2}\left(\frac{L}{M_{seg}}\right)^{3/2} + \left(\frac{\rho\overline{V}^2K_G}{8.8}\right)^{1/2}\left(\frac{f}{M}\right)^{1/2}\right] \qquad (7)$$

where it is assumed that $R_G^3 = K_GM$. K_G is a constant characteristic of each polymer solvent combination. Computation of the drag reduction rate of decrease by use of equation 7 is straightforward but requires evaluations of many terms.

A typical computation of the rate of decrease of drag reduction (increase of drag) may be done using data in Figures 3 and 4 and standard constants for drag calculations. The degrading polymer is assumed to be 1.0% polyisobutylene of $M_w = 4\times 10^6$ in an oil of 2 cp viscosity. The solution viscosity is 13 cp. The flow is at 10 m/s in a pipe of 20 cm inside diameter at 300°K. The Deborah number (N_{De}) is evaluated to be 0.049 $m^{0.8}$. Based on a value of $f/f_{pv} = 0.5$ at $N_{De} = 0.03\ m^{0.8}$, $f/f_{pv} = 0.42$ at $N_{De} = 0.043$ (see Figure 3). M_∞ is evaluated at $\tau_w = 525$ dynes/cm^2 to be 3.49×10^6. τ_{min} is evaluated at $M = 4\times 10^6$ to be 297 dynes/cm^2. Since f_{pv} is calculated to be 0.0025, then f = 0.00105. For the solution in question the values of the constants A,C, and E are -18.0, -44.3, and 2960.

Using these values in equations 5 and 6 yields a rate of drag increase of 0.36% per second. This is a seemingly high rate of degradation, but as the M_∞ value is approached by M, the rate will decrease rapidly. Only with pipe flow degradation rate measurements correlated with such calculations can the final adjustments be made in the model to obtain results useful for design.

CONCLUSION

It is feasible to relate the rate of drag reduction decrease to degradation kinetics as a function of shear stress. The above approximate relationships must be refined through use of more data on final degradation levels and degradation rates both in couette flow and in pipes using the same solutions in both.

REFERENCES

1. Kotliar, A.M., J. Poly. Sci., *A, 1,* 3175 (1963).

2. Yu, J.F.S., J.L. Zakin and G.K. Patterson, J. Appl. Poly. Sci., *23,* 2493 (1979).

3. Shen, A.M., *Rate of Shear Degradation of Polymer Solutions,* M.S. Thesis, Univ. of Missouri-Rolla, 1981.

4. Chen, I.M., *Study of Rate of Mechanical Shear Degradation of Polymer Solutions,* M.S. Thesis, Univ. of Missouri-Rolla, 1983.

5. Virk, P.S., A.I.Ch.E. J., *21,* 625 (1975).

6. Rodriguez, J.M., J.L. Zakin, and G.K. Patterson, Soc. Petrol. Eng. J., *7,* 325 (1967).

7. Patterson, G.K., "Turbulent Drag Reduction", Module C2.5, Transport Volume, A.I.Ch.E. Modular Instruction Series, 1981.

Session 6
Theories I

Turbulence Measurements with Viscoelastic Flow in an Open Channel

H.S. LEE
Hyundai Motor Co.
Seoul, Korea

T.F. IRVINE
SUNY, Stony Brook
Stony Brook, New York

H. USUI
Yamaguchi Univ.
Ube, Japan

ABSTRACT

A semi-emperical turbulent one-equation model is developed for rectangular open channel flows of water and viscoelastic fluids. The model is used to predict friction factor vs. Reynolds number relations, velocity profiles, eddy viscosity distributions and turbulent energy budgets. Comparisons are made between the model and the measured results using a Laser Doppler Anemometer.

In addition, LDA measurements are presented of turbulent intensities for both water and polymer solutions. The latter have smaller peak values than water and the peaks occur at higher values of y^+ for the polymer solutions.

I. Introduction

Intensive research efforts have been directed toward an understanding of the mechanism of drag reduction since the phenomenon was discovered by B. A. Toms (1) in 1948. The flow behavior of the drag reducing fluid has been well summarized by Virk (2) and Poreh and Dimant (3). More recent studies of Mizushina, et al. (4), Donahue, et al. (5) and Oldaker, et al. (6), revealed that drag reducing flows have a lower spatially averaged bursting rate, less turbulent momentum transport, and hence a reduced wall shear stress. Also a decrease of turbulent eddy viscosity and an increase of the integral scale of wall turbulence were confirmed.

For open channel flows, Sellin (7,8), Forester (9), and Petersen (10) investigated the polymer effect and obtained discharge capacity increase in the range of 40 ~ 50%.

Usui and Irvine (11) investigated the drag reduction phenomena in a 90 degree triangular open channel. Velocity distributions were measured by a laser doppler anemometer and the laminarization of the flow was discussed. Velocity distributions and friction factors were calculated by adopting Usui's eddy viscosity model and Krajewski's integral transformation (12). The predictions of the analysis showed reasonable agreements with the measured friction factors at a small slope (s = 0.3%), but failed to explain the flow behavior at large slopes (s = 0.45%, 1%).

The purpose of the present investigation was to study, more fully, drag reduction phenomena in open channels and the complete results are

The Influence of Polymer Additives on Velocity and Temperature Fields
IUTAM Symposium Essen 1984
Editor: B. Gampert

presented in the Ph.D. thesis of the senior author (13). A portion of this thesis material has been presented elsewhere (14) and accordingly these portions will only be summarized here, the unreported turbulence measurements will be emphasized.

II. Analysis

The equations of motion and the energy equation for turbulent flows are:

$$\frac{Du_i}{Dt} = -\frac{1}{\rho}\frac{\partial P}{\partial x_i} + \frac{\partial}{\partial x_j}\left(\nu\frac{\partial u_i}{\partial x_j} - \overline{u_i u_j}\right) \quad (1)$$

$$\frac{Dk}{Dt} = \frac{\partial}{\partial x_j}\left\{\nu\frac{\partial k}{\partial x_j} - \overline{u_j\left(\frac{P}{\rho} + \frac{u_i^2}{2}\right)}\right\} - \overline{u_i u_j}\frac{\partial u_i}{\partial x_j} - \nu\overline{\left(\frac{\partial u_i}{\partial x_j}\right)^2} \quad (2)$$

In Eq. (1) and (2), there are more unknowns than the number of equations which is the basic closure problem. To overcome this difficulty, additional closure equations have to be added in order to solve it.

In this investigation, a one-equation model which is a modification of the Hassid and Poreh (15) scheme by adopting Mizushina and Usui's (4) damping factor is developed. By modifying the dissipation term to account for the polymer effects in the buffer region similar to that of Kubo (16), the eddy diffusivity and dissipation term can be expressed as

$$\varepsilon_m = C_\mu k^{\frac{1}{2}}L\{1-\exp[-A_\mu R_t(-\alpha+(\alpha^2+1)^{\frac{1}{2}})^{\frac{1}{2}}]\} \quad (3)$$

$$\text{Diss.} = \frac{2\nu k}{L^2} + \frac{C_D k^{3/2}}{L}\{1-\exp[-A_\mu R_t(-\alpha+(\alpha^2+1)^{\frac{1}{2}})^{\frac{1}{2}}]\} \quad (4)$$

The value of α is determined in basically the same way as Mizushina and Usui (4) with a slope dependent term.

$$\alpha = \left\{\frac{2\lambda}{\nu}\right\}\left\{\frac{u^{*2}}{26}\right\} f(s) \quad (5)$$

Where λ, ν and u^* are the relaxation time of the polymer solution, the kinematic viscosity and the shear velocity, respectively.

The slope dependant function f(s) was determined empirically from the f-Re relation and can be expressed as $f(s) = \frac{2}{38s+1}$ where s is the slope of the channel. The constant λ is determined from the relaxation time for simple laminar shear flow, λ_ℓ, as

$$\lambda/\lambda_\ell = 3.76\text{x}10^8 We^{1.34} \quad (6)$$

where We is the Weisenberg number, $\lambda_\ell\ \bar{u}/De$, $\bar{u}$ is the mean velocity and De is the flow depth. The relaxation time of the Polymer in a solution

λ_ℓ, was derived by James (17) using linear viscoelastic theory as given by Rouse (18). This relaxation time is only applicable to a very dilute Polymer solution and can be expressed as

$$\lambda_\ell = \frac{2}{5}\frac{\eta_s[\eta]^2 MC}{K_2 T} \tag{7}$$

where η_s is the viscosity of the solvent, $[\eta]$ is the intrinsic viscosity of the solution, M is the molecular weight of the Polymer, C is the Polymer concentration, K_2 is the gas constant and T is the absolute temperature. Thus the basic equations for open channel turbulent flow are

$$(1 + \varepsilon^+)\frac{du^+}{dy^+} = 1 - y^+/d^+ \tag{8}$$

$$L^+ = y^+\sqrt{1 - y^+/d^+} \tag{9}$$

$$\varepsilon^+ = C_\mu k^{+\frac{1}{2}} L^+ \left|1 - \exp\{-A_\mu k^{+\frac{1}{2}} L^+ (-\alpha + (\alpha^2+1)^{\frac{1}{2}})^{\frac{1}{2}}\}\right| \tag{10}$$

$$\frac{d}{dy^+}\left\{\left(1 + \frac{\varepsilon^+}{\sigma_k}\right)\frac{dk^+}{dy^+}\right\} = \frac{\varepsilon^+(1-y^+/d^+)^2}{(1+\varepsilon^+)^2} - \frac{2k^+}{L^{+2}} \tag{11}$$

$$- \frac{C_D k^{+3/2}}{L^+}\left\{1 - \exp\left|-A_\mu k^{+\frac{1}{2}} L^+ \sqrt{-\alpha + (\alpha^2+1)^{\frac{1}{2}}}\right|\right\}$$

with boundary conditions

$$k^+ = 0 \text{ at } y^+ = 0 \tag{12}$$

$$k^+ = 0 \text{ at } y^+ = d^+ \tag{13}$$

where $k^+ = k/u^{*2}$, $L^+ = Lu^*/\nu$ and $\varepsilon^+ = \varepsilon_m/\nu$ (14)

The constants used are almost the same as those used by Hassid (15). $C_D = 0.416$, $C_\mu = 0.22$, $A_\mu = 0.012$, $\sigma_k = 1.0$.
A finite difference method was used to solve the above equations.

III. Experimental Apparatus and Procedure

Flow System

A rectangular open channel (6.35cm x 7.62cm x 556cm) was built from 1.27cm thick plexiglass plate and used as a test section. Two dial calipers were used to measure the fluid depth and hydraulic jacks were adjusted to obtain the desired slopes. A two ton mixing tank was large enough to run the experiment with Reynolds number up to 70,000. The details of flow systems can be found in (13).

Laser Doppler Anemometry System (LDA)

A DISA LDA system was used to measure mean velocity profiles and turbulent intensities at the centerline of the channel. The LDA system was placed on a milling machine bed which provides a 3-dimensional coordinate system with an accuracy up to 2.54×10^{-3} cm. An averaging circuit (TSI 1047) and a digital voltmeter (HP 3465A) were used to measure the analog voltage output from the tracker. For the turbulence intensity measurement, a DISA 55D35 RMS unit was used. For scattering particles 30 wppm of powdered coffee cream was used both for the water and the Polymer solutions.

IV. Results and Discussion

The measurements and predictions of friction factor vs. Reynolds number with 100, 50 and 0 ppm solutions are shown in Fig. 1. The calculated results show good agreements with the measured values in the fully developed flow region. For different concentrations and different slopes, the same trend exists. No significant effect of the Froude number was apparent.

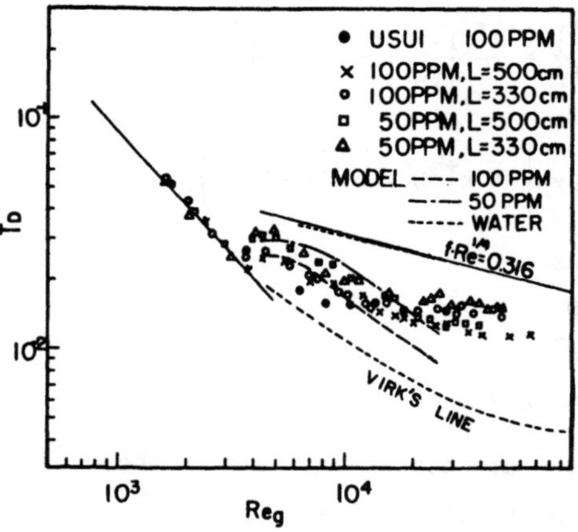

FIG. 1. FRICTION FACTOR MEASUREMENT AT THE DOWN STREAM STATION (L = 500CM)

Velocity Profile Measurements

The LDA measuring station was at 400 cm from the inlet of the channel. The results of the velocity distributions for a slope of 1% is shown in Fig. 2. The letters A, B, and C in the figures mean slopes of 0.2, 0.45 and 1 percent respectively. The following numbers show the concentration of polymer, i.e., 5 means 50 wppm, 7 means 70 wppm and 10 means 100 wppm. The third number indicates the experimental run number. In Fig. 2, Virk's (19) ultimate profile and the logarithmic law region profile does not correlate the data as well as the Dimant and Poreh (20)

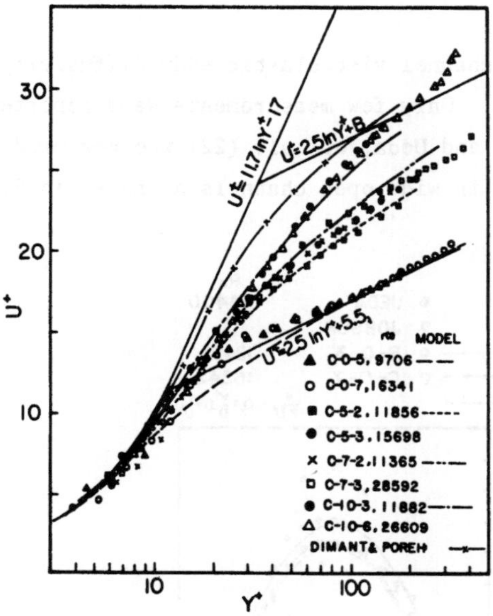

FIG. 2. U^+ - Y^+ MEASUREMENT AT S = 0.01

model or the model developed in this study. The present model predicts the u^+ - y^+ distribution fairly well except near the free surface. The sudden increase of velocity near free surface could be explained by the increase of the turbulence length scale in viscoelastic fluids which increases the free surface damping. Also, undeveloped flow caused by the increased entrance lengths in higher concentrations of Polymer contributes to the acceleration near the free surface.

Eddy Viscosity Measurements

The eddy diffusivity of momentum can be calculated as follows

$$\varepsilon^+ = \frac{\varepsilon_m}{\nu} = (1 - y/d)/(du^+/dy^+) - 1 \tag{15}$$

$$= (1 - \frac{y^+}{d^+})/(du^+/dy^+) - 1 \tag{16}$$

and $$\frac{\varepsilon_m}{\nu d^+} = (1 - y^+/d^+)/d^+(du^+/dy^+) - 1/d^+ \tag{17}$$

Least squares curves in the wall, core and surface regions calculated from the velocity measurement results. The three regions have overlapping zones to prevent discontinuities. The results were substituted in Eqs. (16) and (17) to calculate the eddy diffusivity of momentum or the normalized eddy diffusivity.

Until now, open channel viscoelastic eddy diffusivity measurements have not been available. Only few measurements were reported with water by Jobson, et al. (21) and Ueda, et al. (22) who measured the eddy diffusivity of momentum in wide open channels as shown in Fig. 3.

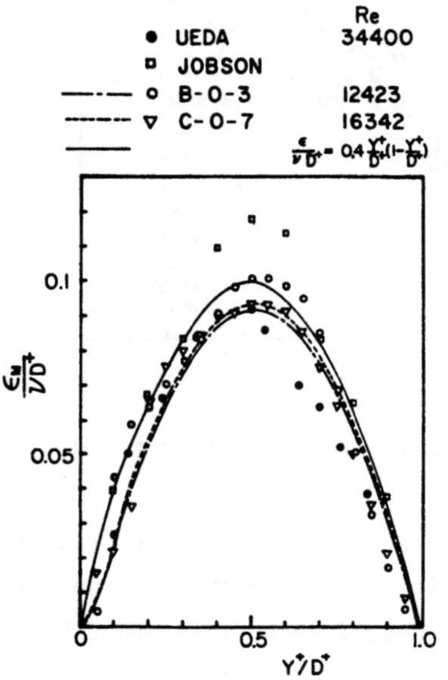

FIG. 3. NORMALIZED EDDY VISCOSITY OF WATER

In Fig. 3, these experimental results agree fairly well with Eq. (15). The eddy diffusivity of momentum results show their maximum at $\frac{y}{d}$ = 0.5 - 0.55 for both slopes (s = 0.0045 and s = 0.01). The numerical calculation results are also shown in Fig. 3 at two different Reynolds numbers. These numberical results underestimate the eddy diffusivity. But if we consider the accuracy of the eddy diffusivity measurements, especially the difficulty in measuring the velocity gradient, then the predictions would seem to be reasonable.

Figure 4 shows the eddy viscosity results without the depth normalization. For 100 wppm, the values are only 25% of the eddy viscosity of water.

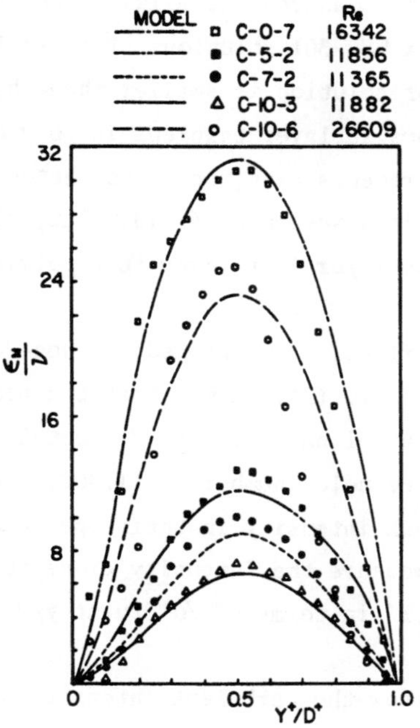

FIG. 4. EDDY VISCOSITIES AT S = 0.01

The Turbulent Intensity Measurements

For the turbulent intensity of a drag reducing fluid in closed ducts, a few experimental results are available (4, 16, 23, 26). For open channel flows, only Newtonian intensity results have been reported (24,25).

Rudd (26) observed a large increase of the axial intensity distribution for a polymer solution in a square duct although no other experimental measurements of polymer solutions in channels and pipes show such a large increase. His 0.01% Separan AP 30 results show that the peak intensity in terms of $\sqrt{u'^2}/u^*$ is near to 8 at y = 20, while the peak intensity of water is 4 at $y^+ = 10$. Logan (23) measured the turbulent intensity in a square duct using 50 ppm Polyox solution. Logan's results show that the turbulent intensities of polymer solutions have smaller peak values at a further location than that of water. Mizushina and Usui (4) measured axial turbulent intensities in a circular pipe with a polymer solution. They proposed that the turbulent intensities in the wall region are damped in proportion to the rate of drag reduction in the polymeric region, and that the damping effect ceases to increase at the maximum reduction asymptote.

Recently, Kubo (16) measured the axial turbulent intensity in a circular pipe with Polyox WSR 301 solution. The result shows that the peak intensity of polymer solution is smaller than that of water. Also, he confirmed that the viscous sublayer thickens in polymer solutions. So far, the peak intensity measurements disagree as to whether they are higher in a polymer solution than in a Newtonian fluid. But, all the measurements agree that the viscous sublayer is thicker in a polymer solution at the peak intensity location.

For open channel flows, as mentioned, no non-Newtonian results are available. Nakagawa, et al. (24) showed that the peak intensity of water in a rectangular open channel has a value of about 2.7 in terms of $\sqrt{\overline{u'^2}}/u^*$ at $y^+ = 15$ at a Reynolds number of 11,000. McQuivey, et al. (25) failed to measure the peak intensity in their open channel measurements because they could not measure the intensity close to the wall. Their highest intensity was 0.16 in terms of $\sqrt{\overline{u'^2}}/\bar{u}$ at $y/d = 0.1$ at a Reynolds number of 11,900.

Figures 5 and 6 show the turbulent intensities in the longitudinal direction at two different slopes; 0.0045 and 0.01 respectively.

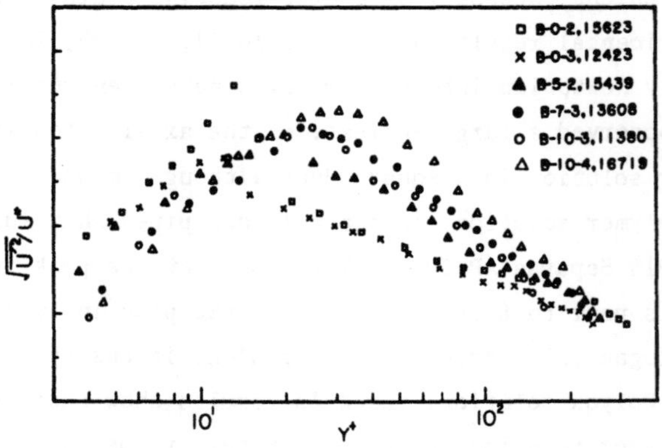

FIG. 5. TURBULENT INTENSITIES AT S = 0.0045

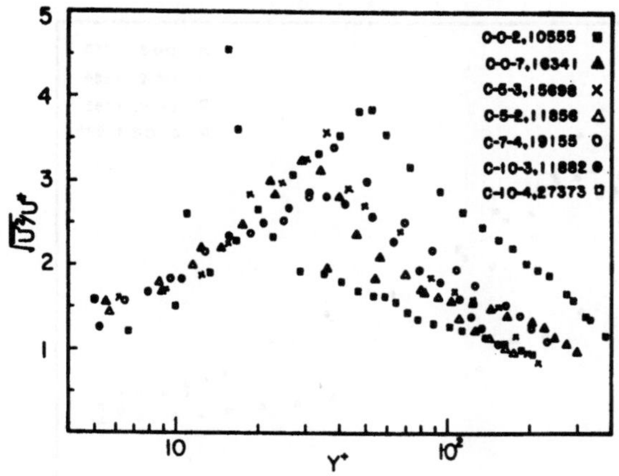

FIG. 6. TURBULENT INTENSITIES AT S = 0.01

The peak intensities occur further from the wall with increasing slope. These effects are more obvious in Figs. 7 and 8.

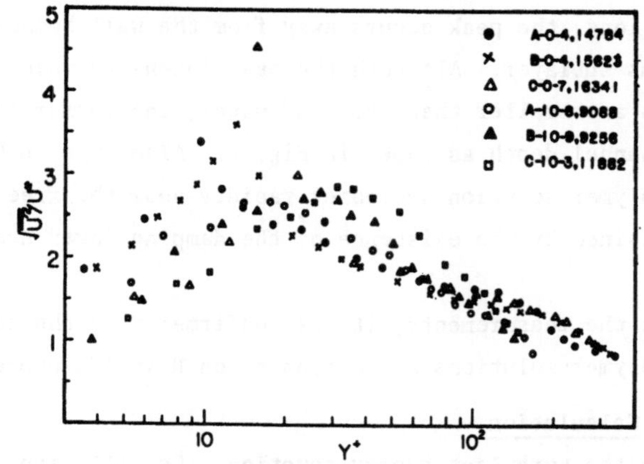

FIG. 7. SLOPE EFFECT ON TURBULENT INTENSITIES

Figure 7 shows the slope effect and Fig. 8 shows the concentration effect. The peak intensities of the polymer solution are smaller than those of water. The water results show sharp peaks, up to 4.5, at y^+ around 15 at the Reynolds number of 16,341. Thus, Mizushina and Usui's (4) and Logan's results (23) are in good agreement with the present results.

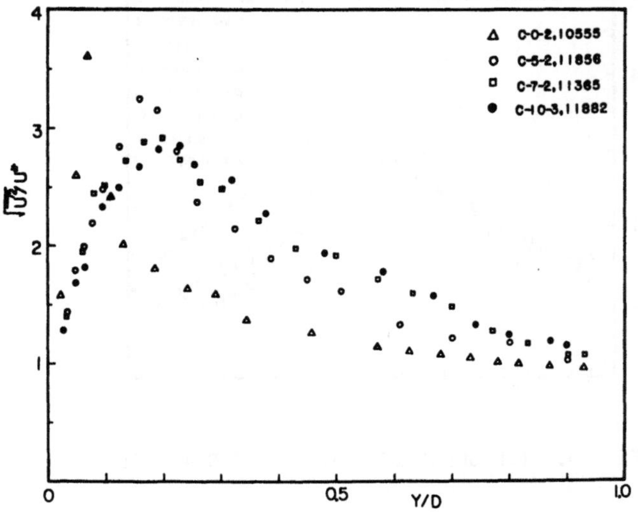

FIG. 8. POLYMER CONCENTRATION EFFECT OF TURBULENT INTENSITIES

For polymer solutions, the peak occurs away from the wall because of the thickening viscous sublayer. Although the peak intensity values of the polymer solutions are smaller than those of water, the former stay higher in most of the channel depth as shown in Fig. 8. Also, the turbulent intensity of a polymer solution decreases rapidly near the free surface which can be explained by the existence of the damping layer near the surface.

Throughout the measurements, it was confirmed that the turbulent intensities of Polymer solutions are dependent on Reynolds number.

Turbulent Energy Calculation

By solving the turbulent energy equation, (Eq. 11), the dimensionless turbulent kinetic energy distribution was obtained as shown in Fig. 9.

For the four different concentrations (water, 50, 70, and 100 wppm solutions), the turbulent kinetic energy, k^+, decreases rapidly near the free surface and must go to zero at the surface because this is a boundary condition for the model.

The Newtonian turbulent kinetic energy results of this calculation compare favorably with that of Hassid and Poreh's (15) in a circular tube. Also, Laufers' (26) measured results in a large square duct agree fairly well with these predictions as shown by square data points in Fig. 8. A definite increase in k^+ is observed for flows with drag reduction. However, measurements of turbulent kinetic energy in flows with drag reduction are

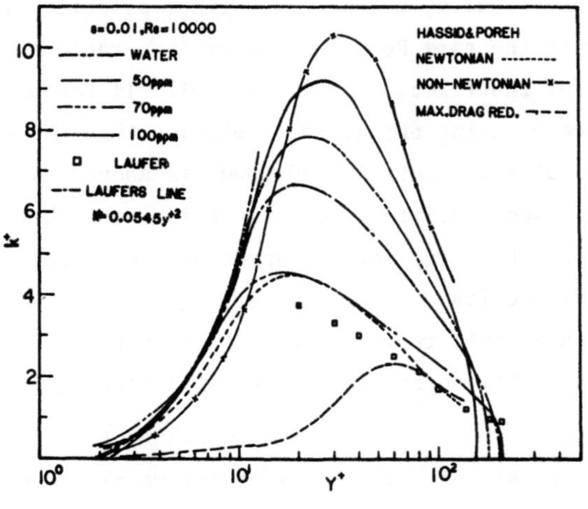

FIG. 9. CALCULATED TURBULENT KINETIC ENERGY DISTRIBUTION

not available to compare with the present model.

Hassid and Poreh's (15) model calculation results for $R^+ = 100$ and $A = 0.001$ show a maximum k^+ value of 10.7. According to their calculations the dimensionless turbulent kinetic energy is increasing in the small drag reduction region and then starts to decrease in the large drag reduction region, and eventually the dimensionless turbulent kinetic energy becomes smaller than that of Newtonian flow in the maximum drag reduction zone as in Fig. 8.

They assumed that the longitudinal velocity fluctuation, u', accounts for most of the turbulent kinetic energy in the wall region. Then using Reishmann (27) and Rudd's (26) measured fluctuation results, the dimensionless turbulent kinetic energy ratio of non-Newtonian to Newtonian $(\frac{k^+}{k_N^+})$ was calculated in terms of the percentage reduction. According to their results, Reishmann's 100 ppm Separan data (27) with 38.9% drag reduction shows about 2.3 times larger turbulent kinetic energy than that of the water of $y^+ > 50$.

In these measurements, the 100 wppm at Re = 10,000 results shows approximately a 45% drag reduction and the turbulent kinetic energy ratio of the maximum values is about two which is quite close to Reishmann's results.

But in any case, the absolute value of the turbulent kinetic energy of the Polymer solutions before normalizing by u^* is not increased compared to that of water. The considerable increase occurs during the normalizing, because the friction velocity, u^*, of the Polymer solutions is much smaller than that of water at the same Reynolds number (the ratio of u^* is the same as the square root of the friction factor ratio). If the turbulent kinetic energy is normalized by using the average velocity then the turbulent kinetic energy of the Polymer solution (100 wppm) is about 15% smaller than that of water at the same average velocity. This shows that, with Polymer solutions, the turbulent energy level changes only moderately compared to the amount of drag reduction.

In the very near wall region, Laufer (26) proposed the turbulent kinetic energy distribution described by the following equation

$$k^+ = 0.0545\, y^{+2} \tag{18}$$

This equation shows excellent agreement with the water and Polymer solution prediction of this work as shown in Fig. 8 at $y^+ < 10$ because even the Polymer solutions behave like water in the viscous-sublayer.

The turbulent energy budgets were calculated for open channel flows with a water and a Polymer solution. As already mentioned, the turbulent dissipation term is divided into two parts. The third term in Eq. (11), $2k^+/L^{+2}$, is interpreted as the dissipation due to the large eddies which partially submerge in the viscous sublayer as Tounsend (28) explains. Since its scale is much larger than the Polymer molecules and it loses energy mainly in the viscous sublayer region where Polymer molecules are not stretched, the dissipation due to the large eddy will not be affected by the Polymer concentration. The last term of Eq. (11),

$$(C_D k^{+3/2}/L^+)\{1-\exp[-A_\mu k^{+\frac{1}{2}} L^+ (-\alpha+(\alpha^2+1)^{\frac{1}{2}})^{\frac{1}{2}}]\}$$

is the dissipation due to small eddies through the energy cascade process. At high turbulent Reynolds numbers, dissipation is mainly due to the small eddies and large eddy dissipation is almost negligible. In the buffer zone, however, the turbulent Reynolds number is not large so that turbulent energy is dissipated even in the smaller wave number region. Also, in the region close to the wall (viscous sublayer), viscous diffusion is balanced by viscous dissipation and the turbulent kinetic energy varies as y^{+2}.

The calculated energy budget of the Newtonian fluid is shown in Fig. 10. The peak of the production term happens at $y^+ = 10$ which is in good agreement with the turbulent intensity measurements where the peak value at around $y^+ = 10$. For the 100 wppm Polymer solution at Re = 10,000, Fig. 11, the peak of the production is approximately 20% smaller than that of water and the location is moved away from the way (about $y^+ = 20$) which again agrees with the intensity measurements.

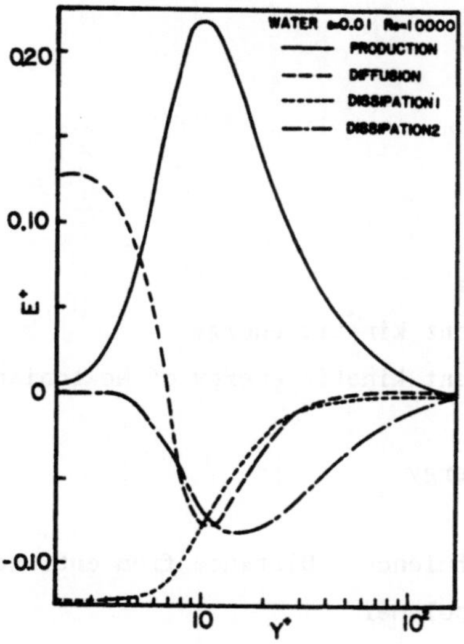

FIG. 10. CALCULATED ENERGY BUDGET OF WATER AT S = 0.01 at RE = 10^4

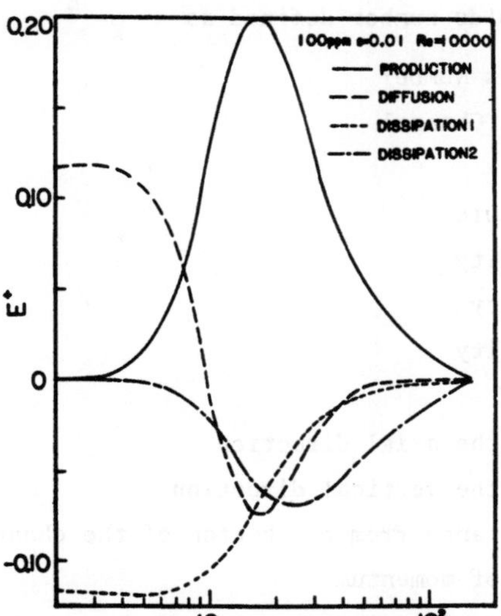

FIG. 11. CALCULATED ENERGY BUDGET OF 100 PPM SOLUTION AT S = 0.01 at Re = 10^4

List of Symbols

A_μ:	Constant
C:	Polymer concentration
C_μ:	Constant
C_D:	Constant
D_e,d:	Channel Flow Depth
d^+:	Dimensionless Depth
f_D:	Darcy Friction Factor
k^+:	Dimensionless turbulent kinetic energy
k_N^+:	Dimensionless turbulent kinetic energy of Newtonian Fluid Flow
K_2	Gas Constant
k:	Turbulent kinetic energy
K:	Fluid consistency
l,L:	Length scales of turbulence: Distance from entrance to measuring static
M:	Molecular weight of polymer
n:	Flow index
P:	Pressure
ppm:	Parts per million by weight
Re:	Reynolds number $\frac{\bar{u}D_h}{\nu}$
Re_g:	Generalized Reynolds number defined as $\frac{\rho\bar{u}^{2-n}D_h^n}{K}$
R_t:	Turbulent Reynolds Number
s:	Slope of the Open Channel
t:	Measuring time
T:	Absolute temperature
u:	Local axial velocity
$\bar{u}$:	Local mean velocity
$\bar{u}'$:	Fluctuation velocity
u^*:	Shear velocity
x:	Coordinate along the axial direction
y:	Coordinate along the vertical direction
y^+:	Dimensionless distance from the bottom of the channel
ε_m,ε:	Eddy diffusivity of momentum
ε_m^+,ε^+:	Dimensionless eddy viscosity
ρ:	Density of the solution
λ_1:	Polymer relaxation time for simple laminar flow
λ:	Polymer relaxation time for turbulent flow
ν:	Kinematic viscosity of solution

α: Parameter

η_s: Solvent viscosity

$[\eta]$: Intrinsic viscosity

σ_k: Turbulent Prandtl number

References

1. Toms, A.B., "Some Observations on the Flow of Linear Polymer Solutions through Straight Tubes at Large Reynolds Numbers", proceedings International Congress on Rheology, 1948, Vol. II, North Holland Publishing Co., Amsterdam, 135 (1949)

2. Virk, P.S., "Drag Reduction Fundamentals", AIChE Journal, 21, 4, 625 (1975)

3. Poreh, M., and Y. Dimant, "Heat Transfer in Flows with Drag Reduction", Advances in Heat Transfer 12, 77, 1976

4. Mizushina, T., and H. Usui, "Reduction of Eddy Diffusivity for Momentum and Heat in Viscoelastic Fluid Flow in a Circular Tube", Phys. Fluids Suppl., 20, S 100 (1977)

5. Donohue, G.L., Tiederman, W.G. and Reischman, M.M., "Flow Visualization of the Near Wall Region in Drag Reducing Channel Flow", J. Fluid Mech. 56, 559 (1972)

6. Oldaker, D.K., and Tiederman, W.G., "Spatial Structure of the Viscous Sublayer in Drag-Reducing Channel Flows", Phy. Fluids, 20, S 133 (1977)

7. Sellin, R.H.J. and B.J.S. Barnard, "Open Channel Applications for Dilute Polymer Solutions", J. of Hydraulic Res., 8, 219 (1970)

8. Sellin, R.H.J., "Experiments with Polymer Additives in a Long Pipeline", 1st Int. Conference on Drag Reduction (B.H. R.A., Cambridge, U.K.) 3 (1975)

9. Forester R.H., R.E. Larson, J.W. Hayden and J.M. Wetzel, "Effects of Polymer Addition on Friction in a 10 inch Diameter Pipe", J. of Hydronautics, 3, No. 1, 59 (1969)

10. Peterson, J.P., "Increased Turbulent Dispersion in Dilute High Polymer Drag Reducing Open Channel Flow", Ph.D. Thesis, Clemson Univ. (1972)

11. Usui, H. and T.F. Irvine, Jr., "Drag Reduction in a 90 degree Triangular Open Channel", J. of Rheology, 24, 525 (1980)

12. Krajewski, B., "Determination of Turbulent Velocity Field in a Recti-linear Duct with Non-Circular Cross Section", Int. J. Heat Mass Transfer, 13, 1819 (1970)

13. Lee, H.S., "Turbulence Measurements and Modeling of Viscoelastic Fluid Flow in a Rectangular Open Channel", Ph.D. Thesis, Mechanical Engineering Dept., State University of New York, Stony Brook, N.Y. (1982)

14. H.S. Lee and T.F. Irvine, Jr., "Turbulence Measurements and Modeling of Viscoelastic Fluid in a Rectangular Open Channel", Presented Energy Sources Technology Conference, New Orleans, Louisiana, Feb. 12-16, 1984, ASME Publication FED. Vol. 11 (1984)

15. Hassid, S. and M. Poreh, "A Turbulent Energy Model for Flows with Drag Reduction", J. Fluids Engr. Trans. ASME, 97, 234 (1975)

16. Kubo, I., "Turbulence in Drag Reducing Polymer Solution", Proc. Symp. Turbulence in Liquids (1979)

17. James, D.F. and A.J. Acosta, "The Laminar Flow of Dilute Polymer Solutions Around Circular Cylinders", J. Fluid Mech., 42, 269 (1970)

18. Rouse, P.E., "A Theory of the Linear Viscoelastic Properties of Dilute Solution of Coiling Polymers", J. Chem. Phys., 21, 1272 (1953)

19. Virk, P.S., "An Elastic Sublayer Model for Drag Reduction by Dilute Polyme: Solutions of Linear Macromolecules", J. Fluid Mech., 45, Part 3, 417 (1971

20. Dimant, Y. and M. Poreh, "Heat Transfer in Flows with Drag Reduction", Advances in Heat Transfer, Academic Press, 12 (1976)

21. Jobson, H.E. and W. W. Sayre, "Vertical Transfer in an Open Channel Flow", J. Hydraulics Div. ASCE, 96, 703 (1970)

22. Ueda, H., R. Möler, S. Komori and T. Mizushina, "Eddy Diffusivity Near the Free Surface of Open Channel Flow", Int. J. Heat Mass Transfer, 20, No. 11 1127 (1977)

23. Logan, E., "Laser Velocimeter Measurement of Reynolds Stress and Turbulenc in Dilute Polymer Solutions", AIAA J., 10, 962 (1972)

24. Nakagawa, H., I. Nezu and H. Ueda, "Turbulence of Open Channel Flow over Smooth and Rough Beds", Proc. of JSCE, No. 241, 155 (1975)

25. McQuivey, R.S. and V. Richardson, "Some Turbulence Measurements in Open Channel Flow", J. Hydraulics Div. Proc. ASCE., 95, 209 (1969)

26. Laufer, J., "The Structure of Turbulence in Fully Developed Pipe Flow", NACA TN, 2954 (1953)

27. Reischman, M.M. and W.G. Tiederman, "Laser Doppler Flow", J. Fluid Mech., 70, part 2, 369 (1975)

28. Tounsend, A.A., "The Structure of Turbulent Shear Flow", Cambridge Univ. Press, Cambridge, Mass. (1954)

A Turbulence Model for Drag Reduction Dilute Polymer Solutions in Boundary Layer on a Flat Plate

A.N.Burov, B.P.Kolobov, A.N.Popkov

Institute of Theoretical and Applied Mechanics
USSR Academy of Sciences, Novosibirsk 630090, USSR

Summary

A turbulence model for small polymer additives in a boundary layer on a flat plate is proposed with an account being taken of optimum polymer concentration. Results of a numerical investigation of the solution flow on a flat semi-infinite plate are presented. These results are in a satisfactory agreement with axperimental data.

Turbulence model

Recently the Tom's effect has attracted a considerable attention of research workers. This is associated with the fact that negligible additives of polymer macromolecules lead to a sharp change in a character of the near-wall turbulent flows which manifests itself in the form of a considerable diminution of the friction drag as well as the convective heat - and mass transfer and of changes in other turbulent characteristics.

The present paper deals with a diffusion model of turbulent boundary layer on a flat plate with small polymer additives.

The basic system of differential equations has the form

$$u\frac{\partial u}{\partial x}+v\frac{\partial u}{\partial y}=u_e\frac{du_e}{dx}+\frac{\partial}{\partial y}\left[(\nu+\nu_T)\frac{\partial u}{\partial y}\right]+D\frac{\partial c}{\partial y}\frac{\partial u}{\partial y},\quad(1)$$

$$u\frac{\partial c}{\partial x}+v\frac{\partial c}{\partial y}=\frac{\partial}{\partial y}\left(D\frac{\partial c}{\partial y}\right),\quad(2)$$

$$\frac{\partial u}{\partial x}+\frac{\partial v}{\partial y}=0.\quad(3)$$

Boundary conditions are as follows

The Influence of Polymer Additives on Velocity and Temperature Fields
IUTAM Symposium Essen 1984
Editor: B. Gampert

$$u(x,0) = u_0(x), v(x,0) = v_0(x),$$
$$a_1(x)\, c(x,0) + b_1(x) \frac{\partial c(x,0)}{\partial y} = d_1(x),\ a_1^2 + b_1^2 \neq 0, \tag{4}$$
$$u(x,\infty) = u_e(x),\ c(x,\infty) = c_e = const.$$

Initial conditions on the line $x = x_0$ have the form

$$u(x,y) = g_1(y),\ c(x,y) = g_2(y), \tag{5}$$

where x, y are coordinates measured along the body generatrix and along the normal to the body surface; u , v are velocity components along the x - and y - axis, respectively; ν , ν_T are coefficients of the kinematic molecular and turbulent viscosity, respectively; $D = \nu/S_C + \nu_T/S_T$ is the coefficient of the polymer diffusion in a solution; S_C , S_T are the laminar and the turbulent Schmidt numbers, respectively; C is the polymer concentration (the relation of polymer mass to the solution mass in a unit volume). The subscript "E" refers to the quantities calculated on the outer edge of the boundary layer.

The equations (1)-(3) were obtained under the condition that $C << 1$.

In the present work a turbulence model for turbulent viscosity is used which is based on the mixing length

$$\nu_T = \ell^2 |\partial u/\partial y|. \tag{6}$$

The mixing length ℓ in a near-wall layer is determined on the basis of the turbulence model modification with small polymer additives [3] :

$$\ell = \psi \kappa y (1 - exp(-y\sqrt{\tau/\rho}\, F/26\nu)), \tag{7}$$

where the function F has the form

$$F = 1 - f_1(M,C) f_2(u^*/\hat{u}^*_{nop}), \tag{8}$$

$$f_2 = \begin{cases} 0, if\ u^* \leq \hat{u}^*_{nop}, \\ \frac{2}{\pi} \text{arctg}\, (1{,}7(u^*/\hat{u}^*_{nop} - 1)),\ if\ u^* > \hat{u}^*_{nop}, \end{cases} \tag{9}$$

where u^* is the dynamic velocity.
$\hat{u}^*_{\text{пор}} = E_0 M^{-.89}$ is the threshold dynamic velocity at the wall. $E_0 = const$, M is the moleoular polymer weight.

The function $f_1 = f_1(M, C, C_{opт})$ is determined from the comparison with experimental data with regard for the presence of the optimum solution concentration $C_{opт}$ at which the effect of the polymer influence is the greatest. Experimental data [3] show a nonmonotonic influence of polymer additives on the reduction in the friction drag. This may be taken into account as follows

$$f_1 = \begin{cases} 58 \text{ arctg } (1.2M^{.85}C), & \text{for } C \leq C_{opт}, \\ 58 \text{ arctg } (1,2M^{.85}C^2_{opт}/C), & \text{for } C > C_{opт}. \end{cases} \quad (10)$$

From the analysis of experimental data it can be assumed in the first approximation that

$$C_{opт} = C_{opт}(M) = \gamma / M^{.85} \quad (11)$$

where γ is the model parameter determined experimentally. Outside the near-wall region the mixing length is determined by a dependency [2]

$$\ell = K_1 \delta^{**} = 1.18756 \delta^{**} \quad (12)$$

where δ^{**} is the momentum loss thickness.

<u>Solution algorithm and the results</u>

Let us introduce the new variables

$$\begin{aligned} &\bar{x} = \int_0^x \sigma(x)\,dx, \quad \bar{y} = \int_0^y Q(x,z)\,dz, \\ &u = u_e \varphi(\bar{x},\bar{y}) = \frac{\partial}{\partial \bar{y}}\left(\sqrt{2\nu\xi}\,\phi(\bar{x},\bar{y})\right), \\ &v = -\frac{\partial}{\partial x}\left(\sqrt{2\nu\xi}\,\phi(\bar{x},\bar{y})\right), \end{aligned} \quad (13)$$

where

$$\begin{aligned} &\sigma(x) = f(x),\ Q(x,y) = \frac{u_e}{\sqrt{2\nu\xi}}\,\frac{f(x)}{1+\nu_T/\nu},\ \xi = \int_0^x f u_e\,dx, \\ &\phi(\bar{x},\bar{y}) = \phi(\bar{x},0) + \int_0^y (1+\nu_T/\nu)/f(\bar{x})\,\varphi(\bar{x},\bar{y})\,d\bar{y}, \\ &\phi(\bar{x},0) = -\int_0^{x(\bar{x})} v_0(x)\,dx/\sqrt{2\nu\xi} \end{aligned} \quad (14)$$

the function $f(x)$ is defined in what follows.

In the new variables the problem (1)-(5) may be written in the form

$$\frac{\partial^2 \varphi}{\partial \bar{y}^2} + \left(\phi + \alpha\frac{\partial \phi}{\partial \bar{x}} + D_1\frac{\partial c}{\partial \bar{y}}\right)\frac{\partial \varphi}{\partial \bar{y}} + BN(1-\varphi^2) = \alpha N \varphi \frac{\partial \varphi}{\partial \bar{x}}, \quad (15)$$

$$\frac{\partial}{\partial \bar{y}}\left(D_1 \frac{\partial c}{\partial \bar{y}}\right)+\left(\phi+\alpha \frac{\partial \phi}{\partial \bar{x}}\right)\frac{\partial c}{\partial \bar{y}}=\alpha N \varphi \frac{\partial \varphi}{\partial x} \tag{16}$$

with boundary conditions

$$\varphi(\bar{x},0)=u_0(\bar{x})/u_e(\bar{x}),\quad \varphi(\bar{x},\infty)=1,$$
$$\alpha_1(\bar{x})\,c(\bar{x},0)+b_1(\bar{x})\,Q(\bar{x},0)\,\frac{\partial c(\bar{x},0)}{\partial \bar{y}}=d_1(\bar{x}), \tag{17}$$
$$c(\bar{x},\infty)=c_e.$$

On the line $\bar{x}=\bar{x}_0$ the system (15)-(16) reverts to ordinary differential equations whose solutions give initial conditions. In (15)-(16) the following notations are introduced

$$N=(1+\nu_T/\nu)/f(\bar{x}),\quad D_1=D/(\nu+\nu_T),$$
$$\alpha=2\xi/u_e,\quad B=\frac{2\xi}{u_e^2}\cdot\frac{du_e}{d\bar{x}}.$$

The equations (15)-(16) were integrated numerically by the modified technique of [2].

In the computations the normalizing function $f(x)$ was taken to be equal to

$$f(x)=1+\psi\bar{C}_1\,(u_e x/\nu)^{\bar{C}_2},$$

where $\bar{C}_2=.8$, $\bar{C}_1=(.77\div 2{,}5)\cdot 10^{-4}$ for the laminar flow case, $\psi=1$ for the turbulent flow.

Basic computations were carried out for the flow of the solution of the polymer WSR-301 ($M=4\cdot 10^6$) at $S_c=S_T=1$.

Results of a comparison of the computed universal velocity profile on a plate with experimental data of [5] are presented in Fig. 1 for different values of the constant polymer concentration ($\hat{u}^*_{nop}/u_e$) = $3.644\cdot 10^{-4}$.

A comparison with the experiment of [6] on a plate for the comlete friction coefficient C_f is presented in Fig. 2 at $\hat{u}^*_{nop}/u_e = 3.644\cdot 10^{-4}$. Fig 3 shows the computations of diffusion boundary layer with homogeneous polymer injection at the beginning of a plate $0 \leq x/L \leq 0.005$ under the conditions $v_0/u_e = 10^{-4}$, $\hat{u}^*_{nop}/u_e = 3.644\cdot 10^{-4}$.

Experimental data of [7] and computed data on the reduction in the friction drag are presented in Fig. 4 for the flow of polyethulene oxide solutions depending on the polymer concentration.

References

1. Ignatiev,V.N.; Kuznetsov,B.G. : Diffusion model of a turbulent boundary layer with polymer. Chislennye Metody Mekhaniki Sploshnoi Sredy, Novosibirsk, 1973, 4, № 4, p. 78-87.
2. Kolobov,B.P.; Popkov,A.N.: Numerical investigation of turbulent flows with polymers in a two- dimensional boundary layer. Chislennoe Modelirovaniye v Dinamike Zhidkosti, Novosibirsk, ITPM SO AN SSSR, 1983, p. 41-50.
3. Sedov,L.I.; Vasetskaya,N.G.; Ioselevich,V.A., Pilipenko,V.N.: On the reduction in hydrodynamic drag by polymer additives. Mekhanika Turbulentnyh Potokov. Moscow, Nauka, 1980, p.7-28.
4. Alexin,V.A.; Mikhailu,A.G.; Pilipenko,V.N.: Effect of distributed polymer solution injection on the characteristics of a turbulent boundary layer. Mekhanika Zhidkosti i Gasa, 1983, №5, p.58-64.
5. Ivanyuta,Yu.F.; Chekalova,L.A.: Investigation of velocity profile of turbulent flows of weak polymer solutions in a pipe. Inzhenerno-Fizicheskii Zhurnal, 1974, v. 26, 5, p. 799-806.
6. Fruman D, Sulmont P.: Reduction de la resistance de frottement d'une plaque plane dans les solutions de polymeres. C.R.Acad. Sci., Paris, 1969, t.268, serie A., p. 1493-1496.
7. Wu, J.:Drag Reduction in External Flows of Additive Solutions. Viscous Drag Reduction, ed. by Welta, C.S. Plenym Press, New Jork, 1969, p. 331-350.

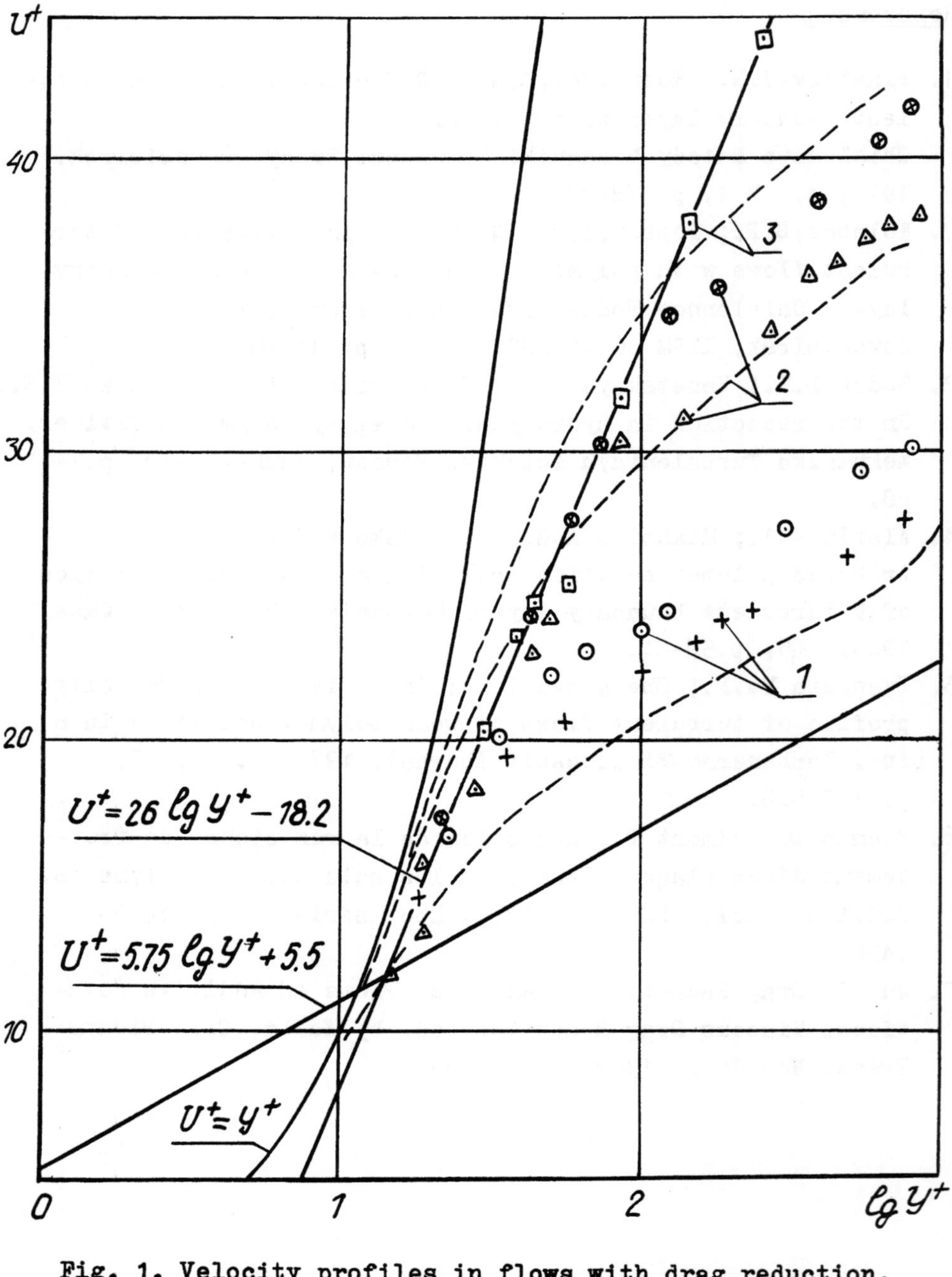

Fig. 1. Velocity profiles in flows with drag reduction. 1 - C = 2 × 10^6, 2 - C = 10^{-5}, 3 - C = 2.5 × 10^{-5}.

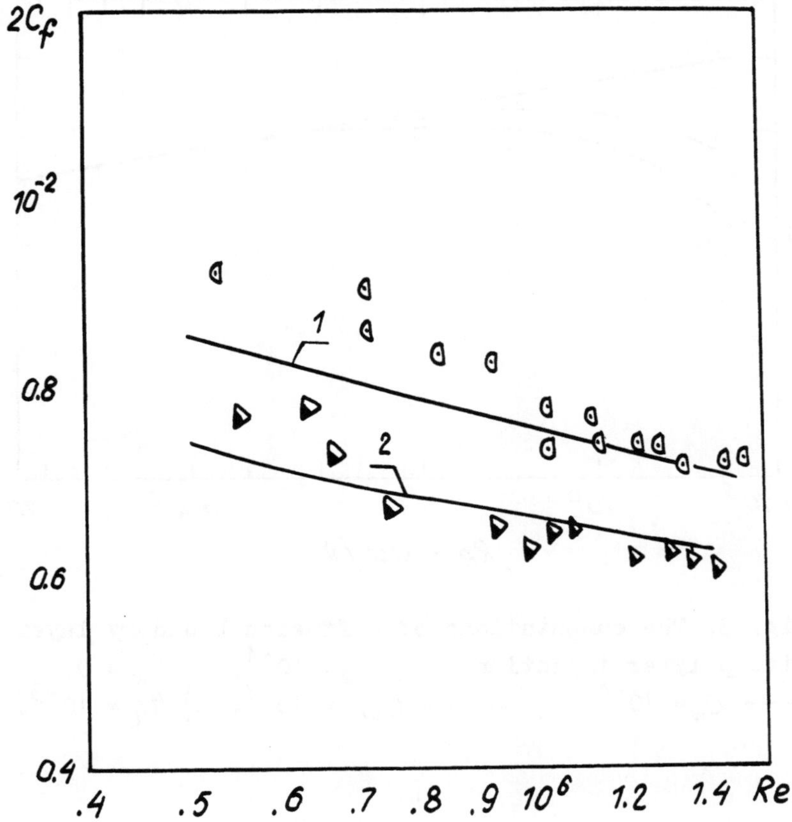

Fig. 2. Predicted and experimental data.
1 - $C = 10^{-6}$, 2 - $C = 2 \times 10^{-6}$.

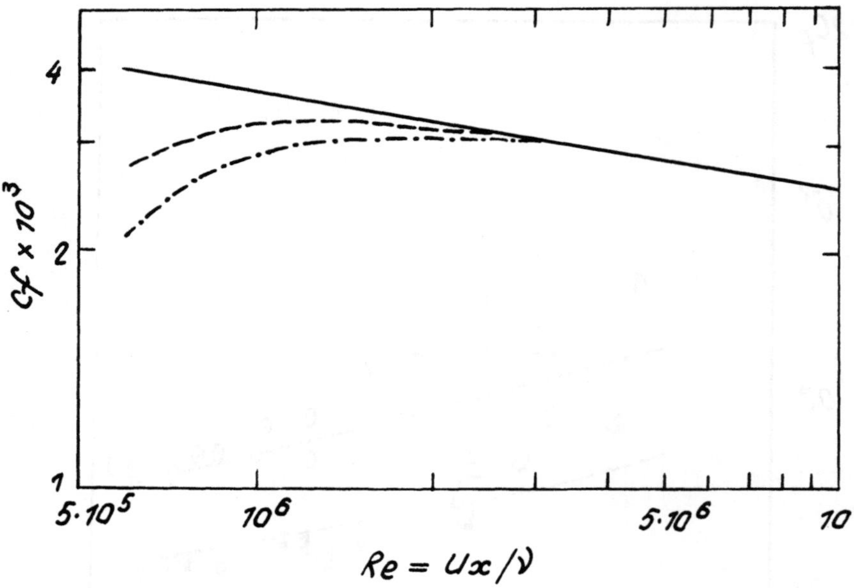

Fig. 3. The computations of diffusion boundary layer with polymer injection $v_0/u_e = 10^{-4}$, — $C_w = 0$, – – – $C_w = 10^{-4}$, — · — $C_w = 10^{-3}$, $v_0/u_e = 10^{-3}$.

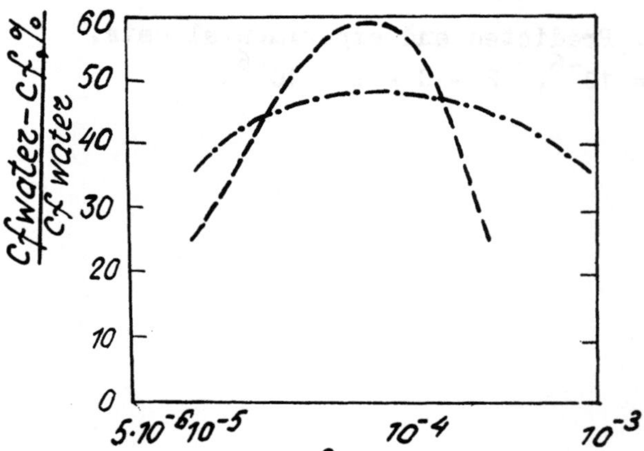

Fig. 4. The efficiency of polymer solutions as a function of concentration. – – experimental data, — · — predicted.

Session 7
Structures in Turbulence I

The Structure near the Wall in Turbulent Shear Flow

HELMUT ECKELMANN

Max-Planck-Institut für Strömungsforschung
D 3400 Göttingen, Federal Republic of Germany

Summary

A survey concerning coherent structures which can play an important part in the phenomenon of drag reduction by polymer additives is given. Most of these structures have long been known. Recent measurements have shown that both the scaling of low speed streaks and the bursting frequency remain constant when made dimensionless with inner variables. Streamwise vortices have been visually verified and their intensities anemometrically determined. Various authors have independently shown that too little attention has been paid to the problem of small-scale structure in turbulence measurements.

Introduction

In the year 1949, Toms (1) observed that the pressure drop in a turbulent pipe flow becomes substantially smaller when a minute amount of a polymer is added to the flow medium. One has since attempted to understand this effect. Known as Toms phenomenon, this effect is found in all wall-bounded turbulent shear flows.

Experimental investigations have shown that a region close to the wall is mainly responsible for the drag reduction produced by polymer additives. This region, known as the inner region, extends out to about 100 viscous lengths from the wall. Here the viscous length is defined as the ratio of kinematic viscosity ν to friction velocity $u_\tau = \sqrt{\tau_w/\rho}$ (τ_w: wall shear stress, ρ: fluid density). The inner region of a wall-bounded turbulent flow can be regarded as the center of turbulent activity. Here coherent structures of diverse types are observed. This region is the most strongly altered by polymer additives. The inner region can also be influenced directly by the wall, e.g. blowing, suction or surface alteration. The latter can lead either to an increase in drag (wall roughness) or to a drag reduction (i.e. wall grooves in streamwise direction). This paper will consider only smooth surfaces and only the coherent structures ob-

The Influence of Polymer Additives
on Velocity and Temperature Fields
IUTAM Symposium Essen 1984
Editor: B. Gampert

served close to the wall. It is worthy of mention that just at transition the boundary layer thickness measures roughly 100 viscous lengths. After transition, only the outer or logarithmic region of the boundary layer further increases with increasing Reynolds number, $Re = \overline{U} R/\nu$ ($\overline{U}$: either centre line velocity U_c in pipe or channel, or U_∞ in boundary layer flow; R: either radius, channel half-width b, or boundary layer thickness δ). The inner region then remains constant as measured in viscous lengths. This is a strong indication that the inner region plays an important part in wall bounded turbulence.

Information concerning coherent structures in wall vicinity was supplied especially by visual investigations (2-10). Coherent structures observed in the inner region were further investigated quantitatively with the help of hot wire or hot film techniques and employing the method of conditional sampling or conditional analysis (11-22). Particularly the region close to the wall has also been investigated with the help of electrochemical (23, 24) and optical (25) methods. Finally, a number of general reviews should be mentioned: Davies and Yule (26), Antonia (27), Cantwell (28), Wallace and Brodkey (29) and Hussain (30).

The structure of wall-bounded turbulent flow

Wall-bounded turbulent flows are generally subdivided in the direction normal to the wall (y coordinate). To elucidate this, the velocity profile measured in a turbulent channel flow is shown in Fig. 1 with non-dimensional coordinates (inner coordinates). The non-dimensional local mean velocity $u^+ = \overline{U}/u_\tau$ in-

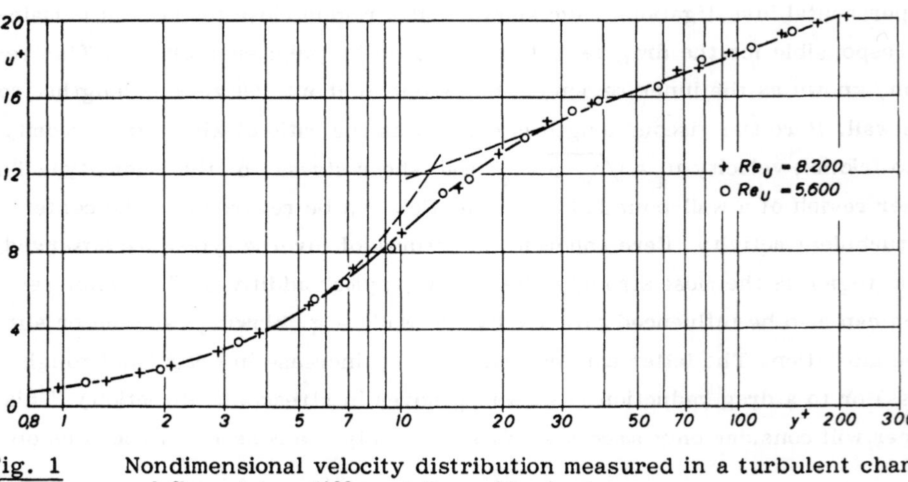

Fig. 1 Nondimensional velocity distribution measured in a turbulent channel flow at two different Reynolds numbers

creases linearly up to a wall distance of approximately $y^+ = y\, u_\tau/\nu = 5$. This region is known as the viscous sublayer. Adjacent to the sublayer and extending to about $y^+ = 50$ the buffer layer represents the transition from the linear to the logarithmic velocity law. Only the logarithmic region becomes larger with increasing Reynolds number. The other two regions actually become smaller with increasing Re but remain constant when measured in non-dimensional coordinates. The velocity profile presented here was measured at a relatively small Re, and thus the logarithmic region is small with an outer edge at $y^+_{max} = 140$ and 210, respectively. Turbulent flows at higher Re typically exhibit values for y^+_{max} of 10^3 to 10^4.

The subdivision introduced here is valid for all wall-bounded turbulent flows, i.e. channel, pipe and boundary layer flows. In turbulent boundary layers, part of the logarithmic region is influenced by the outer flow. In this case an intermittency region develops with an inner edge at about $y = 0.4\,\delta$. The region up to a non-dimensional wall distance of about $y^+ = 100$ is denoted as the inner region. The inner region thus compresses the viscous sublayer, the buffer layer and a part of the logarithmic region. The centre of turbulent activity is located here. The outer region, with which an intensive interaction occurs, plays a basically passive role.

Low speed streaks

The streaky structure of the viscous sublayer is the structure known longest and can be easily verified either by visual, anemometric, or electrochemical means. The streaky structure is characterized by thin, elongated regions with alternatingly high and low velocities which are randomly distributed in space and time. Low speed streaks appear most pronounced in visual investigations; their spacing decreases with increasing Re but remains constant when expressed in non-dimensional lengths. According to recent measurements by Smith and Metzler (9), the streak spacing remains constant at about $\lambda_z^+ = 100$ over a large range of Reynolds numbers. For $y^+ > 5$ however, this spacing increases with wall distance due to a merging and intermittency process. Oldaker and Tiedeman (7) report that in the presence of polymer additives, the average spacing of the streaks increases as the amount of drag reduction increases. The low speed streaks were first observed by Ferrell, Richardson and Beatty (31) and Hama (see Corrsin 32) and subsequently studied systematically at Stanford University by Kline and co-workers. The low speed streaks are of a passive nature. They are evidently produced by structures (e.g. sweeps, streamwise vortices) appearing further out from the wall. Low speed streaks become important in turbulence production only after lift up from the wall.

Bursts

According to Kline et al. (2), a sequence of events consisting of three phases described as lift-up, oscillation, and break-up of a low speed streak is designated as burst. Bursts appear randomly in immediate wall vicinity and can be made visible with either dye or hydrogen bubbles. Their detection by anemometric means becomes much more difficult. The VITA technique (variable interval time averaging) developed by Kaplan and Laufer (33) is most suited for this purpose. However, the use of this method as a burst detector is not without problems. Johansson and Alfredsson (34) have shown that the VITA technique has properties of a band-pass-filter.

It is generally assumed that the bursting process plays an important part in turbulence production as well as in momentum, heat, and mass transfer. However, the relation between the bursts and the outer flow region remains unclear. It has long been accepted that the bursting frequency assumes a constant value independent of Re when this frequency has been made dimensionless with outer variables such as U_c and R, or U_∞ (the velocity at the outer edge of the boundary layer) and δ. Recent measurements by Blackwelder and Haritonidis (35) have shown, however, that when scaled with the variables appropriate for the wall region (u_τ, ν), the non-dimensional bursting frequency is constant independent of the Reynolds number. Measurements have been carried out in the past with unsuitably large sensors leading to spatial-averaging effects. The measurements carried out by Blackwelder and Haritonidis were based on sensors of a spatial scale always less than twenty viscous length scales. According to measurements by Oldaker and Tiedeman (7), bursting is more energetic and less frequent in the presence of polymer additives.

Pockets

A new flow structure first observed by Falco (36) was designated pocket by that author. Pockets appear in immediate wall vicinity and measure approximately 100 x 100 viscous lengths. They are apparently produced by high speed fluid which has been displaced from the outer region towards the wall. However, the cause of this inflow of high speed fluid remains unclear to the present date. The hypothesis originating from Falco that pockets are the result of vortex rings which have formed in the outer region and been transported towards the wall by high speed fronts seems highly improbable to the author of the present paper. It is still unclear which process could be responsible for generating the vortex rings.

Ejections

Ejections were observed by Corino and Brodkey (3) in a turbulent pipe flow. They can be described as abrupt movements of fluid away from the wall and occur especially frequently near the wall up to about $y^+ = 30$. Ejections are local events and occur randomly in both space and time. The average ejection angle measured from the axis of the pipe is 8.5°. The origin of ejections is not known, but it is assumed that they are closely associated with the lift-up phase during bursting. The ejections transport fluid from the wall region into a region of much greater velocity and must therefore be regarded as part of a turbulence generating mechanism. Ejections can be described in Eulerian coordinates by $v > 0$ (motion away from the wall) and $u < 0$ (decelerated fluid), thus yielding a positive contribution to the Reynolds stress $-\rho\overline{uv}$. The contribution of ejections to Reynolds stress (Fig. 2) has been determined by Willmarth and Lu (37) and Wallace et al. (14). Ejections are part of a sequence of fluid motions observed by Corino and Brodkey which is terminated by the sweep.

Sweeps

The sweep can be described as a high speed ($u > 0$) motion of fluid directed towards the wall ($v < 0$). The leading edge of a sweep is called the front. This front generates a high-shear layer as it interacts with lower-speed fluid. Sweeps also yield a positive contribution to the Reynolds stress (Fig. 2).

The front of a sweep is curved. In the viscous sublayer, the mean angle between the front and the wall is 4.7° according to measurements by Kreplin and Eckelmann (21). Thomas and Bull (38) determined this angle as approximately 45° at greater distances from the wall. According to the visual investigations of Nychas, Hershey and Brodkey (39), a sweep running parallel to the wall is also possible. The lateral form of the front is unknown. The width of a sweep is well known only for the viscous sublayer, where it equals the width

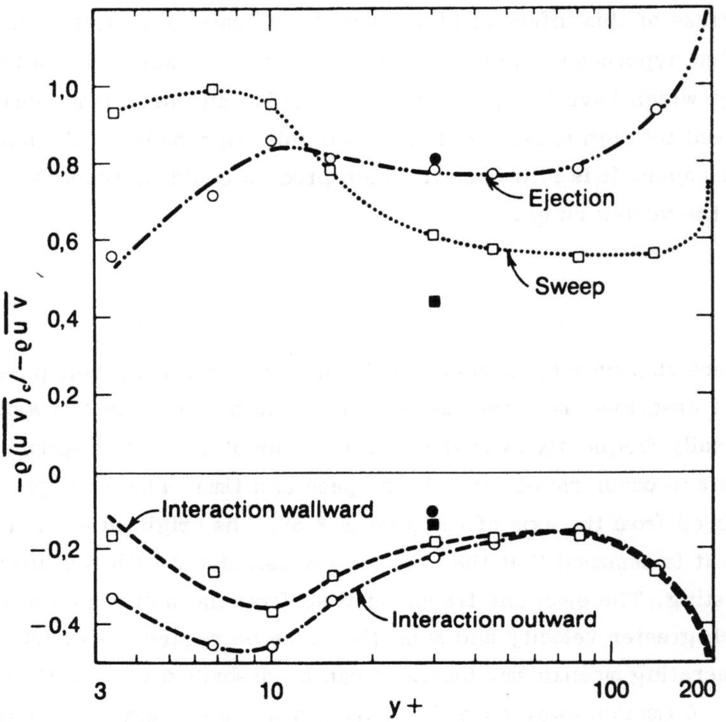

Fig. 2 Classified Reynolds stress normalized with the local average Reynolds stress. ●, ■ results of Willmarth and Lu (37). From (15)

of a high speed streak, or roughly $\Delta s^+ = 80$. The convective velocity of the sweep front was also determined as a function of distance from the wall by Kreplin and Eckelmann. This velocity increases from about 12 u_τ at the wall to 17 u_τ farther out, corresponding to approximately 60 to 90 % of the centre line velocity.

Streamwise vortices

Anemometric investigations of Bakewell and Lumley (11) and Blackwelder and Eckelmann (20) as well as visual investigations carried out recently by Smith and Schwartz (10) give evidence for the existence of counter-rotating streamwise vortex pairs. It is generally assumed that the vortex pairs are responsible for the lift up of fluid from the wall. Fig. 3 shows the vortex model proposed by Blackwelder and Eckelmann. The sense of rotation of the two vortices is such that they "pump" low speed fluid away from the wall and simultaneously

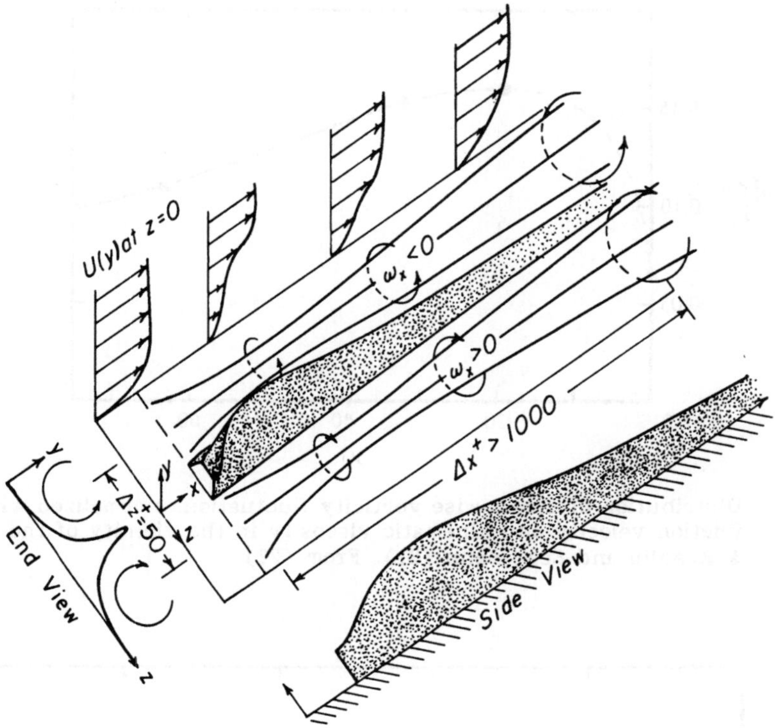

Fig. 3 Model of counter-rotating streamwise vortices together with the resulting low speed streak. The velocity profiles are from Blackwelder and Kaplan (17). From (20)

form a low speed streak. The vortex centre is located at about $y^+ = 20$. Kastrinakis and Eckelmann (22) also found a maximum in the distribution of the streamwise vorticity fluctuations (Fig. 4) at this location. The distribution of streamwise vorticity fluctuations over the channel half-width is shown in Fig. 5. A connection between the streamwise vorticity component and ejections was given by Kastrinakis, Nychas and Eckelmann (40). It was found that the ejections provide the major contribution to the streamwise vorticity. Also Kim et al. (4) found streamwise vortices in the wall region associated with the lift-up of the low speed streaks. Praturi and Brodkey (41), who used a stereoscopic flow viszalization technique, observed most of the streamwise vortex motion around $y^+ = 50$. In the visual studies of Kim et al. and Praturi and Brodkey, most of the vortices did not make more than one complete revolution. This is in accordance with an estimate which can be made from the anemometric measurements of Kastrinakis and Eckelmann. The strong rise of the flatness factor for the streamwise vorticity fluctuations up to the channel centre line (Fig. 6)

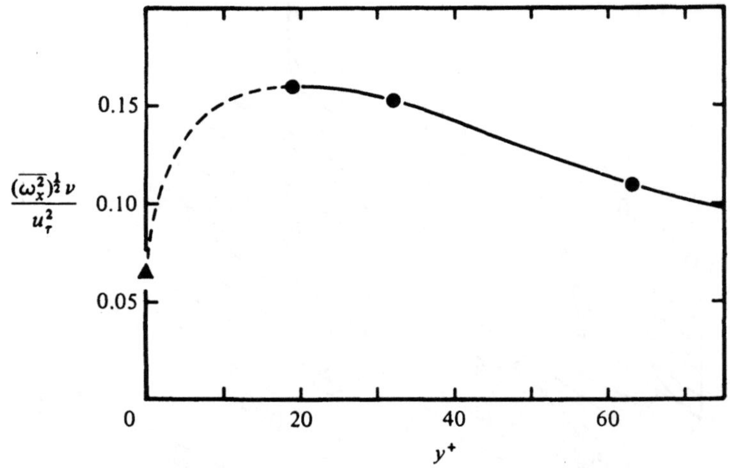

Fig. 4 Distribution of streamwise vorticity fluctuations normalized with the friction velocity and kinematic viscosity in the vicinity of the wall. ▲ Kreplin and Eckelmann (21). From (22)

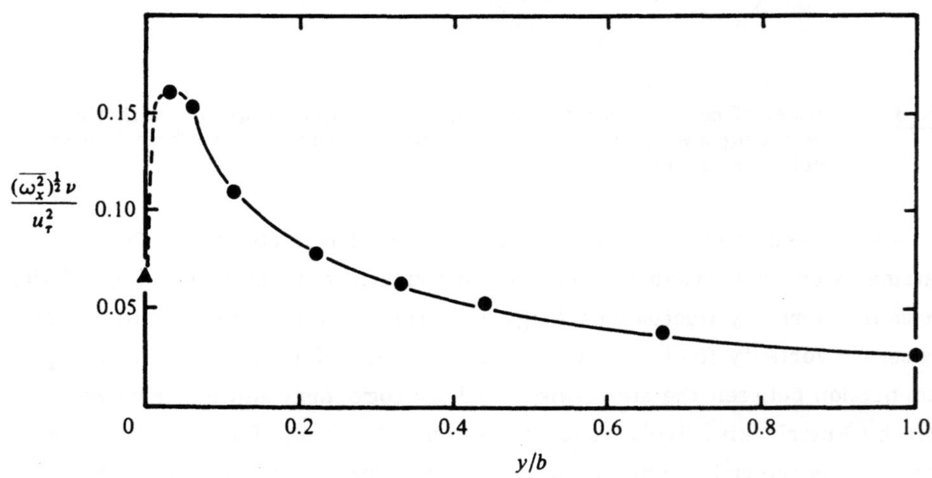

Fig. 5 Same as Fig. 4 for the channel half-width b. From (22)

implies that streamwise vortices become less frequent as the distance from the wall increases. The relationship between the streamwise vortices observed near the wall and the horseshoe or hairpin vortices described in the literature remains unclear to the present date. Similarly, it is unclear whether the wall region vortical motion causes the burst sequence, or whether the vortices are simply a by-product of other structures.

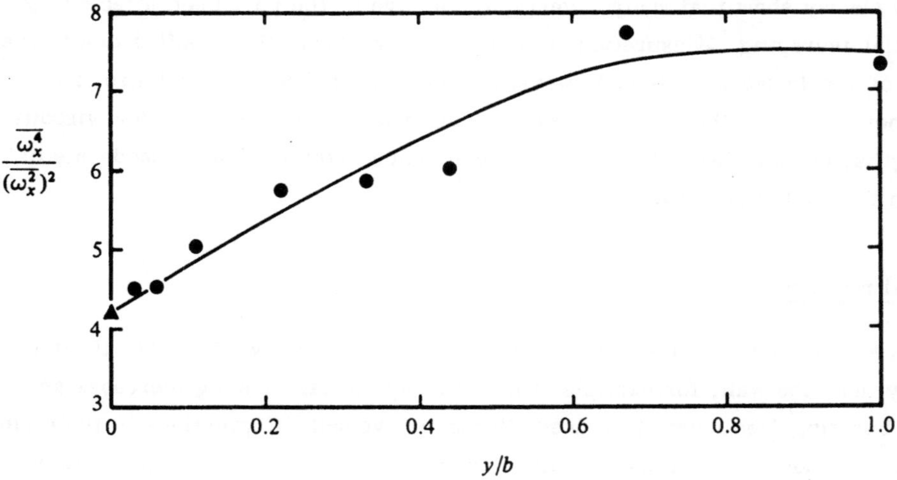

Fig. 6 Distribution of skewness and flatness factors for the streamwise vorticity fluctuations. ▲ Kreplin and Eckelmann (21). From (22)

Small-scale structure

Observations of Corino and Brodkey (3) in a turbulent pipe flow showed that a small-scale structure is quite common during the observed sequence of events (ejections and sweeps). In their visual investigation, motion pictures were produced of the path lines of small suspended particles in a thin illuminated plane of fluid. This plane was oriented parallel to the flow and normal to the wall and was approximately 20 viscous lengths thick. Particles within this plane and originating at the same location in the film were seen to move in different directions. This observed two-layer phenomenon is direct evidence that severe velocity differences exist within a distance of less than 20 viscous lengths. Also measurements in a turbulent boundary layer by Willmarth and Bogar (42) with an extremely small hot wire X probe gave evidence for a small-scale structure which previously had remained undetected. However, hot film measurements by Johnson and Eckelmann (43) could not confirm the small-scale structure observed by Willmarth and Bogar. These later measurements were carried out in the Göttingen oil channel with hot film probes having nearly the same non-dimensional scaling (sensor length l^+ about 2) as in the work of Willmarth and Bogar. Nevertheless, a small-scale strucutre in turbulent flows must certainly play an important part in the mechanism of drag reduction. Hot wire measurements by Blackwelder and Haritonidis (35), hot film measurements by Johansson and Alfredsson (44), and wall-pressure measurements by Schewe

(45) clearly show that many investigations have in the past been affected by spatial averaging. Measurement results were systematically falsified due to the use of overly large probes. However, probes of sufficiently small size, i.e. probes for which the non-dimensional scaling is on the order of a few viscous lengths, cannot always be found for turbulence investigations at moderate or high Reynolds numbers.

Final remarks

It is known that the structure of turbulence is altered by additives, particularly near the wall; for example, the low speed streak spacing increases and the bursting frequency decreases. Moreover, velocity fluctuations in the mean flow direction become more violent, whereas the intensity of the fluctuations normal to the wall and the correlation between the two velocity fluctuations both decrease. With the help of additives the experimenter thus has a knob in hand with which he can control the structure of turbulence. In spite of this tool, it has still not been possible to better understanding of the mechanism of turbulence to the present date. On the contrary, drag reduction by additives poses additional problems. The measurement, for example, of turbulent velocity fluctuations is made much more difficult. Since the drag reduction phenomenon can only be observed when the wall shear stress measures at least about 7 N/m^2 (46) measurements at small turbulent Reynolds numbers become almost impossible, and thus the region near the wall is practically inaccessible to anemometric measurements (either laser or hot film). It should therefore also prove very difficult to directly verify the existence of streamwise vortices in flows with polymer additives and thus answer the question posed by Willmarth and Bogar (42) as to whether the small-scale vortical structure near the wall is inhibited by polymer additives.

The author wishes to thank Mr. F.D. Johnson for help with the translation and Frau C. Köneke for typing the manuscript.

References

1. Toms, B.A., Proceedings of the 1st International Congress on Rheology, Vol. 2, p. 135. North-Holland, Amsterdam 1949.

2. Kline, S.J., Reynolds, W.C., Schraub, F.A. and Runstadler, P.W., The Structure of Turbulent Boundary Layers, J. Fluid Mech. 30, (1967), 741.

3. Corino, E.R. and Brodkey, R.S., A visual Investigation of the Wall Region in Turbulent Flow, J. Fluid Mech. 37, (1969), 1.

4. Kim, H.T., Kline, S.J. and Reynolds, W.C., The Production of Turbulence Near a Smooth Wall in a Turbulent Boundary Layer, J. Fluid Mech. 50, (1971), 133.

5. Falco, R.E., Some Comments on Turbulent Boundary Layer Structure Inferred from the Movements of a Passive Contaminant, AIAA Paper 74-99, 1974.

6. Falco, R.E., The Production of Turbulence Near a Wall, AIAA Paper 80-1356, 1980.

7. Oldaker, D.K. and Tiederman, W.G., Spatial Structure of the Viscous Sublayer in Drag Reducting Channel Flows, Phys. Fluids 20, (1977), S133.

8. Smith, C.R., Visualization of Turbulent Boundary Layer Structure Using a Moving Hydrogen Bubble-Wire Probe, in Workshop on Coherent Structure of Turbulent Boundary Layer, Smith & Abbot Ed., AFOSR/Lehigh, 1978.

9. Smith, C.R. and Metzler, S.P., The Characteristics of Low-Speed Streaks in the Near-Wall Region of Turbulent Boundary Layer, J. Fluid Mech. 129, (1983), 27.

10. Smith, C.R. and Schwartz, S.P., Observations of Streamwise Rotation in the Near-Wall Region of a Turbulent Boundary Layer, Phys. Fluids 26, (1983), 641.

11. Bakewell, H. and Lumley, J.L., Viscous Sublayer and Adjacent Wall Region in Turbulent Pipe Flow, Phys. Fluids 10, (1967), 1880.

12. Gupta, A.K., Laufer, J. and Kaplan, R.E., Spatial Structure in the Viscous Sublayer, J. Fluid Mech. 50, (1971), 493.

13. Rao, K.N., Narashimha, R. and Badri Narayanan, M.A., The "Bursting" Phenomenon in a Turbulent Boundary Layer, J. Fluid Mech. 48, (1971), 339.

14. Wallace, J.M., Eckelmann, H. and Brodkey, R.S., The Wall Region in Turbulent Shear Flow, J. Fluid Mech. 54, (1972), 39.

15. Brodkey, R.S., Wallace, J.M. and Eckelmann, H., Some Properties of Truncated Turbulence Signals in Bounded Shear Flows, J. Fluid Mech. 63, (1974), 209.

16. Ueda, H. and Hinze, T.O., Fine Structure Turbulence in the Wall region of a Turbulent Boundary Layer, J. Fluid Mech. 67, (1975), 125.

17. Blackwelder, R.F. and Kaplan, R.E., Wall Structure of the Turbulent Boundary Layer, J. Fluid Mech. 76, (1976), 89.

18. Wallace, J.M., Brodkey, R.S. and Eckelmann, H., Pattern-Recognized Structures in Bounded Turbulent Shear Flows, J. Fluid Mech. 83, (1977), 673.

19. Eckelmann, H., Nychas, S.G., Brodkey, R.S. and Wallace, J.M., Vorticity and Turbulence Production in Pattern Recognized Turbulent Flow Structures, Phys. Fluids 20, (1977), S225.

20. Blackwelder, R.F. and Eckelmann, H., Streamwise Vortices Associated with the Bursting Phenomenon, J. Fluid Mech. 94, (1979), 577.

21. Kreplin, H.-P. and Eckelmann H., Propagation of Perturbations in the Viscous Sublayer and Adjacent Wall Region, J. Fluid Mech. 95, (1979), 305.

22. Kastrinakis, E.G. and Eckelmann, H., Measurement of Streamwise Vorticity Fluctuations in a Turbulent Channel Flow, J. Fluid Mech 137, (1983), 165.

23. Sirkar, K.K. and Hanratty, T.J., Relation of Turbulent Mass Transfer to a Wall at High Schmidt Numbers to the Velocity Field, J. Fluid Mech. 44, (1970), 589.

24. Lee, M.K., Eckelman, L.D. and Hanratty, T.J., Identification of Turbulent Wall Eddies through the Phase Relation of the Components of the Fluctuating Velocity Gradient, J. Fluid Mech. 66, (1974), 17.

25. Emmerling, R., Meier, G.E.A. and Dinkelacker, A., Investigation of the Instantaneous Structure of the Wall Pressure under a Turbulent Boundary Layer Flow, AGARD Conf. Proc. No. 131, Belgium Meeting 1973.

26. Davies, P.O.A.L. and Yule, A.J., Coherent Structures in Turbulence, J. Fluid Mech. 69, (1975), 513.

27. Antonia, R.A., Conditional Sampling in Turbulence Measurement, Annual Reviews of Fluid Mechanics 13, (1981), 131.

28. Cantwell, B.J., Organized Motion in Turbulent Flow, Annual Review of Fluid Mechanics 13, (1981), 457.

29. Wallace, J.M. and Brodkey, R.S., The Delta Conferences Columbia, Ohio March 1979, 1980 and 1981. The Ohio State University Research Foundation, Columbus, Ohio 1982.

30. Hussain, A.K.M.F., Coherent Structures - Reality and Myth, Phys. Fluids 26, (1983), 2816.

31. Ferrell, T.K., Richardson, F.R. and Beatty, K.O., Jr., Dye Displacement Techniques for Velocity Distribution Measurements, Ind. Eng. Chem. 47, (1955), 29.

32. Corrsin, S., Symp. Naval Hydrodyn., Publ. 15, NAS NRC 373, 1957.

33. Kaplan, R.E. and Laufer, J., The Intermittently Turbulent Region of the Boundary Layer, Proc. 12th Int. Cong. Appl. Mech., Stanford University, p. 236, Springer 1969.

34. Johansson, A.V. and Alfredsson, P.H., On the Structure of Turbulent Channel Flow, J. Fluid Mech. 122, (1982), 295.

35. Blackwelder, R.F. and Haritonidis, J.H., Scaling of the Bursting Frequency in Turbulent Boundary Layers, J. Fluid Mech. 132, (1983), 87.

36. Falco, R.E., Coherent Motions in the Outer Region of Turbulent Boundary Layers, Phys. Fluids 20, (1977), S124.

37. Willmarth, W.W. and Lu, S.S., Structure of the Reynolds Stress Near the Wall, J. Fluid Mech. 55, (1972), 65.

38. Thomas, A.S.W. and Bull, M.K., On the Role of Wall-Pressure Fluctuations in Deterministic Motions in the Turbulent Boundary Layer, J. Fluid Mech. 128, (1983), 283.

39. Nychas, S.G., Hershey, H.C. and Brodkey, R.S., A Visual Study of Turbulent Shear Flow, J. Fluid Mech. 61, (1973), 513.

40. Kastrinakis, E.G., Nychas, S.G. and Eckelmann, H., Some Streamwise Vorticity Characteristics of Coherent Structures, IUTAM Symposium on Structure of Complex Turbulent Shear Flows, R. Dumas and L. Fulachier eds. Springer 1983, 31.

41. Praturi, A.K. and Brodkey, R.S., A Stereoscopic Visual Study of Coherent Structures in Turbulent Shear Flow, J. Fluid Mech. 89, (1978), 251.

42. Willmarth, W.W. and Bogar, T.J., Survey and New Measurements of Turbulent Structure Near the Wall, Phys. Fluids 20, (1977), S9.

43. Johnson, F.D. and Eckelmann, H., Has a Small-Scale Structure in Turbulence been Experimentally Verified?, Phys. Fluids 26, (1983), 2408.

44. Johansson, A.V. and Alfredsson, P.H., Effects of Imperfect Spatial Resolution on Measurements of Wall-Bounded Turbulent Shear Flows, J. Fluid Mech. 137, (1983), 409.

45. Schewe, G., On the Structure and Resolution of Wall-Pressure Fluctuations Associated with Turbulent Boundary-Layer Flow, J. Fluid Mech. 134, (1983), 311.

46. Virk, P.S., Drag Reduction Fundamentals, AIChE J. 21, (1975), 625.

Statistical Measurements in the Turbulent Flow of Dilute Drag-Reducing Polymer Solutions in a Round Tube

Hiromoto Usui, Noriyuki Iwata and Yuji Sano

Department of Chemical Engineering
Yamaguchi University
2557 Tokiwadai
Ube, 755 JAPAN

Abstract

The drag reduction phenomenon in a fully developed turbulent pipe flow was examined by measuring the longitudinal velocity by means of a laser-Doppler anemometer. Statistical analysis of the fluctuating velocity showed that the most significant effect of the polymer additive on the organized structure of wall turbulence was detected near the outer edge of a buffer layer. The typical ejection process was detected as the remarkable negative fluctuating velocity both through the buffer region and the inner part of turbulent core.

1. Introduction

In spite of the general agreement for the usefulness in the practical applications of drag reduction phenomenon, there is no generally acceptable theory that is able to explain the interaction between the organized turbulence structure and the polymer additive. Most previous investigations in the tube flow imply that the effects of polymer additives on turbulence are limited to the region near the wall where the bursting phenomenon is responsible for the major part of turbulent energy production and dissipation. However, the detailed mechanism of drag reduction has been left unclarified because of the difficulty of turbulence measurements in polymer solutions.

It was intended, in this work, to give a possible explanation for the drag reduction phenomenon using a statistical analysis of the turbulent fluctuating velocity. For this purpose, a laser-Doppler velocimeter was used to measure the longitudinal local velocity in a round tube flow. The results support the previous experimental evidence in drag

The Influence of Polymer Additives
on Velocity and Temperature Fields
IUTAM Symposium Essen 1984
Editor: B. Gampert

reducing turbulent flow, i.e. the thickening of laminar sublayer and the buffer region, the enlargement both of the bursting time and the turbulence integral scale.

2. Experiments

The experimental setup was just the same as reported in the previous work (Usui and Sano (15)) except for the test tube. A once-through flow system including a large storage tank (1.5 m^3) was employed to minimize the mechanical degradation of polymer additives. A PVC round tube (4.5 m in length and 0.0254 m in inner diameter) was used in this work. The test tube was constructed of a transparent polyester film (40μm thickness), and a compensator attached around the pipe was filled with the test solution. The same type of test section was employed by Mizushina and Usui (10) and the other several investigators, Berman (2) and Kunen and Ooms (8).

A laser-Doppler anemometer, with a He-Ne 5 mW laser (NEO-5M), similar to that previously reported (Mizushina and Usui (10)) was used to measure the longitudinal velocity in a fully developed turbulent flow of drag reducing polymer solutions. A reference beam mode of Doppler system, with the cross angle, θ, of 6.77 degree. was used in this work. The laser beam diameter, σ, at the measuring point is, for convenience, estimated by the following equation given elsewhere (e.g. George and Lumley (6)),

$$\sigma = \frac{2^{1/2}}{\pi} \frac{f_0 \lambda}{d_0} \tag{1}$$

In this experiment, the focal length of the lens, f_0, wave length, λ, and the diameter of laser beam at the lens, d_0, were 250 mm, 0.6328μm and 1 mm, respectively. The scattering volume is usually approximated by an ellipsoid of which major axis length is σ_1, and the lengths of shorter axes are σ_2 and σ_3. Using the above-mentioned values, σ_1, σ_2 and σ_3 were estimated as 1.22, 0.072 and 0.072 mm, respectively. The major axis of scattering volume was oriented in the spanwize direction of the test pipe. The instantaneous longitudinal velocity was recorded

Table 1 Experimental condition for LDV measurements

c (ppm)	Re	ν (cm^2/s)	f	u* (cm/s)	$\frac{f_N^{1)} - f_P^{2)}}{f_N}$ x 100(%)	T_B $^{3)}$	$\bar{t}$ $^{4)}$
0	10370	0.0110	0.0076	2.76	-	0.316	-
30	9850	0.0113	0.0063	2.46	20.7	0.411	-
50	8920	0.0122	0.0052	2.18	36.1	0.61	-
100	9630	0.0133	0.0037	2.17	53.7	0.75	0.65
200	8500	0.0155	0.0035	2.17	57.5	0.81	0.65

1) f_N is the Newtonian frictionfactor estimated by $f=0.0791Re^{-0.25}$.
2) f_P is the friction factor for a polymer solution.
3) T_B is the bursting period determined by short time sampling auto-correlation method.
4) $\bar{t}$ is the mean time interval of two remarkable nagative peaks appeared in the longitudinal fluctuating velocity of polymer solutions.

by a data recorder (TEAC,R-61). The voltage outputs of the instrument were digitized and stored on magnetic tape. The sampling interval and the data size were 0.005 second and 8000, respectively. Statistical processing of the digitized data was made with ACOS-800 computer system.

Polyethylene oxide (grade Alcox, E-160), supplied by Meisei Chemical Corp., was used as the polymer additive. The intrinsic viscosity of Alcox, E-160 in water solution was determined experimentally as 1200 cm^3/g. The test fluids for the laser-Doppler measurements contained a fresh milk of 30 ppm.

3. Results and discussion

The experimental conditions for the velocity measurements are shown in Table 1.

The present experimental results of the time smoothed velocity distribution are in good agreement with those reported by several authors (Reishman and Tiederman (13), Mizushina and Usui (10)), showing the increase both in laminar sublayer and buffer layer thickness.

The turbulent intensity of a Newtonian fluid (water, c = 0 ppm) was considerably lower than the previously well established value throughout the whole radial positions. This discrepancy was interpreted as the results of larger sampling volume of the present LDV system compared with the microscale of turbulence. The correction of measured turbulent intensity based on the spatial resolution analysis by George and Lumley (6) was made. The detailed correction method has been described in the previous paper (Usui and Sano (15)). The corrected data was in reasonable agreement with the measurements by Pennel et al. (12).

In the case of polymer solutions, it is well accepted that the flow stabilizing effect of the polymer additives causes the enlargement of the turbulence scale at lower frequency and causes the suppression of the microscale eddies (e.g. see Fortuna and Hanratty (5), Berner and Scrivener (3)). The present turbulent intensity profiles of polymer solutions confirms the previous data of Reishman and Tiederman (13) and Mizushina and Usui (10). A shift of the peak toward the axis is observed near the wall.

It is desirable to obtain the result of statistical analysis of the digitized data for more detailed understanding of turbulence mechanism in a drag reducing pipe flow. Thus, in this work, we attempted to give a more quantitative analysis including the skewness factor and the flatness factor.

The skewness factor and the flatness factor of longitudinal fluctuating velocity, u, are defined as, $S(u)=\overline{u^3}/(\overline{u^2})^{3/2}$ and $F(u)=\overline{u^4}/(\overline{u^2})^2$, respectively. Since the skewness factor is an odd-order moment, it is an indication of the degree of asymmetry of the probability density distribution, and the flatness factor is the measure of the skirt of the probability density distribution. The calculation of higher order moments of fluctuating velocity is very sensitive to the spike signal caused by a signal dropout. Thus, in this study, a great deal of care (e.g. adjustments of scattering particle concentration, sampling volume and optical arrangements) was paid to avoid the signal dropout. The authors believe that the effect of signal dropout on the final results of this paper is negligible.

As mentioned in the previous section, the length of the

major axis of the scattering ellipsoid, σ_1, was estimated as 1.22 mm. This value is nondimensionalized as $\sigma_1 u^*/\nu$=30.2 for a Newtonian fluid and 18.9 for 200 ppm polymer solution at Re=10,000. On the other hand, the previous experiments (Donohue et al.(4), Fortuna and Hanratty (5) and Achia and Thompson (1)) have demonstrated that the non-dimensional wall-layer streak spacing had a constant value of λ^+=100 in the fluid flow. In drag reducing flows, the trends of these investigations show an increase in λ^+ with increasing drag reduction. Thus it might be concluded that the present finite scattering volume is sufficiently small to detect the large-scale motion of wall turbulence both in a Newtonian fluid and polymer solutions.

The distribution of skewness factor and flatness factor are presented in Figures 1 and 2. The results of water are in reasonable agreement with the data of Laufer (9), Ueda and Hinze (14) and Nakagawa and Nezu(11). Although the results of polymer solutions show almost the same distribution as Newtonian fluids very near the wall, the skewness factor decreases anomalously in buffer layer or at the inner edge of turbulent core, and the flatness factor shows a remarkable increase at the same position. In addition, these distributions approach to the Newtonian value at the center of a pipe flow. Berner and Scrivener (3) have reported that the probability density distribution of longitudinal fluctuating velocity is very much skewed in the direction of negative fluctuating velocity in buffer layer. However, it is impossible to read the skewness factor out of the data presented in their paper. Usui and Sano (15) have reported the same tendency of the skewness factor and the flatness factor in a drag reducing open channel flow, although the deviation from the Newtonian profile was not so large as those observed in Figures 1 and 2. The geometrical difference of flow cross sections seems to be the reason of their discrepancy.

It is well accepted that the ejection process in the organized motion of wall turbulence causes a negative value of the skewness factor. The experimental evidence shown in Figures 1 and 2 suggests that the fluctuating velocity caused by the

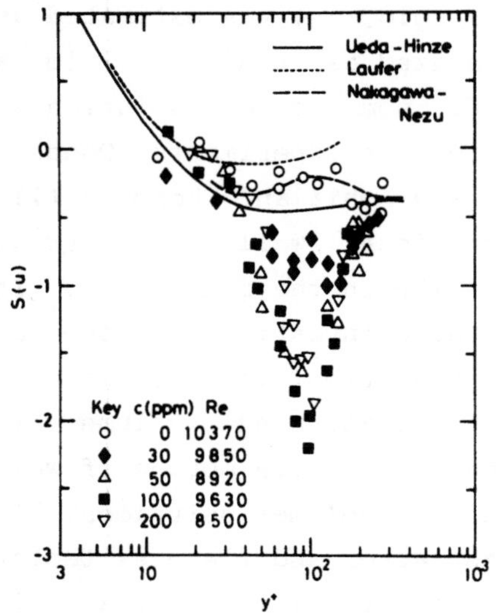

Fig.1. Distribution of the skewness factor of u.

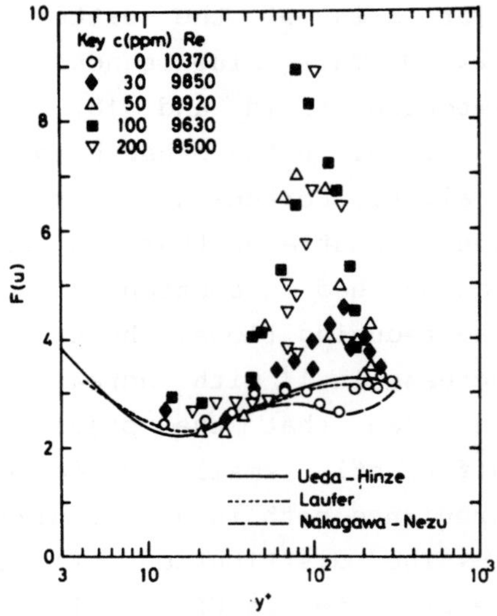

Fig.2. Distribution of the flatness factor of u.

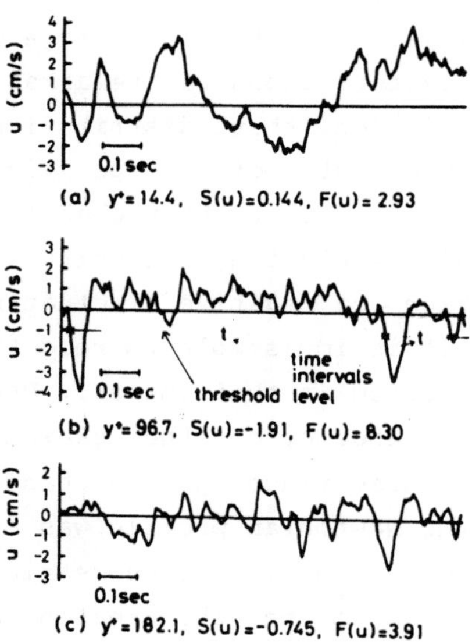

Fig.3. Fluctuation of the longitudinal velocity component for a 100 ppm polymer solution at Re=9,630.

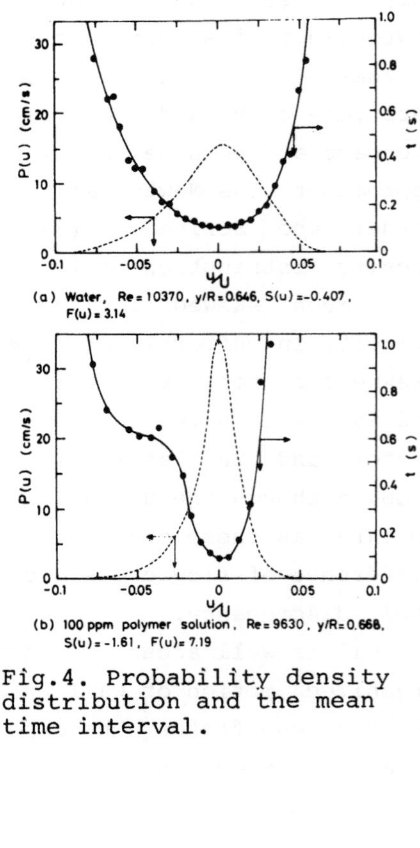

Fig.4. Probability density distribution and the mean time interval.

ejection process may be very cleary detected. The typical fluctuating velocity is illustrated in Figure 3 for the case of 100 ppm polymer solutions. The measuring positions were located at $y^+ = 14.4$ (just above the laminar sublayer), at $y^+ = 96.7$ (at the outer edge of the buffer layer where the minimum skewness factor was detected), and at $y^+ = 182.1$ (in the turbulent core), respectively. Although no remarkable difference from that of a Newtonian fluid flow is observed in Figures 3-(a) and 3-(c), we notice that unusually big negative peaks of longitudinal fluctuating velocity appear occasionally in Figure 3-(b). It is obvious that these remarkable peaks cause the negative skew of probability density function. In a drag reducing flow, the organized structure of wall turbulence is considerably enlarged both in space and time. Combined with this enlargement in flow structure, an appropriate selection of sampling volume might cause a clear detection of ejection process.

The negative peak in Figure 3-(b) is so remarkable that it is possible to determine the time interval between two ejection processes. This means that the bursting period is possibly determined. For this purpose, a threshold level shown in Figure 3-(b) was changed. The time interval was defined as the time to cross twice the same threshold level with the same sign of (du/dt). The mean value of this time interval, $\bar{t}$, may be meaningless if the velocity fluctuation is random. However, as shown in Figure 3-(b), the determination of mean time interval of large negative peaks seems to be possible as the negative peaks are prominent. The mean time interval as a function of threshold level is shown in Figure 4 with the result of probability density distribution. The data of a 100 ppm polymer solution is very much skewed in the negative direction of the fluctuating velocity. The distribution of mean time interval of negative peaks shows, in the case of a polymer solution (Figure 4-(b)), almost constant value at $u/U = -0.003 \sim -0.006$, while in the case of a Newtonian fluid (Figure 4-(a)) no plateau is observed. This value ($\bar{t} \fallingdotseq 0.62$ s) should correspond to the mean time interval of the ejection processes.

The results of the mean time interval are compared in

Figure 5 with the measurements of bursting period using auto-correlation method with short sampling time. The holographic data of Achia and Thompson (1) is also compared in this diagram. As pointed out by Berman (2), the bursting period determined by auto-correlation method depends on the sampling volume and the concentration of scattering particle. The present data of a Newtonian fluid is a little larger than the Stanford data (Kim et al.(7)) and those of Achia and Thompson. The present Newtonian data seems to correspond to "the high concentration data" of scattering particles shown in the paper of Berman (2). Comparing with the results of a Newtonian fluid, the bursting period in polymer solutions is slightly higher than the Newtonian values, and the same tendency is observed in the data of Achia and Thompson. Although the mean time interval of negative peak, which was interpreted as the time interval between ejection processes, is a little lower than the results of auto-correlation method, it may be concluded that the detection of mean time interval is possibly adopted as a measuring technique of sublayer period. This detection method is restricted only to the highly drag-reducing system. Although it was confirmed implicitly that the remarkable negative peak observed in Figure 3-(b) was caused by the ejection process in wall turbulence, more direct verification should be done, in the future work, combining with the flow visualization technique or the simultaneous measurements of multi-velocity components.

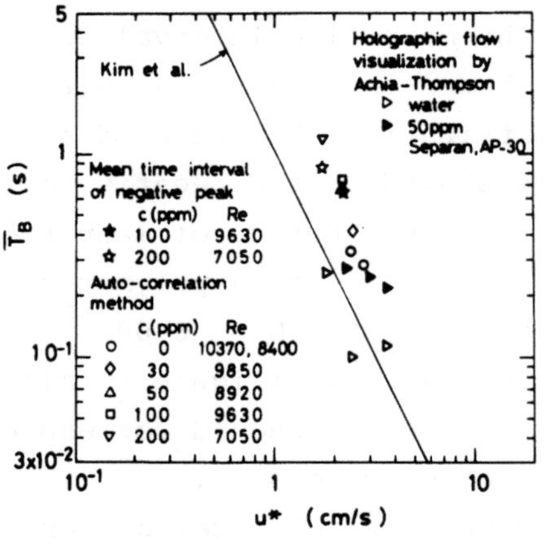

Fig.5. Sublayer-period data in Newtonian and drag-reducing flows.

4. Conclusion

The effects of polymer additives on the organized structure of wall turbulence in a fully developed pipe flow were discussed by means of the measurements of longitudinal fluctuating velocity detected by a laser Doppler velocimeter. The skewness factor of polymer solutions showed anomalously high negative value at the outer edge of buffer layer. The corresponding flatness factor was much larger than that of water. Direct observation of fluctuating velocity showed that the anomalous profiles both of skewness and flatness factors were caused by the remarkable negative peak. This phenomenon was interpreted as an increased scale of the ejection process in polymer solutions. The detection of the time interval between the two negative peaks gave almost the same sublayer period as that measured by the auto-correlation method.

References

1. Achia, B. U. and D. W. Thompson, "Structure of Turbulent Boundary in Drag-Reducing Pipe Flow", J. Fluid Mech., 81, (1977) 439-464

2. Berman, N. S., "The Effect of Sample Probe Size on Sublayer Period in Turbulent Boundary Layers", Chem. Eng. Commun., 5, (1980) 337-345

3. Berner, C. and O. Scrivener, "Drag Reduction and Structure of Turbulence in Dilute Polymer Solutions", Prog. Astronut. Aeronaut., 72, (1980) 290-299

4. Donohue, G. L., W. G. Tiederman and M. M. Reishman, "Flow Visualization of the Near-Wall Region in a Drag-Reducing Channel Flow", J. Fluid Mech., 56,(1972) 559-575

5. Fortuna, G. and T. J. Hanratty, "The Infulence of Drag-Reducing Polymers on Turbulence in the Viscous Sublayer", J. Fluid Mech., 53, (1972) 575-586

6. George, W. K. and J. L. Lumley, "The Laser-Doppler Velocimeter and its application to the Measurements of Turbulence", J. Fluid Mech., 60, (1973) 321-362

7. Kim, H. T., S. J. Kline and W. C. Reynolds, "Production of Turbulence near a Smooth Wall in a Turbulent Boundary Layer", J. Fluid Mech., 50, (1971) 133-160

8. Kunen, J. M. G. and G. Ooms, "On the Detection Methods for Coherent Structure in Turbulent Flows", Proceedings of Eighth Biennial Symposium on Turbulence, J. L. Zakin and G. K. Patterson ed., Sience Press, N. J. (1983)

9. Laufer, J., "The Structure of Turbulence in Fully Developed Pipe Flow", NACA TA2954 (1953)

10. Mizushina, T. and H. Usui, "Reduction of Eddy Diffusion for Momentum and Heat in Viscoelastic Fluid Flow in a Circular Tube", Phys. Fluids, 20, (1977) S100-108

11. Nakagawa, H. and I. Nezu,"Prediction of the Contributions to the Reynolds Stress from Bursting Events in Open-Channel Flows", J. Fluid Mech., 80, (1977) 99-128

12. Pennell, W. T., E. M. Sparrow and E. R. G. Eckert, "Turbulence Intensity and Time Mean Velocity Distributions in Low Reynolds Number Turbulent Pipe Flows", Int. J. Heat Mass Transfer, 15, (1972) 1067-1074

13. Reishman, M. M. and W. G. Tiederman,"Laser-Doppler Anemometer Measurements in Drag-Reducing Channel Flows", J. Fluid Mech., 70,(1975) 369-392

14. Ueda, H. and J. O. Hinze, "Fine-Structure Turbulence in the Wall Region of a Turbulent Boundary Layer", J. Fluid Mech., 67, (1975) 125-143

15. Usui, H. and Y. Sano, "Turbulence Structure of Dilute Drag-Reducing Polymer Solutions in a Rectangular Open Channel Flow", AIChE J., 29, (1983) 611-617

The Vorticity Diffusion of Λ-Vortices in Drag Reducing Solutions

A. GYR

Federal Institute of Technology ETHZ
Institute of Hydromechanics and
Water Resources Management
CH-8093 Zurich-Hönggerberg
Switzerland

Abstract

The vorticity diffusion in a Λ -vortex attached to the wall in a complex eigenvalue critical point is investigated. It is shown that extensional viscosity is produced near the critical point. The internal strainfield of the vortex rods is damped and an accelerated vorticity diffusion occurs due to polymer additives.
Some ideas on drag reduction based on such a mechanism are discussed.

Introduction

It is a well known hypothesis that drag reduction has to do with the elongation of the long chain molecules by turbulent flow field, and with the local change in rheology which results from this process. It is further assumed that this change in rheology reacts with the flow in redistributing vorticity so that the mean drag of the flow will be reduced. These hypotheses are supported by experimental evidences.

First evidence: the effect sets on at a specific wall shear stress. The onset can be explained as that state of the flow which is able to stretch the molecules in areas of high strain, producing a dramatic local change in the rheology of the fluid.

Secondly, the effect is restricted to boundary layer flows, the interaction with the wall must therefore be essential. In other words, the flow close to the wall must have the ability of stretching the molecules.

In the zone close to the wall the momentum exchange with the outer regions is mainly sustained by ejection and sweeps, the events which are thought to be responsible for the stretching.

These events have been modeled by a variety of different flow structures. In this paper we make use of a model proposed by PERRY & CHONG (1982). The reasons why this -vortex model is prefered are given in chapter I.

In chapter II the interaction of the flow field in such events with the mole-

The Influence of Polymer Additives
on Velocity and Temperature Fields
IUTAM Symposium Essen 1984
Editor: B. Gampert

cules is described. It is shown that there are indeed local areas where the molecules can be stretched and due to the associated change in rheology the diffusion of vorticity is highly accelerated in a much larger region. This result is consistent with the few visualisation of flow events in dilute polymer solutions.
In chapter III some ideas are scetched which could be used to relate the above results to the drag reducing effect but no final statements on this problem is made.

I. The Λ-vortex model of ejections

It is now well established that in a turbulent boundary layer flow the transport of momentum to the wall is intermittent. The events responsible for this locally large instantaneous momentum transport are called sweeps and ejections depending on whether the transport is towards or away from the wall. A variety of models describing the flow field in such events as well as their overall organisation in the main flow have been proposed.
Visualisation of ejections show that they are associated with a jet-like and a vortex-like flow field. The relation between the strengthes of these fields are not well known since in order to measure the vorticity in such events, one would need probes which are not available to-day.
The jet-like or elongational flow field in these features can be understood as a local separation from the wall. An analysis of the vorticity field however poses a fundamental problem related to the non-slip condition at the wall since the Helmholtz laws are incompatible with this condition. One of Helmholtz's law states that a vortex line has to be closed or that it has to end at a boundary. Since at the wall (xy-plane) the velocities u and v parallel to the wall are identical zero, the same must hold for the vorticity component ζ normal to the wall. No vortex line can hence be attached to the wall. The main aspects of this problem are described by LIGHTHILL (1963).
It is evident that for a description of such flows the viscosity of the fluid has to be introduced. This has been done by LIM, CHONG & PERRY (1980) who showed that there exists a linearised solution of the Navier-Stokes equations which they called a viscous tornado or a complex eigenvalue critical point flow. This solution permits a vortex line to be quasi attached to the wall under non-slip conditions in the sense that a versatil spiral flow around a separation streamline arrises. The concept be scetched as follows:
The flow near a separation point at the wall can be classified by the form

of the solutions for the streamlines at the critical point, i.e. the point where the vorticity is zero and instantaneous streamline solutions have indeterminate slopes; (OSWATITSCH (1958), LIGHTHILL (1963)). By substituting a Taylor expansion of the velocities into the Navier-Stokes and continuity equations, the relations for the trajectories of a fluid particle contain terms of "measurable" quantities such as surface pressure and surface vorticity and their various derivatives in the xy-plane:

$$\begin{bmatrix} \dot{x} \\ \dot{y} \\ \dot{z} \end{bmatrix} = \begin{bmatrix} \eta,_x & \eta,_y & p,_x/2 \\ -\xi,_x & -\xi,_y & p,_y/2 \\ 0 & 0 & 1/2(\xi,_y - \eta,_x) \end{bmatrix} \begin{bmatrix} x \\ y \\ z \end{bmatrix} \tag{1.1}$$

For solutions of the form

$$x = R_x e^{\lambda\tau} \quad , \quad y = R_y e^{\lambda\tau} \quad , \quad z = R_z e^{\lambda\tau} \tag{1.2}$$

LIM et al. showed that for

$$4\eta,_y \xi,_x > (\eta,_x - \xi,_y)^2$$

The trajectories in canonical and real time form are found as

$$x = K(\sigma Ht + 1) \cos(M \ln(\sigma Ht + 1) + \Phi) \tag{1.3a}$$

$$y = K(\sigma Ht + 1) \sin(M \ln(\sigma Ht + 1) + \Phi) \tag{1.3b}$$

$$z = \frac{H}{(\sigma Ht + 1)} \tag{1.3c}$$

$$\sigma = -\frac{(\xi,_y - \eta,_x)}{2}$$

$$M = \omega/\delta \text{ and } \omega = 1/2\left\{4\eta,_y \xi,_x - (\eta,_x + \xi,_y)^2\right\}^{1/2}$$

H, K and Φ are constants depending on the initial condition.

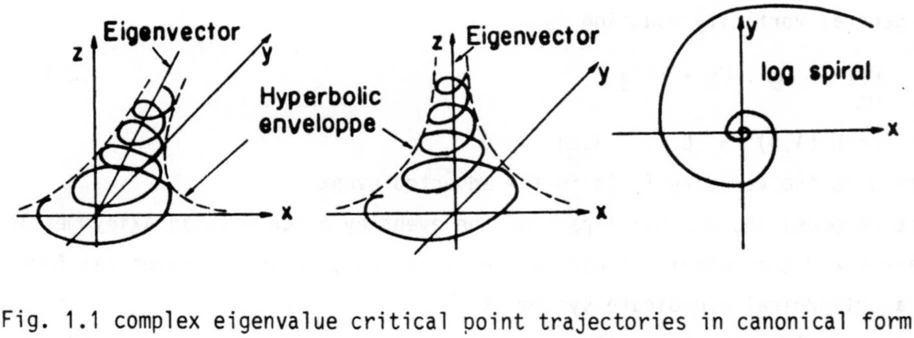

Fig. 1.1 complex eigenvalue critical point trajectories in canonical form

This means that near the wall this events have a jetlike flow field with axisymmetric strain, on which a weak vorticity is superimposed which by stretching of the flow field will be concentrated.
Such a concept is compatible with a comprehensive model of ejections which has been proposed by PERRY & CHONG (1982) who call them Λ-vortices. Λ-vortices are simplified picture of events which are called vortex loops, horseshoe or hairpin vortices, all of which are topologically equivalent but are different stage of stretching.
The authors postulate that a Λ-vortex consists of viscous sublayer material and that it formed from a sheet of such material which rolls up at the edges into rods. This model is close to the wall combined with the model of a viscous tornado. Using such a Λ-vortex model GYR & SCHMID (1984) show that the onset of the drag reducing effect can be explained by events just able to stretch the molecules. The local rheology in these events is changed by the stretched molecules and it is therefore of interest in which way the internal flow in these events is altered. Of main interest is the interaction of the vorticity stretching and the diffusion under these new material conditions.

II. The interaction of polymer molecules with Λ-vortices

The main part of the simplified events are the vortex rods of the Λ-vortices with an internal flow field which can be described as an axysymmetric strain field with a weak vorticity field superimposed on it. The boundaries of this jetlike event are shear layers since the events have convectional velocities less than the velocity of the mean flow at their positions.
To study the behavior of the vorticity in these events under the influence of the molecules one has to solve the vorticity equation in the advected flow field with changing rheology.
The general vorticity equation is

$$\frac{D\underline{\omega}}{Dt} = \underline{\omega} \cdot \nabla\underline{u} + \nu\nabla^2\underline{\omega} \qquad (2.1)$$

with $(\nabla,\underline{u}) = 0 \; ; \; (\nabla,\underline{\omega}) = 0$

where $\underline{u}$ is the velocity field in the advected event.
Since we model the essentiel part of the event by a rod with an axisymmetric strain field the velocity field in the event can be given in canonical form for a cylindrical coordinate system as

$$\underline{u} = (u_z, u_r, u_\varphi) = (2\beta z, -\beta r, 0) \qquad (2.2)$$

with β the strain parameter.

Equation (2.1) can now be written as

$$\frac{\delta\zeta}{\delta t} = \frac{\beta}{r}\frac{\delta(\zeta r^2)}{\delta r} + \nu\left(\frac{\delta^2\zeta}{\delta r^2} + \frac{1}{r}\frac{\delta\zeta}{\delta r}\right) \tag{2.3}$$

and a steady-state solution for eq. (2.3) can be found in BATCHELOR (1967):

$$\zeta(r) = \zeta_1 \exp\left(-\frac{\beta}{2\nu} r^2\right) \tag{2.4}$$

It can be shown that eq. (2.4) is in fact the distribution to which ω tends, as $t \to \infty$, for an arbitrary initial distribution of ζ with regret to r, subject only to the condition that $\zeta \to 0$ is faster than r^{-2} as $r \to \infty$ and that $\int_0^\infty \zeta 2\pi r \, dr$ is finite and non-zero. This latter integral represents the flux of vorticity across a plane normal to the z-axis and is invariant thereby determing the constant ζ_1 in terms of the initial conditions

$$\zeta_1 = \frac{\beta}{2\nu} r_o^2(0)\zeta(0) \tag{2.5}$$

r_o is the typical width scale of the problem with $r_o(0)$ its initial condition. $r_o(0)$ is the initial radius of the vortex core in which we have a uniformely distributed vorticity $\zeta(0)$.

We obtained these results by assuming that β and ν are constant or only function of time and $\underline{\omega}(0) = (\zeta(0), 0, 0)$ is weak in the sense that its feedback influence on the flow field $\underline{u}$ can be neglected.

These calculations consider the case in which the vorticity initially concentrated in a vortex rod tries to diffuse, thereby opposing the straining flow field. The solution shows that in this kind of flow where β and ν are constant an equilibrium between the two effects will be established at

$$r_o(\infty) = (\nu/\beta)^{1/2} \tag{2.6}$$

Since the flow field in an event cannot be described by a timeindependent strain rate - otherwise the flow would accelerate to an infinite velocity - we have to assume that $\beta = \beta(t)$. Under the hypothesis that in such events the molecules are stretched, the viscosity changes and we have to introduce $\nu = \nu(t)$. In order to solve eq. (2.3) we make use of the assymptotic solution by introducing the time dependence in the form of $r_o(t)$ and $\zeta_1(t)$ as

$$\zeta = \zeta_1(t) \exp\left(\frac{-r^2}{2r_o^2(t)}\right) \tag{2.7}$$

For this type of solution eq. (2.3) becomes an equation of Bernoulli type

$$\frac{\partial r_0(t)}{\partial t} = -\beta(t) r_0(t) + \frac{\nu(t)}{r_0(t)} \tag{2.8}$$

By substituting $s = r_0^2(t)$ with the initial condition $s(0) = r_0^2(0)$ eq. (2.8) transforms into the linear differential equation with the solution

$$r_0^2(t) = \exp(-2\int_{t_0}^{t}\beta(\tau)d\tau)[r^2(0) + 2\int_{t_0}^{t}\nu(\vartheta)\exp(2\int_{t_0}^{\vartheta}\beta(\tau)d\tau)d\vartheta] \tag{2.9}$$

With given $\beta(t)$ and $\nu(t)$, $r_0(t)$ can be calculated and $\zeta(t)$ can be found by eq. (2.7) using again the circulation theorem to evaluate ζ_1, as

$$\zeta_1(t) = \frac{r^2(0)}{2r_0^2(t)}\zeta(0) \tag{2.10}$$

By solution (2.9) it is easy to see that for $\beta = ¢$, $\nu = ¢$ the time behaviour of such a flow would be for $t_0 = 0$

$$r_0(t) = \sqrt{\frac{\nu}{\beta}}\left\{1 - \exp(-2\beta t) + \frac{\beta}{\nu}r_0^2(0)\exp(-2\beta t)\right\}^{1/2} \tag{2.11}$$

As time scale for this process we introduce the time needed for the viscous diffusion and the stretching process to be of the same order of magnitude. Since the first two terms on the right hand side of eq. (2.11) are viscous effects and the third one is the contribution from the stretching we obtain.

$$\hat{t} = \frac{1}{2\beta}\ln\left(\frac{\beta r_0^2(0)}{\nu} + 1\right) \quad \text{at} \quad r_0(\hat{t}) = \sqrt{\frac{\nu}{\beta}}\, r_0(0)\sqrt{\frac{2}{r_0^2(0) + \nu/\beta}} \tag{2.12}$$

Under this flow condition the stretching wins insofar as the vortex gets restricted to an assymtotic core given by eq. (2.6) (Fig. 2.1a).

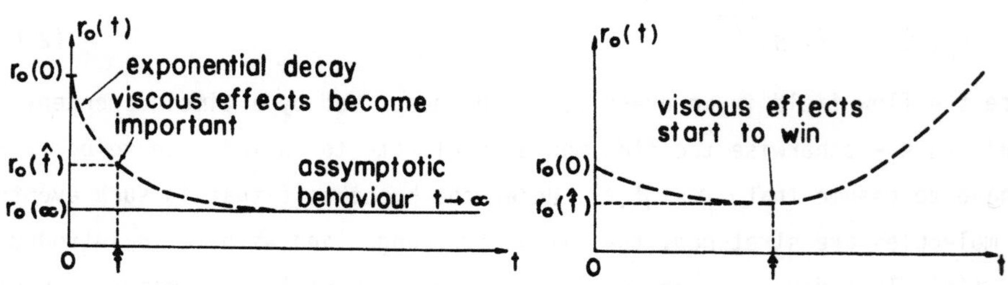

Fig. 2.1a) Typical behaviour of $r_0(t)$ for $\beta = ¢$, $\nu = ¢$; b) Typical behaviour of $r_0(t)$ for $\beta = \frac{v}{2(z_0+vt)}$.

As mentioned earlier, the flow field in the event cannot be described by eq.(2.11) because β decays with time. From experimental results it appears that the flow in a rod may be assumed to have a constant velocity in axial direction. We obtain from eq.(2.2) for a time in the order of the life span of the event.

$$u_z = \frac{dz}{dt} = 2\beta(t)z = v \; ; \; z = z_o + vt \; ; \; \beta(t) = \frac{v}{2(z_o+vt)} \quad (2.13)$$

By introducing this function in solution (2.9), we have

$$r_o(t) = \left\{ \frac{2\nu(z_o t + 1/2vt^2) + z_o r_o^2(0)}{z_o + vt} \right\}^{1/2} \quad (2.14)$$

With $t \rightarrow \infty$ the diffusion process wins always but this process is a fairly slow one with the behaviour which in practically all cases has the form as shown in fig. 2.1b.

Using the time $\hat{t}$ as a time scale r_o should become a minimum at

$$\hat{t}_{1,2} = \frac{1}{v}\left(-z_o \pm r_o(o)\sqrt{\frac{z_o v}{\nu} - \frac{z_o^2}{r_o^2(0)}}\right) > 0 \quad (2.15)$$

With this model for β we run into a problem for $t \rightarrow 0$, since for $z_o \approx 0$ would then become infinite. We therefore have to find a more realistic model. The singularities can be avoided by a two layer model composed of an inertial zone $z_o < z \leqslant z_i$ and an equilibrium layer $z_1 > z \geqslant z_i$.

In the inertial zone $\beta = \beta_o$ where as in the equilibrium zone $\beta = \beta(t)$ as given by eq.(2.13).

Since at z_i all quantities have to match, i.e. u_z, β and ζ, we replace the singularity problem by a steadiness problem. Moreover β has to be only time dependent.

The steadiness in the velocity field is described by eq.(2.14) and fulfilled if β is steady in z_i. Since the time behaviour of β is different in both zones, the steadiness of β at a fixed level z_i can only be fulfilled at a certain time. A two layer model of this kind would therefore be unappropriate unless additional assumptions based on the observed behaviour of the events are used to take care of this difficulty.

For example, we may argue, that it takes time until the events reachethe hight z_i and that this time t_i can be used as the initial time for describing the flow in the outer region. In order to secure its steadiness at z_i we also use a time depending description of β in the inner region after the initial time t_i. That means that after the initial time the flow starts to decay also in the inertial region. This is consistent with visualisations. Formaly we subdivide the period of existence of an event into a starting phase an equilibrium phase characterized by a constant axial velocity and a decay period. Only the first two will be discussed and are at one instant equivalent to the spacial description. What we need therefore is the time t_i which the event needs to reach the level z_i. Since the direct measurement of this quantity is quite difficult one has to find a way to describe the events by other means.

In the description of the Λ-vortex as it has been used in eq.(2.2) and the following equations we did not take care of the vorticity problem which occurs at walls with non-slip conditions as treated in chapter I.

We know that PERRY & CHONG's Λ-vortex model is based on a wrapping process as shown in fig. 2.2

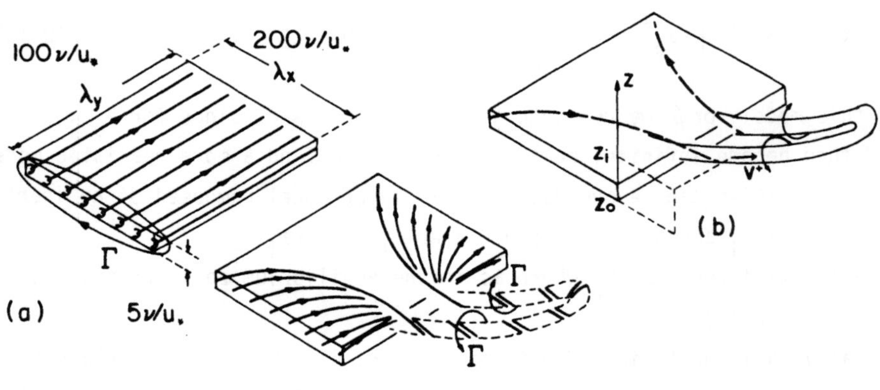

Fig. 2.2 Schematic of the Λ-vortex formation and definition scetch
Initially the vorticity is organized spanwise at it is uniformly distributed in flow direction. By an instability process, the vorticity starts to wrap up into a vortex which by stretching takes the idealized form of a Λ composed mainly of two side-vortex-rods with an internal flow behaviour as described by eq.(2.2). This process ends when the vorticity cannot be concentrated anymore and viscous diffusion processes start to dominate the flow field in the event, which starts from thereon to decay. Since this model also has to account for the non-slip condition, the wall near flow can be described by a viscous tornado. Therefore the question arises whether by incorporating the model of the viscous tornado into the Λ-vortex model it would be possible to describe the flow field completely.

We encounter two essential difficulties. The first one stems from the fact that the so called measurable quantities in eq. (1.1) are mesurable only in a theoretical sense but not in practice.
The second problem is a theoretical one, since this model cannot be introduced by a superposition of a viscous tornado and a Λ-vortex. The tornado is a wrapping process whereas the vortex rod dynamics as described in eq. (2.3) presumes that at a level z_0 the vorticity is uniformly concentrated in a rod of radius r_0. The development of this rod is given by the interaction of a strain field with a diffusion process of vorticity. The difference of the two mechanisms is evident since the wrapping process implies a continuous feeding of the core area of the vortex with vorticity from the outer areas.
In the outer flow with respect to the wall, the Λ-vortex model will be dominant. Since in this region the vorticity equation is tractable in the form of eq. (2.3) we approximate the outer flow field of the event by this model. The tornado like wall-near flow then represents a boundary condition for the Λ-vortices.
This approximation is also supported by the very successful calculation of the momentum transfer made by PERRY & CHONG, which are based on the Λ-vortex model and on a similarity assumption according to which the relevant vorticity processes close to the wall scale with its inner parameter (ν, u_*). This assumption was made by KLINE et al. (1967) based on their famous observations. The main profit of such a model is the conservation of the circulation. The level at which the vorticity is wrapped completely into the rods can be estimated from the visual study by PERRY et al. (1981). We find from this very small set of pictures a mean value of $\overline{r_0^+}|_{P_{yz}} = 6.5$ where P_{yz} is the projection on the yz-plane at the level the wrapping process ends up. With a mean inclination $\overline{\alpha} = 22^0$ of the events BERTSCHLER et al.(1979) this gives in the canonical form a value of $\overline{r_0^+} = \overline{r_0^+}|_{P_{yz}} \cos \alpha = 6$. If we assume that the wrapping process can be explained by a complex eigenvalue singular point behaviour, then the wall flow is dominated by a flow as given by eq. (1.3a-c). From this equation it follows that in the xy-projection the trajectories in canonical form are simple logarithmic spirals, fig. 1.1, and the solution trajectories intersect the xz and yz planes along curves given by

$$z = \frac{HK}{x} = \frac{HK}{y} \tag{2.16}$$

Hence trajectories wrap themselves on to rectangular hyperbolae of revolution. In the more general non-canonical form, the pattern is obtained by distorting the canonical form by affine transformation.
With the estimate that $r_o^+ \simeq 50$ at $z_o^+ = 5$, the sublayer thickness, one would obtain

$$HK = z_o^+ r_o^+ \frac{\nu^2}{u_*^2} = 250 \frac{\nu^2}{u_*^2} = (HK)^+ \frac{\nu^2}{u_*^2}$$

if z_o^+ can be set close to the wall equal $\tilde{z}_o^+$.
With the Λ-vortex we postulated that the velocity in axial direction in the core is limited and a constant value equal to the velocity v at the end of the inertial flow period. This value can be measured fairly easily since in the equilibrium zone, which we find at a level far enough away from the walls that its influence is not disturbing the measurements.
Since we postulated that the behaviour of the envelope of the trajectories close to the wall is more or less similar to the behaviour in a viscous tornado hyperbolic envelopes would suggest that z_i^+, the level at which the wrapping ends is given by

$$\tilde{z}_i^+ = (HK)^+/r_i^+ \approx 250/6 \text{ or } z_i^+ = \tilde{z}_i^+ \sin 22 \approx 15.6 \qquad (2.17)$$

Since at z_i^+ the inertial and the equilibrium solutions habe to match, there results from eq (2.13)

$$\beta_o^+ = \frac{v^+}{2z_i^+} \text{ with } \beta_o^+ = \beta_o / u_*^2 \text{ and } z_i^+ \approx 15.6 \qquad (2.18)$$

If needed this result can always be transformed in a transient one by integration of eq. (2.14)

$$t_i^+ = \frac{1}{2\beta_o^+} \ln \frac{z_i^+}{z_o^+} \qquad (2.19)$$

With a measured v^+ we can calculate β_o^+ from eq. (2.18) and would now have all needed parameter for describing the flow field of this vortex completely in the frame of this model.
After the Λ-vortex has reached z_i, its further development is given by eq. (2.14).
With measurements of v^+ at the distance $z^+ \approx 30$ are able to calculate β_o from eq. (2.18). With β_o the flow field is given and the ejection event can be described as a stagnation point flow with a superimposed vorticity which is weak

close to the wall and increasing with distance from the wall. The fluid particles are spiralling along hyperbolic surfaces with an increasing shear. Since the stretching of the molecules is associated with a high strain and a low shear field the molecules can only be stretched or prestretched close to the stagnation point. This means that this happens at a distance from the wall of the order of the viscous sublayer in a field described by the given parameter β_0.

GYR & SCHMID (1984) show that the onset of drag reduction is associated with a $\beta_0 > \beta_{0\ crit}$ where $\beta_{0\ crit}$ is the strain parameter needed to stretch the molecules in a pure strain field.

Based on a paper of PETERLIN (1966), LUMLEY (1969) argued that the stretching of the molecules is associated with the flow field in which the strain dominates the shear significantly. He attributed the stretching of the molecules to a statistical behaviour of the turbulent boundary layer by excluding the viscous sublayer. In some contradiction to this we state here that the stretching occurs in typical ejection events and the molecules start to stretch in the sublayer but are not stretched before they reach the edge of the sublayer.

When the molecules have reached a certain distance from the wall the shear reaches a level at which the molecules dont get elongated any more but where they will align. The direction of the aligned molecules is at a specific angle to the shear profile. This has been shown by HINCH (1967) for molecules behaving as flexible inextensible, nearly straight threads in a shearing flow. In other words they will follow the fluid trajectories but stay at a defined angle to them.

In direction of the elongation of the aligned molecules an extreme elongational viscosity will be found. In absence of any vorticity these molecules would align in strain direction and therefore damp the axial flow in the event. This mechanism could be one reason for drag reduction.

Since the direction of the elongated particles does not coincide with the axial direction as a consequence of the spiraling of the streamlines as well as of the perpendicular shear, we find that extensional viscosity can act in an anisotropic way in the plain perpendicular to the axial movement. As long as the molecules are stretched the viscosity used in eq. (2.1) therefore becomes time dependent and its value increases with wall distance. It can easily be seen from the integration (2.9) of the vorticity equation (2.3) that

an increase in viscosity causes a hyperdiffusion of the vorticity. This together with damping could give rise to drag reduction.
This description is based on the assumption that the overall behaviour can be studied in a Newtonian fluid, and that a locally higher viscosity can be introduced to evaluate the vorticity diffusion. Since the main mechanism is an intertial one and the restoring time of the molecules rather small this simplification appears to be permissible.
The simplified model is characterized by an inertial range and an equilibrium range. In the inertial range the molecules are stretched in a strong strain field. The axial flow field is damped but also "frozen". This means that the core of the vortex rod is similar to a solid body motion at least as long as the molecules do not relax.
In the equilibrium range, the diffusion process of the vorticity is significant and as a pecularity of the polymer solutions it will be highly accelerated by the new rheology which becomes important as soon as the shear in the events is dominant.
As a conclusion we would expect ejection in polymer solutions which are damped in axial direction and in which the core size increases. This is exactly what ACHIA & THOMPSON (1974 or 1977) have found in their visualisation by using holographic fringe interferograms. It seems therefore obvious that the interaction of the new rheology with the vorticity in the events must be of importance to explain the drag reduction in such flows.

III. Drag reduction

The aim of this investigation was to make a step towards the explanation of drag reduction by means of an altered vorticity diffusion behaviour in the ejections. We were able to show that we presumably will have a damping of the strain and an acceleration of the vorticity diffusion in the ejection type of events.
These behaviours could be integrated in several theories explaining drag reduction, but we would like to renounce here from offering one explanation, since this bit of information seems too small to give an overall explanation. Instead we would like to show that the described flow behaviour fits well into some common ideas about drag reduction.
All models based on the extensional viscosity arrising from stretched molecules have to show that this viscosity acts selectively in such a way that the total energy dissipation is decreasing in spite of the fact that a higher

viscosity in general would result in a higher dissipation rate.
The simplest mechanism for drag reduction implies that turbulent shear flows are comprised of many flow structures. The high extensional viscosity then selectively dissipates those structures which are more efficient at transporting momentum. It is shown here that these efficient structures are not suppressed but altered.
An other common explanation for drag reduction is that the extensional viscosity stabilizes the flow of the viscous sublayer or the organisation of the different features in a grand structure. It would also be possible to incorporate the enhanced vorticity diffusion discussed here into such an explanation. Without knowing the grand structure in Newtonian fluids this is naturally highly speculative.
In this model the momentum transport in the ejections will be reduced by damping the axial flow component. Moreover the dissipation is reduced due to a reduction of the internal shear by a vorticiy diffusion process. Finally the vortex core increases in width and the axial momentum is smeered over larger areas. Therefore it is thinkable that the competition between vorticity diffusion and the cascade process, which both are not dissipative processes, would result in a reduction of the energy transport since the diffusion transports the energy in the opposite direction than the cascade process: This means towards vortices of larger scale.
LUMLEY (1977) gave purely statistical arguments for the drag reduction mecha nism. He starts from a model of turbulence in which at high Reynolds numbers the vorticity and the strain rate are uncorrelated with each other and can therefore occur in virtually any combination. Since both vorticity as well as the strain have distribution with long tails, so that large and small values of each are more probable than for Gaussian distribution, there is substantial probability of finding regions of relatively large strain rate and relatively small vorticity.
He argues that for the viscous sublayer vorticity is equal to the strain rate and, since molecules from the outer flow region would relax before being diffused into this layer, the viscous sublayer is therefore uneffected by the polymers. This means that the extensional viscosity would cause substantial increase in viscosity in the turbulent part of the flow and virtually no increase in the viscous sublayer. The higher viscosity would thus supress the eddies which carry the Reynolds stress in the buffer zone resulting in a

thickening of the sublayer, and a reduction of drag.

The mechanism described here contradicts some of these views. First of all, in the area where the ejections are "attached" to the wall, i.e. in the sublayer, we have the most effective strain fields, but the consequences of this stretching can only be found in the buffer zone: The viscous sublayer is presumably also in this model enlarged by the damping process. This is also consistent with the finding that the large eddies are essentially the same as in a Newtonian flow except that their scale is expanded to fit the altered sublayer. This result agrees with LUMLEY's (1971) prediction that the large eddy structures are parasitic and determined primarily by the mean velocity profile. The parasitic character of large eddies is insofar essential for our study as we treated the Λ-vortices as inbeded in Newtonian fluid. Therefore the mechanism is quite different from that of Lumley although its conclusion hold for this kind of model.

With these remarks the limits of such models of drag reduction are given. It seems therefore reasonable to investigate this flow behaviour in much more detail before conclusive statements on the drag reducing mechanism can be made. However most of the main feature of the drag reduction can be explained by this vorticity diffusion mechanism.

References

ACHIA B.U. & THOMPSON D.W. (1974): Laser holographic measurements of wall turbulence structures in drag reducing pipe flow. Proc. Int. Conf. on Drag Reduction Cambridge A2, 23-40 (1977) Structure of the turbulent boundary in drag reducing pipe flow JFM 81, 439-64

BATCHELOR G.K. (1967): An introduction to Fluid Dynamics. Cambridge Univ. Press.

BERTSCHLER M., DRACOS T. & GYR A. (1979): On the flow structures giving rise to high Reynolds stresses in a turbulent boundary layer. Turbulence in liquids. Dept. Chem. Eng. Univ. of Missouri-Rolla (Proc. 6th Symp. on tub. in liquids Oct. 1979)

HINCH E.J. (1976): The distortion of a flexible inextensible thread in a shearing flow. JFM 74, 317-33

KLINE S.J., REYNOLDS W.C., SCHRAUB F.A. & RUNDSTADLER P.W. (1967): The structures of turbulent boundary layers JFM 30, 741-73

LIGHTHILL M.J. (1963): Laminar boundary layers (Rosenhead) 48-88, Oxford Univ. Press., Claredon Press.

LIM T.T, CHONG M.S. & PERRY A.E. (1980): The viscous tornado. Proc. 7th Austr. Conf. on Hydraulics and Fluid Mechanics, Brisbane, 250-3

LUMLEY J.L. (1969): Drag reduction by additives. Ann. Rev. of Fluid Mech. I, 367-84 (1971) Developments in Mechanics (Univ. of Notre Dame, Notre Dame Indiana 6, 63), (1977) Drag reduction in two phase and polymer flows. Phys. fluids 20, S 64-71

OSWATITSCH K. (1958): Die Ablösungsbedingung von Grenzschichten. Grenzschicht Forschung (H. Görtler) 357 Springer

PETERLIN A. (1966): Hydromechanics of linear macromolecules. Pure & Appl. Chem. 12, 273-86

PERRY A.E., LIM T.T. & TEH E.W. (1981): A visual study of turbulent spots. JFM 104, 387-405

PERRY A.E. & CHONG M.S. (1982): On the mechanics of wall turbulence. JFM 119, 173-217

Session 8
Structures in Turbulence II

Influence of Polymer Additives on Neutral Stability of Görtler Vortices

Y. TOMITA

Department of Mechanical Engineering
Kyushu Institute of Technology, Kitakyushu 804, Japan

Summary

By towing a concave wall of 500mm radius of curvature at a constant speed in a tank filled with an aqueous solution of polythylene oxide of about 4×10^6 molecular weight, the neutral stability of Görtler vortices was determined for the 2 and 5wppm solutions with the hydrogen bubble technique and the streamwise velocity of basic flow in boundary layer along the wall was also measured.

Introduction

There is a hypothesis that the drag reduction by polymer additives is caused by a large extensional viscosity of the solution and by the increased viscosity flows are stabilized. Many experiments suggest that flows in which drag reduction is observed are in most cases accompanied by extensional type of secondary flows in one form or another. In flows along concave walls, instability with respect to three-dimensional disturbances, represented by counter-rotating vortex pairs, occurs before actual transition to turbulence. These disturbances, called Görtler vortices, also represent extensional flows. Thus, it is inferred that by addition of a drag reducing polymer the vortices will be stabilized. In this paper, we experimentally test this hypothesis.

Background of Görtler vortices

Centrifugal instability in boundary layer flows along concave walls was first examined theoretically by Görtler [1] in 1940. Since then many investigators have been studying this problem. Several assumptions are necessary to solve the problem. Among those the parallel flow assumption is important. Görtler assumed that the basic flow in boundary layer is parallel and that the disturbance are temporally growing. Smith [2] introduced a non-parallel flow model which includes the transverse component of the basic flow, and considered spatially growing disturbances. The neutral disturbances of this model do not develop streamwise. Floryan and Saric [3] showed that the streamwise variation of

The Influence of Polymer Additives
on Velocity and Temperature Fields
IUTAM Symposium Essen 1984
Editor: B. Gampert

the transverse component of the basic flow is also important. On the neutral disturbances Hall [4] discussed that if the basic flow develops streamwise the disturbances should also develop streamwise. In this paper, we show our computations of the neutral stability of Görtler model, Smith model and modified Smith model for Newtonian fluids by using body-oriented coordinates. We assume the neutral disturbance velocities in the dimensionless form as

$$u' = u(\eta)\cos\sigma\zeta \tag{1a}$$

$$v' = v(\eta)\cos\sigma\zeta \tag{1b}$$

$$w' = w(\eta)\sin\sigma\zeta \tag{1c}$$

The equations governing the neutral stability of the Görtler model in the dimensionless form are

$$\frac{d^2u}{d\eta^2} - \sigma^2 u = \frac{dU}{d\eta}\, v \tag{2}$$

$$\frac{d^4v}{d\eta^4} - 2\sigma^2\frac{d^2v}{d\eta^2} + \sigma^4 v = -2G^2\sigma^2 Uu \tag{3}$$

where u', v' and w' are the disturbance velocities in the streamwise, transverse and spanwise directions, U is the streamwise velocity of the basic flow, σ is the wave number of disturbance in the spanwise direction, η and ζ are the transverse and spanwise coordinates and G is the Görtler parameter. When U_∞ is the free stream velocity, r is the radius of curvature of the concave wall, λ is the wave length of disturbance, x is the distance from the leading edge and ν is the kinematic viscosity, G and σ are defined as

$$G = \left(\frac{x}{r}\right)^{1/2}(R_x)^{1/4} \tag{4}$$

$$\sigma = \frac{2\pi x}{\lambda}\,(R_x)^{-1/2} \tag{5}$$

and R_x being the Reynolds number defined by $U_\infty x/\nu$. The streamwise velocities are nondimensionalized by using U_∞ and the transverse velocities are by using $U_\infty/\sqrt{R_x}$, and η is $y\sqrt{U_\infty/\nu x}$, where y is the distance from the wall.
The equations for the modified Smith model are

$$\frac{d^2u}{d\eta^2} - \sigma^2 u = \frac{\partial U}{\partial\eta}\, v \tag{6}$$

$$\frac{d^4v}{d\eta^4} - (V+2k)\frac{d^3v}{d\eta^3} - \left(\frac{\partial V}{\partial\eta} + 2\sigma^2 - kV\right)\frac{d^2v}{d\eta^2} + \left(\sigma^2 V + k\frac{\partial V}{\partial\eta} + 2k\sigma^2\right)\frac{dv}{d\eta}$$
$$+ \sigma^2\left(\frac{\partial V}{\partial\eta} + \sigma^2\right)v + \sigma^2\left\{2G^2U + (1 + k\eta)\frac{\partial V}{\partial\xi}\right\}u = 0 \tag{7}$$

where k is the dimensionless curvature defined by $(x/r)/\sqrt{R_x}$, V is the transverse velocity of the basic flow and ξ is the streamwise coordinates. The equations for the Smith model are the same as those for the modified Smith model except that $\partial V/\partial\xi$ is omitted. The boundary conditions are the same for all models and given by

$$u = v = \frac{dv}{d\eta} = 0 \qquad \text{at } \eta = 0, \infty \tag{8}$$

For a given basic flow each model forms an eigenvalue problem for the parameters. The characteristics equation for the Görtler model is $F(G,\sigma) = 0$ for a given U, and for the Smith and modified Smith models those are $F(G,\sigma,k) = 0$ for a given U and V. In this paper we apply the Blasius flow to the basic flow.

On the other hand available experimental results are few on the neutral stability of the Görtler vortices. Wortmann [5] obtained a neutral point which corresponds to the minimum Görtler parameter according to Görtler's computations with the tellurium method. Bippes [6] examined the high wave number range with the hydrogen bubble technique and found out that the neutral line obtained slightly shifts to the higher wave number than that given by the computations. In this paper we lay emphasis on measurement in the low wave number range.

Experimental apparatus and procedure

Our experiment was carried out along the lines of the one adopted by Bippes. The concave wall was towed at a given speed in a liquid-filled tank. The behaviour of a given disturbance which was produced by a row of wings of finite span was observed by time lines obtained with the hydrogen bubble method. The tank was made of acryl and was 4500mm long with a cross-section of 400 by 400mm. The concave wall whose radius of curvature was 500mm was made of casting of aluminum alloy. Figure 1 shows the main dimensions of the wall. The surface was painted black and was drawn white lines in a checkered pattern having 20mm pitch. The upper curved wall was used to keep the free stream velocity constant and the flat plate above the upper curved wall was used to shut out the diffused reflection of light caused by surface waves. The wings as a disturbance producer had symmetrical section whose chord length and maximum thickness were 10 and 2mm. The used lengths of span were 10, 15, and 20mm which were assumed to produce the disturbances having 20, 30 and 40mm wave lengths. The row of wings was set at the angle of attack of about zero degree and was fixed at 7mm above the wall level and 20mm ahead of the leading edge of the wall. To produce the hydrogen bubbles we used tungsten wires of 60μm O.D. for stability experiment and stretched those in the spanwise direction about 1 to 2mm above the wall at several places between 5 and 260mm from the leading edge. When we measured the streamwise velocity distribution of the basic flow in boundary layer, we used tungsten wires of 10μm

O.D. and stretched those in the transverse direction at 80 and 180mm from the leading edge. We used a pulse generator to make the time lines. In the hydrogen bubble technique a duty ratio is an important factor which controls the size of bubbles and the interval of time lines. This is a ratio of time which a current is being applied to the wires to the pulse period. If we make it large, the bubble size increases, which results in the high rising velocity of bubble. If we make it small it becomes difficult to get a better view of the time lines. We adjusted the ratio according to the free stream velocity so as to obtain clear lines. Furthermore, salt was dissolved about 50wppm to promote the bubble generation. It was confirmed by a simple pipe flow experiment using a solution after the stability experiment that the addition of salt does not lower the drag reducing ability of polymer solution.
The polymer used was polyethylene oxide (Alkox E-160, Meisei Chemical Works) whose molecular weight is more than 4×10^6. We used the 2 and 5wppm solutions. When the concentrantion of polymer was made higher, it became difficult to generate the bubbles. We also used a solution stabilizer (Sandex-C, Meisei Chemical Works) at the same concentration with the added polymer.
Before starting each run we waited at least about one hour to calm the solution in the tank. Once the solution was made the same solution was used for about 20 runs.

Stability criterion

It was a delicate problem to determine visually whether the disturbance was amplified or decayed. If we assume that the neutral disturbance is given by Eq.(1) and that the basic flow is parallel, the time lines of neutral disturbance generated by pulse of constant period will produce a pattern as shown in Fig. 2a, and those of amplified disturbance will produce a pattern as shown in Fig. 2b, where the interval between the time lines for a given ζ will decrease downstream in the roll up region of the vortices and increase downstream in the roll down region, thus forming longitudinal streaks made of bubbles. In this work we determined the stability of a given disturbance according to the above criterion, that is, the time lines which represent other than the pattern as ahown in Fig.2b were stable and the time lines as shown in Fig.2b were unstable.

Results and discussion

Figure 3 shows examples of determination of the stability. When the polymer was added, the period of pulse was increased since the time lines were not so clear as compared with those in water. When the disturbance producer was not used, the time lines were parallel to the cathod wires and disappeared downstream, which means that the wires do not disturb the basic flow. It was observed that the time lines lagged behind at the place where a large bubble or polymer blob attached to the wires.

Figure 5 shows the stability diagram for water in which the experimental results due to Wortmann and Bippes as well as our computations of neutral stability for three models are included. As for the Görtler model those well correspond to the computations due to Hämmerlin [7], Herbert [8], Ragab and Nayfeh [9] and Floryan and Saric [3]. There is no minimum Görtler parameter for the Görtler model in the finite wave number. As for the Smith and modified Smith models our computations well correspond to those due to Ragab and Nayfeh and Floryan and Saric. There is also no minimum Görtler parameter.

It appears that our measurement follows the Smith model. But it is noted that the Reynolds number of our data for low Görtler parameter is considerably small. Then, it seems that the Blasius flow no longer holds good for those data. In connection with this problem Ragab and Nayfeh computed the neutral stability curves taking into account of the effect of displacement thickness of boundary layer. According to their results the curves approach to that of the Smith model departing from that of modified Smith model at low wave number when R_x becomes small.

Figures 6 and 7 show the stability diagrams for the 2 and 5wppm solutions, where we used for the viscosity of solution as that of water at the same temperature. It is seen that by adding a small amount of polymer the stable region of Görtler voritces is extended as compared with that of water alone. And the shift of neutral lines to the low wave number against that of water alone is discernible when the polymer was added. If we accept this shift of neutral lines, it follows that for a given Görtler parameter the wave number of neutral stability is lowered by addition of polymer, that is, the size of neutral vortices is increased against that of water alone. This increase in voritces must be caused by an increase in viscosity. However, it is wellknown that the shear viscosity of solution is practically equal to that of water. If we regard that the flow pattern produced by the disturbance represents a kind of extensional flow, it is possible to infer that the extensional viscosity of polymer solution acts on the Görtler vortices. Thus by adding the polymer a large extensional viscosity acts on the Görtler vortices and the stable size of vortices is increased for a given flow field.

Figure 8 shows the streamwise velocity distribution of basic flow for water and the 5wppm solution with R_x as a parameter. The solid lines in the figure are the Blasius profile. From this figure it is seen that the Blasius profile holds good in such a low Reynolds number. It is noted that the second order approximation of boundary layer for semi-infinite plate is the same as that of first order concerning the streamwise velocity.

Conclusions

From this experimental work it is found out that the addition of drag reducing polymer stabilizes the Görtler vortices. If one accept that the Görtler vortices and sublayer eddies in turbulent boundary layer are analogous structure, it is inferred that the addition of polymer in turbulent boundary layer flows would also increase the size of the sublayer eddies and suppress the production of energy dissipating eddies.

The author acknowledges the useful cooperation on this work made by K. Funatsu, K. Miyazaki and S. Fujino.

References

1. Görtler, H.:Über eine dreidimensionale instabilität laminarer Grenzschichten an konkaven Wänden. Nachr. Ges. Wiss. Göttingen, Fachgruppe I, Neue Folge 2, 1 (1940)1-26.

2. Smith, A.M.O.:On the growth of Taylor-Görtler vortices along highly concave walls. Quart. Appl. Math. 13 (1955) 233-262.

3. Floryan, J.M.; Saric, W.S.:Stability of Görtler vortices in boundary layers. AIAA JOURNAL, 20, 3 (1982) 310-324.

4. Hall, P.:Taylor-Görtler vortices in fully developed or boundary-layer flows:linear theory. J. Fluid Mech. 124 (1982) 233-494.

5. Wortmann, F.X.:Experimentelle Untersuchungen laminarer Grenzschichten bei instabiler Schichtung. Proc. XI Int. Cong. Appl. Mech. (1964) 815-825.

6. Bippes, H.:Experimentelle Untersuchung des laminar-turbulenten Umschlags an einer parallel angestromten konkaven Wand. Sitzungsberichte der Heidelberger Akademie der Wissenschaften Mathematisch-Naturwissenschaftliche Klasse, 3 (1972) 103-180.

7. Hämmerlin, G.:Über das Eigenwertproblem der dreidimensionalen Instabilität lamimarer Grenzschichten an konkaven Wänden. J. Rat. Mech. Anal. 4 (1955) 279-321.

8. Herbert, Th.:On the stability of the boundary layer along a concave wall. Arch. Mech. Stos. 28 (1976) 1039-1055.

9. Ragab, S.A.; Nayfeh, A.H.:Effect of pressure gradients on Görtler instability. AIAA Paper No.80-1377, July (1980).

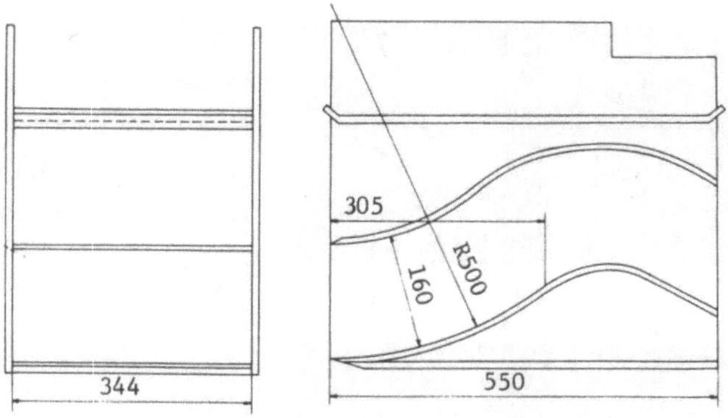

Fig.1. Diagrammatical sketch of concave wall.

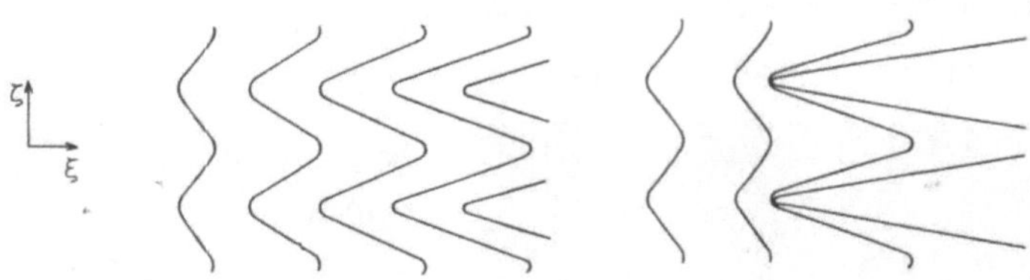

a. Neutral disturbance   b. Amplified disturbance

Fig.2. Approximate pattern of time lines by the Görtler vortices.

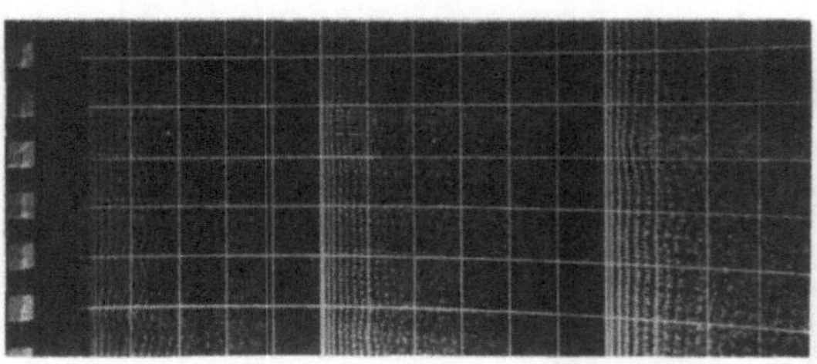

Water

λ=20mm

1. σ=0.323, G=0.220 o
2. σ=1.445, G=2.085 o
3. σ=2.143, G=3.767 o

1 2 3

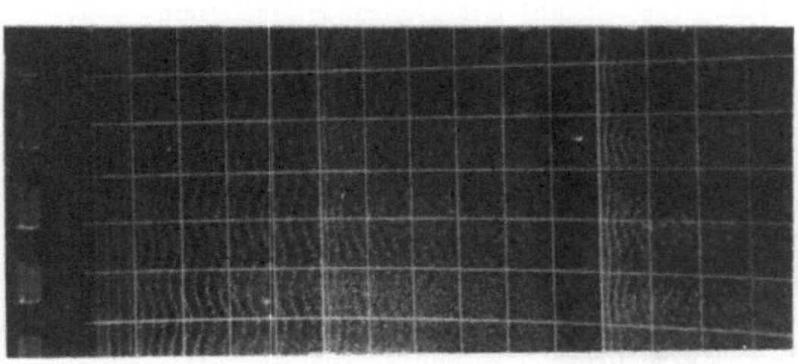

Water

λ=30mm

4. σ=0.189, G=0.235 o
5. σ=0.846, G=2.226 o
6. σ=1.254, G=4.021 ×

4 5 6

Fig.3. Time lines by disturbance producer. Continued

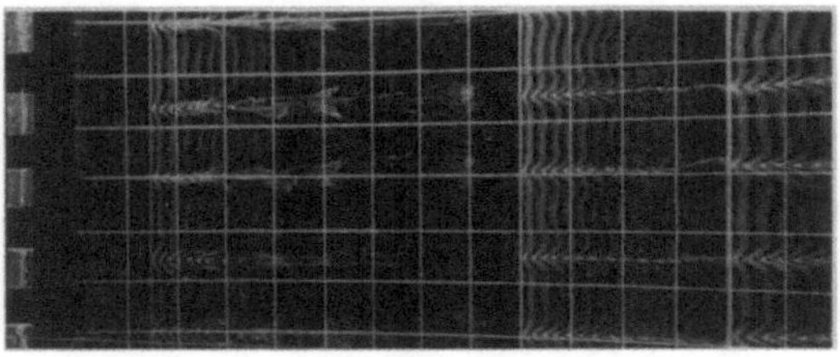

7 8 9

Water

$\lambda=30$mm

7. $\sigma=0.208$, G=1.345 ×
8. $\sigma=0.510$, G=5.157 ×
9. $\sigma=0.613$, G=6.794 ×

10 11 12 13

2wppm solution

$\lambda=30$mm

10. $\sigma=0.631$, G=0.773 o
11. $\sigma=1.152$, G=1.907 o
12. $\sigma=1.546$, G=2.963 o
13. $\sigma=1.858$, G=3.904 o

14 15 16 17

2wppm solution

$\lambda=40$mm

14. $\sigma=0.123$, G=0.253 ×
15. $\sigma=0.388$, G=1.422 ×
16. $\sigma=0.650$, G=3.078 ×
17. $\sigma=0.815$, G=4.320 ×

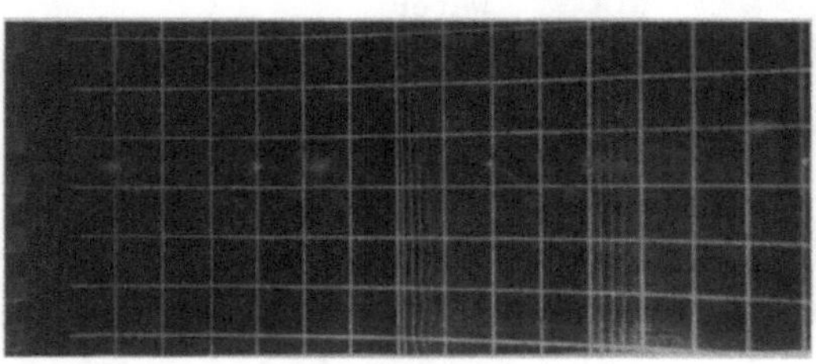

18 19 20

5wppm solution

$\lambda=20$mm

18. $\sigma=0.557$, G=0.672 o
19. $\sigma=1.473$, G=2.892 o
20. $\sigma=1.846$, G=4.059 o

21 22 23

5wppm solution

$\lambda=20$mm

21. $\sigma=0.724$, G=1.473 o
22. $\sigma=1.211$, G=3.189 ×
23. $\sigma=1.651$, G=5.073 ×

Fig.3. Time lines by disturbance producer.

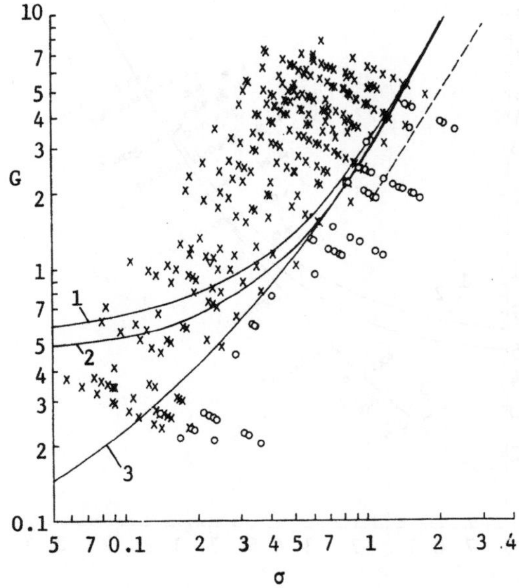

Fig.4. Stability diagram of Görtler vortices for water:Neutral stability curves;1)Görtler model; 2)Modified Smith model for $k=10^{-8}$; 3)Smith model for $k=10^{-8}$:Experimental points; o, stable disturbances, ×, unstable disturbances in this work; ∇,Wortmann's point;------, Bippes' neutral line.

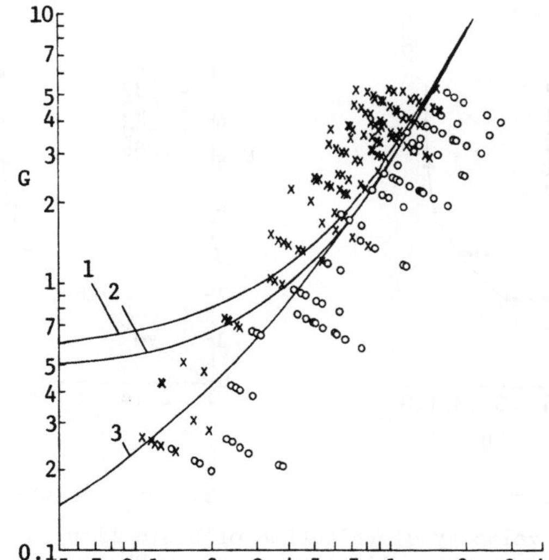

Fig.5. Stability diagram of Görtler vortices for 2wppm solution;symbols same as Fig.4.

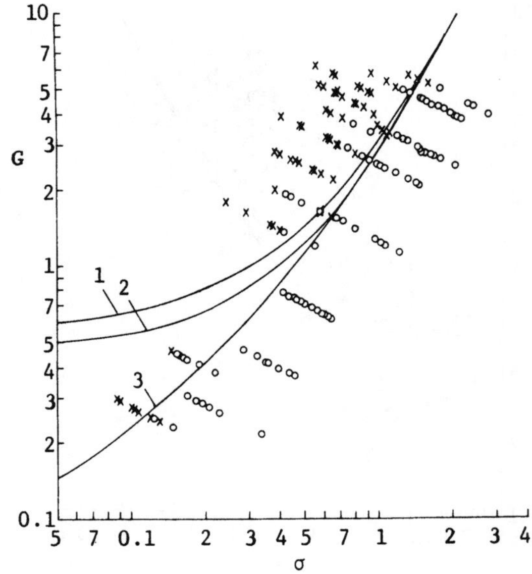

Fig.6. Stability diagram of Görtler vortices for 5wppm solution;symbols same as Fig.4

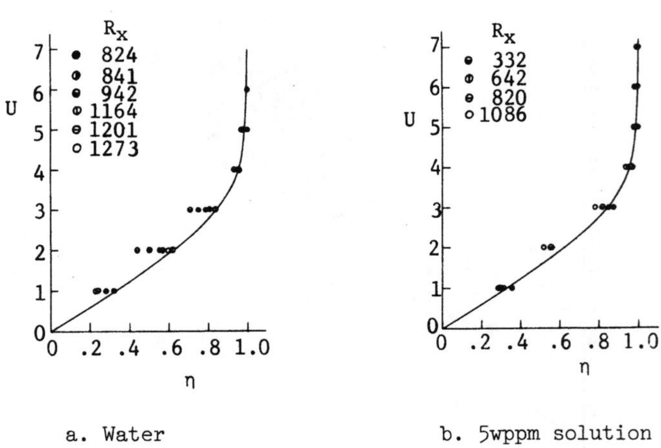

Fig.7. Streamwise velocity distribution of basic flow in boundary layer along the concave wall;the solid line shows the Blasius profile.

Laser-Doppler Measurements of Vortex-Shedding Past Small Circular Cylinders in Dilute Polymer Solutions

E. KIT

Faculty of Engineering, Tel-Aviv University, Ramat-Aviv, Tel-Aviv, Israel.

Z. ROSENTRAUB

Israel Oceanographic & Limnological Research, LTD., Haifa, Israel

Abstract

Measurements of the vortex-shedding frequency in the flow past small circular cylinders (d ~ 0.2÷2mm) in a wide range of Reynolds numbers (Re ~ 10÷1000) are presented. The measurements were performed both in water and in dilute drag-reducing polymer solutions by means of Laser-Doppler anemometry. The data measured in water were in good agreement with similar data obtained by others (e.g. Roshko [15], Berger & Wille [2]). In the presence of polymer additives the onset of the vortex-shedding occurred at smaller Reynolds number (Re~13) and the Strouhal number decreased considerably with increasing of polymer concentration and/or decreasing cylinder diameter. No influence of polymer additives was observed for cylinder diameters larger than 2mm. These results agree with previous investigations [4,8,9] for very thin cylinders and with [16] for much larger cylinders.

Introduction

The flows of dilute polymer solution with strong longitudinal deformation of the polymer molecula (for example, past cylinders of small diameter [6, 7,10]) are of special interest. The main reason for this is that such deformation is followed by a significant increase of energy dissipation of the flow [14]. These effects occurring in the region of the upstream critiçal point of the cylinder [10] were found to be responsible for the increase of the drag coefficient.

On the other hand, the flow in the region downstream of the cylinder and in the wake, where vortex formation occurs has only been scetchily investigated. Measurement of the detailed flow structure for this region is extremely difficult since the diameter of the cylinder should be very small (0.1-0.5 mm). Even measurements of Strouhal number, St=fd/V were accomplished in a small number of works [4,8,9], and in very limited range of Reynolds numbers, Re=dV/ν, where f is the vortex shedding frequency, d-cylinder, diameter, V-velocity of the undisturbed flow, ν-kinematic viscosity. The

The Influence of Polymer Additives on Velocity and Temperature Fields
IUTAM Symposium Essen 1984
Editor: B. Gampert

purpose of this work was a more detailed investigation of the influence of the polymer additives on the Strouhal number using a Laser-Doppler anemometer.

Experimental Set-up.

The experimental set-up (Fig. 1) was arranged in order to measure vortex shedding frequencies from very thin circular cylinders, perpendicular to the flow. A 500 liter constant-head tank was used to supply a steady flow of water or of polymer solution. The flow was controlled by a valve at the extreme downstream end of the conduit and measured by an accurate magnetic flowmeter.

The conduit was made up of 10 mm thick plexiglass and was composed of two square sections. The first section $3\times3\times110$ cm^3 (inside) was utilized to create fully developed flow conditions in the test section. Along the centerline of the test section a row of vertical stainless steel cylinders were fixed, diameters varying from 0.25 mm to 1.0 mm. Additionally one location was provided for insertion of a plexiglass cylinder with diameters varying from 2 to 5 mm.

Following the initial experimental runs a smaller rectangular section was also built in order to produce higher velocities, thereby providing a higher Reynolds number. This section ($3\times1\times105$ cm^3) was gradually expanded to connect with the original cross section of the conduit (3×3 cm^2). The cylinders with diameters varying from 0.1 to 1.0 mm. were put along this conduit.

A Thermo-Systems (TSI) Laser-Doppler Anemometer was used to measure the resulting vortex shedding frequencies. This was accomplished by measuring the velocity fluctuations in the direction of flow at some distance downstream from the cylinders. This type of measurements, being inherently optical in nature, did not disturb the flow field and was not influenced by the presence of the polymer.

Because of the small cylinder diameters used in these experiments, high vortex shedding frequencies resulted, up to 2 Khz. Thus, in order to obtain a continuous signal of the velocity fluctuations, it was necessary to increase the number of particles in the flow. This was accomplished by mixing a small amount, 200 gr. of milk into the head tank.

Experimental Results

Measurements of shedding frequency were conducted separately for each cylinder for various flow rates. For every flow rate during each run, the temperature of water was measured for viscosity calculations purposes. The value of an undisturbed velocity for St and Re numbers calculations was obtained by averaging of 10 velocity readings, measured at 20 diameters upstream of the cylinders.

The LDA measuring volume was located at a distance of 4-10 cylinder diameters downstream from a cylinder and at 2-5 diameters aside (Y-direction). The final position of the measuring volume was found by varying its location until the best possible sinusoidal shape of velocity fluctuations with the highest amplitude was obtained. The velocity fluctuations were recorded on a magnetic tape for further spectral analysis. The quality of the signal was good enough and it was possible to measure the frequency manually by direct recording on the U.V. recorder. The discrepancy between manual measurements and spectral analysis did not exceed 3%.
The polymeric solutions were made from two batches of master solution at 2% concentration. The polymer solution in this tank was allowed to remain at rest for 30 minutes prior to beginning a run in order to achieve a more homogeneous solution. The solution quality was checked during the experiments with an accurate turbulent rheometer. This instrument checked the friction reduction efficiency of the polymer in controlled measurements. The intrinsic viscosity $[\eta]$ of the polymer solution was measured with a capillary viscometer. A value of $[\eta]$=15dl/gr was obtained, i.e., the relative increase in viscosity, $[\eta]c$, was only 15% for the most concentrated solution of 100 ppm. The kinematic viscosity for calculation of Re was corrected according to the polymer concentration and changes in solution temperature.

The Strouhal numbers in water flow past cyclinders of diameters 0.17, 0.25, and 0.7 mm measured in the rectangular conduit for Reynolds numbers up to 1300 (Fig. 2) were found to be in good agreement with the measurements of Roshko [15] although they were about 5% lower. For cylinders of 1 and 2mm diameter in the square conduit the Strouhal numbers obtained were 10% and and 15% respectively lower compared to the results of Roshko. The increasing departure for the larger cylinder diameters could be attributed to the smaller aspect ratio. The higher Strouhal number obtained by Roshko can be attributed also to the very low turbulent intensity (0.03%) in the flow of his experiment as compared to that in the present work (4%).

In Figures 3-5 the results are presented for three cylinder diameters 0.17, 0.25 and 0.7mm, separately, for various polymer concentrations. Each plot contains all the points resulting from several repetative runs. It should be noted that the vortex shedding frequencies were sensitive to the method of preparation of the polymer solution in the tank.

All the plots clearly demonstrate a decrease in vortex shedding frequency resulting from the use of polymer additives as compared to the Newtonian case. This decrease became greater with increasing polymer solution concentrations or with decreasing cylinder diameters. At concentrations of 100 ppm (Fig. 6) no significant decrease in vortex shedding frequency was measured for cylinder of 2 mm in diameter, however, a small decrease was observed for a cylinder of diameter 0.7 mm.

At this concentration a drastic change in the character of the correlation function between the Strouhal and the Reynolds numbers was observed for the thinnest of the cylinders, 0.17 mm. In this case contrary to the Newtonian case at low Reynolds numbers, the Strouhal number became approximately constant. The measured value of St=0.05 is a drastic reduction of almost 400% as compared to the Newtonian case. At smaller polymer concentrations (50 and 20 ppm) the polymers influence rapidly decreased. The Strouhal number increased with the Reynolds number as in the Newtonian case but to lower values. This type of decrease was observed for Re < 100. For Re > 100 the Strouhal number tends to return to values observed for the Newtonian case and sometimes even higher.

The marked influence of the polymer solution at 100 ppm concentration, as compared to the less-concentrated solutions, was accompanied by two other phenomena. The first of these was an earlier appearance of vortices in the wake. For cylinders of 0.17 and 0.25 mm diameter, the vortices appeared at Re=13 (as compared with first appearance at Re=40 for the Newtonian case) The second was the disappearance of the velocity fluctuation signals (Fig.6) indicating vortex shedding from the cylinder. The disappearance occurred at Re=150 for the 0.25 mm cylinder and at Re=100 for the 0.17 mm cylinder, although the shedding frequency before the signal disappearance was the same for both cases.

For the cylinder of 0.7 mm in diameter only small departures from the Newtonian vortex shedding frequency was observed. No early appearance of vortices in the wake of their disappearance was detected in this case.

Discussion and Conclusive Remarks

The phenomenon of the reduction of vortex shedding frequency may be explained on the basis of the linear stability theory. Nishoika & Sato [13] showed that the Strouhal number corresponds (at least for small Reynolds numbers) to the frequency of flcutuations with maximal growth rate as predicted by the linear stability theory. It follows from their measurements and Nakaya's [12] calculations that the non-dimensional frequency, $\beta=2\cdot fb/U$ is about 0.5, whenever f is the frequency of maximum linear growth. In this equation b is the half-width of the wake and U is the free-strain velocity. Measurements [6,7,10] showed that there is a considerable increase in the drag on small diameter cylinders in a flow with polymer additives relative to the Newtonian drag. This increase in drag results in the increase of the wake width and a corresponding decrease in the frequency of vortex shedding as was observed in the present work, using the same polymer additive.

Additional confirmation for the increase of the wake width in polymer solution may be obtained from James and Acosta [6] photographic recordings for thin cylinders and Reynolds numbers up to 35. Their photographs of dye traces show a marked departure of the flow characteristics from those of Newtonian flow within a zone of several diameters around the cylinders and, in particular, a large increase in wake width.

The use of Nishoika & Sato's results for a flow with polymer additives is justified since the behavior of a fluid with polymer additives in free-shear turbulent flows (jets, wakes, etc.) differs very little from the behavior of an ordinary Newtonian fluid (e.g., Barker [1] for a round turbulent jet). Additional confirmation to this fact stems from the result obtained in the current investigation that the known relations between the Strouhal number (St) and drag coefficient (C_d), for the Newtonian fluid $St=0.21/C_d^{3/4}$ is valid for polymer solutions (at least for a concentration of 100 ppm).

The check of the above formula was done by comparing the drag coefficient data from the measurements [7] in Polyox coagulant solutions with an intrinsic viscosity of 17.8 dl/gr with present measurements of shedding frequency (with a similar type of polymer, possessing a measured intrinsic viscosity 15 dl/gr) and also with the shedding frequency measurements [8] with polyox WSR-301. All these experiments were carried out at a concentration of 100 ppm (Table 1).

Table 1

d-cylinder dia.-mm	C_d [7]	d- mm.	St [8]	St present exp.	$St=0.21/C_d^{3/4}$
0.15	7	0.17	0.05	0.05	0.048
0.25	4.5	0.25	0.07	0.08	0.068
0.31	4	0.31	0.076		0.075
0.35	3.5				

This comparison is limited since only a few similar experiments have ever been published. However, the high degree of agreement between the Strouhal numbers, calculated according to the above formula, and from the experiments, indicates the validity of the above relation for polymer solutions.

The phenomenon of the disappearance of downstream vortices can be explained supposing that, at this concentration the longitudinal strain rates in the flow around the cylinder are restricted to the critical longitudinal strain rate due to the presence of the polymer. According to Lumley [11] the critical strain rate is achieved when $2\tau\varepsilon=1$, where ε is the longitudinal strain rate and τ is the relaxation time of the polymer solution.

As the vortex shedding frequency increases, higher local accelerations occur on the downstream side of the cylinder, these being caused by the instantaneous entrainment of fluid through the shear layer into the formation region during the process of vortex shedding (Gerrard [5]). The characteristic longitudinal strain rate caused by this entrainment is on the order of the vortex shedding frequency f. In the present experiment with the cylinders of 0.17 and 0.25 mm in diameter the maximum frequency which existed before the disappearance of the signals was 200 Hz. According to the model in the work [3] the relaxation time was 3 msec, provided $[\eta]=15$dl/gr and c=100ppm. It may be seen that for these values of f and τ $2\tau f \approx 1$, which signifies the achievement of the critical strain rate.

References

1. Barker, S.J.: Laser-Doppler measurements on a round turbulent jet in dilute polymer solution. J. Fluid Mech. Vol. 60, part 4,(1973) 721-731.

2. Berger, E.; Wille, R. Periodic flow phenomena. Anu. Rev. of Fluid Mech. Vol. 4, (1972) 314-340.

3. Everage, A.E.; Gordon, R.J. On the stretching of dilute polymer solutions. AIChE Journal, Vol. 17, No. 5 (1971) 1257.

4. Gadd, G.E.: Effects of long chain molecular additives in water on vortex streets. Nature, Vol. 211, (1966) 169-170.

5. Gerrard, J.H.: The mechanics of formation region of vortices behind bluff bodies. J. Fluid Mech. Vol. 25, part 2. (1966) 401-413.

6. James, D.F.; Acosta, A.J. The laminar flow of dilute solutions around circular cylinders. J. Fluid Mech. Vol. 42, (1970) 269.

7. James, D.F.; Gupta, O.P. Drag on circular cylinders in dilute polymer solutions. Chem. Eng. Progr. Symposium Ser. No. 111, Vol. 67, (1971) 62.

8. Kalashnikov, V.N.; Kudin, A.M. Karman vortices in flow of solutions of friction-drag reducing polymers. DISA Information, No. 10 1979.

9. Kalashnikov, V.N.; Kudin, A.M.; Ordinartsev, S.A. Influence of polymeric additives on the vortex generation in the wakes of thin cylinders. DISA Information, No. 19 1976.

10. Kit, E.; Poreh, M. Drag of the cylinders in dilute polymer solutions. Paper F2, Proc. 2nd Int. Conference on drag reduction, BHRA Fluid Eng. Granfield, U.K., 1977.

11. Lumley, J.L.: Drag reduction in turbulent flow by polymer additives. J. Polymer Sci. : Macromolecular Rev. Vol. 7, (1973) 263.

12. Nakaya, C.: Instability of the near wake behind a circular cylinder. J. Phys. Soc. Japan, Vol. 41, (1976) 1887.

13. Nishioka, M.; Sato, H. Mechanism of determination of the shedding frequency of cortices behind a cylinder at low Reynolds numbers. J. Fluid Mech. Vol. 89, part 1, (1978) 49-60.

14. Petrlin, A.: Hydrodynamics of linear macromolecules, Pure Appl. Chem. Vol. 12, (1966) 563-586.

15. Roshko, A.: On the drag and shedding frequency of two dimensional bluff bodies. NACA. Tech. Note. No. 3169, 1954.

16. Sarpkaya, T.; Rainey, P.G.; Kell, R.E. Flow of dilute polymer solutions about cicular cylinders. J. Fluid Mech.Vol. 57 (1973) 177-208.

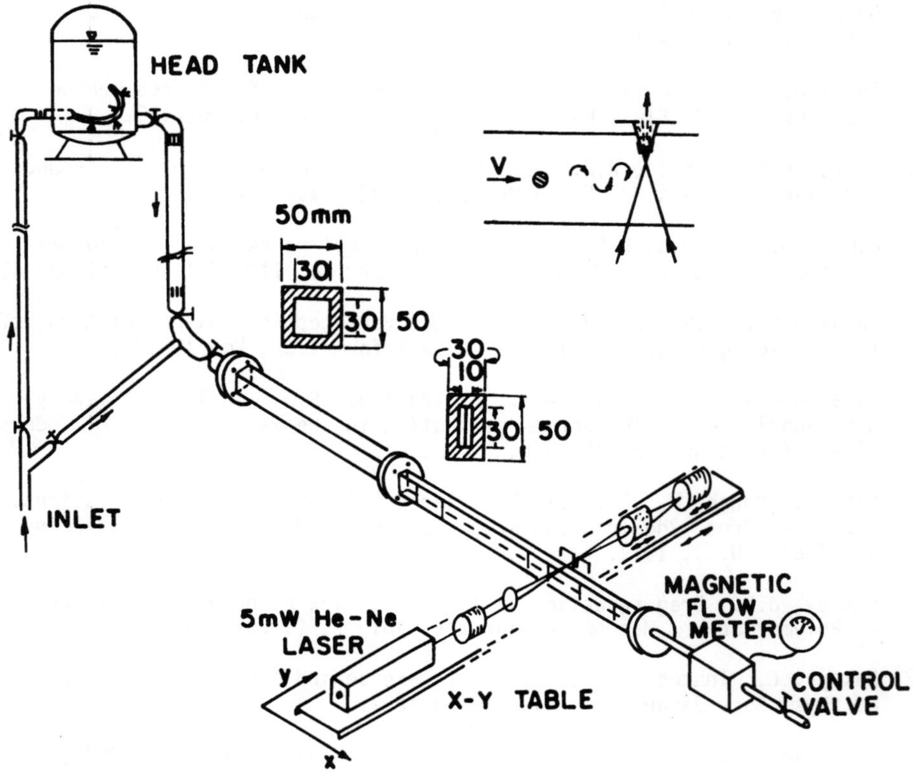

Fig. 1 The experimental system

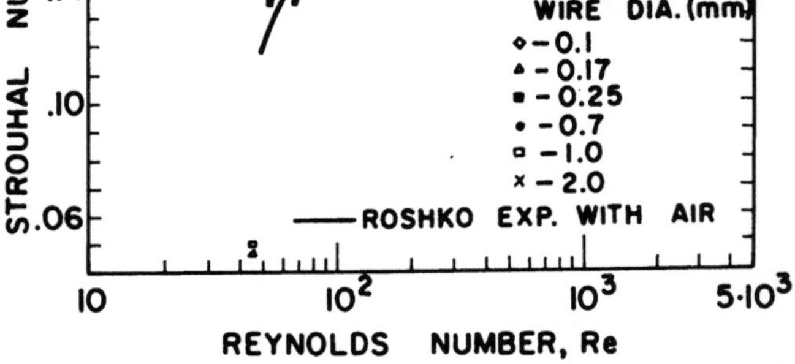

Fig. 2 Non-dimensional vortex shedding frequency from various cylinders in water

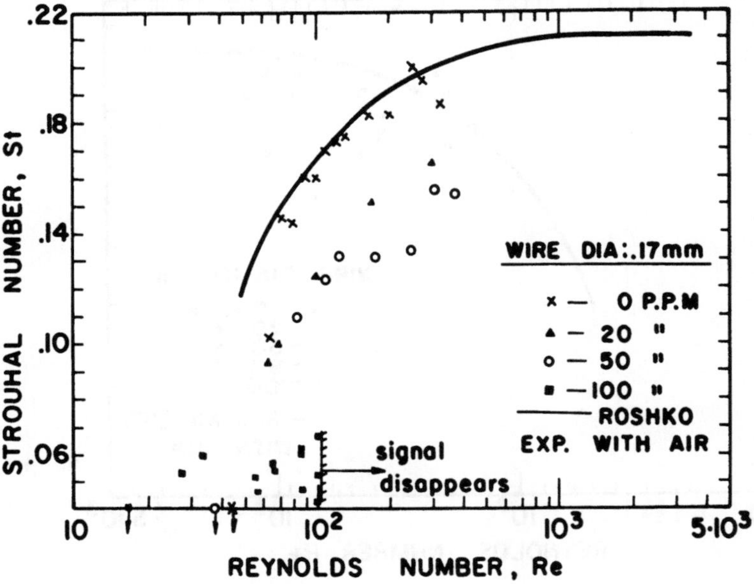

Fig. 3 Non-dimensional vortex frequency from cylinder of 0.17mm diameter at various polyox coagulant concentrations

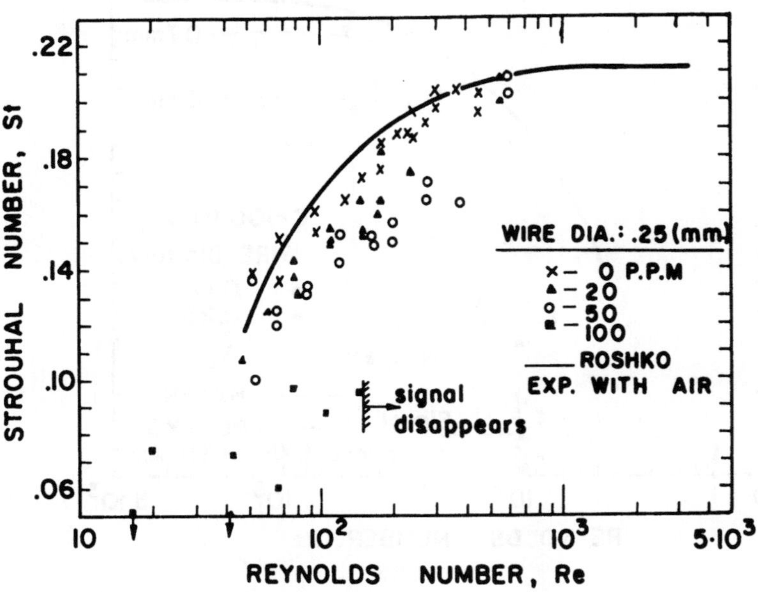

Fig. 4. Non-dimensional vortex shedding frequency from cylinder of 0.25mm diameter at various polyox coagulant concentrations.

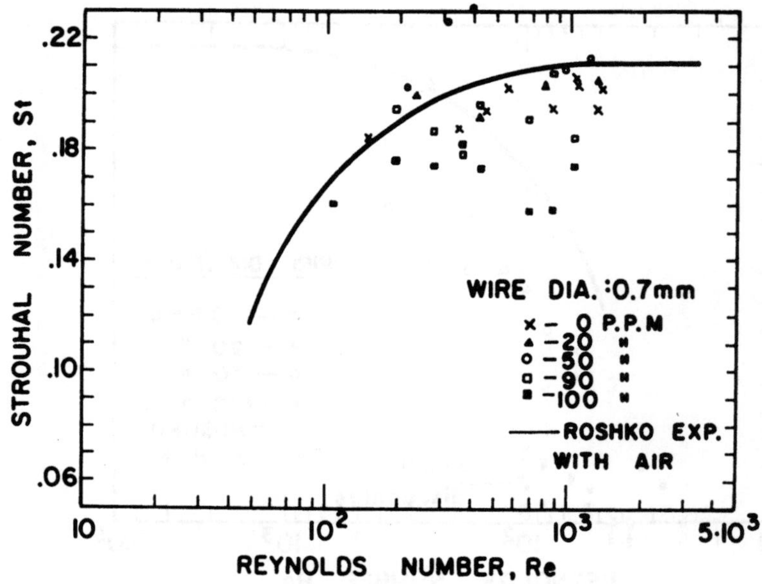

Fig. 5. Non-dimensional vortex shedding frequency from cylinder of 0.7mm diameter at various polyox coagulant concentrations.

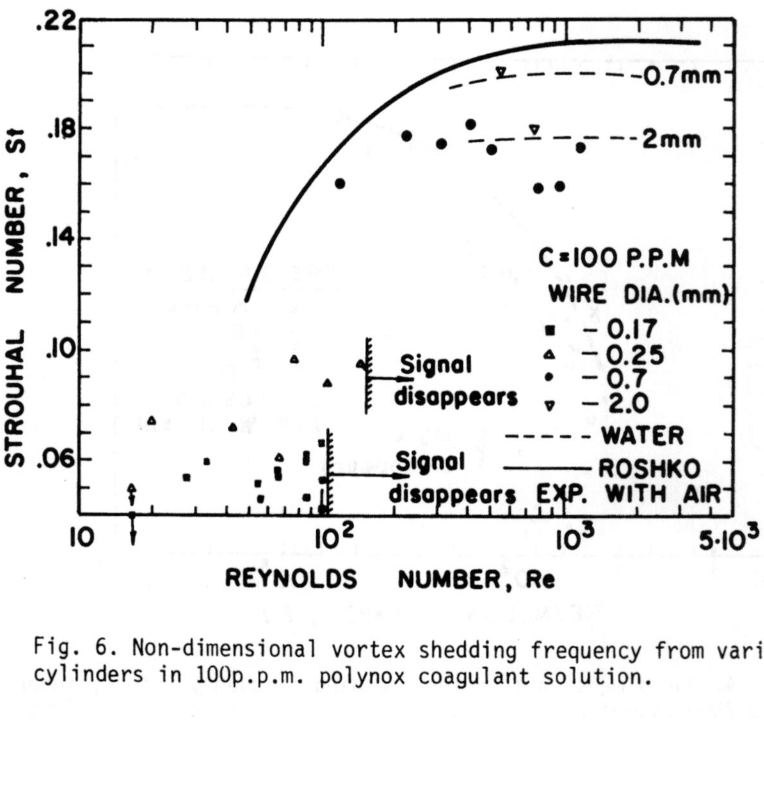

Fig. 6. Non-dimensional vortex shedding frequency from various cylinders in 100p.p.m. polynox coagulant solution.

Influence of Polymer Additives on Turbulence in a Mixing Layer

M.F. HIBBERD

Lehrstuhl für Strömungsmechanik
Abteilung Chemietechnik
Universität Dortmund
Postfach 50 05 00
D-4600 Dortmund 50

Summary

Mixing layers provide an opportunity for investigating the influence of polymer additives on the turbulent momentum transfer processes that are important in all turbulent flows. In the investigations described here both flow visualization and laser-Doppler anemometry were employed to study the differences between a plane mixing layer in water and in a 50 ppm aqueous polyacrylamide (Separan AP30) solution. The polymer additive was found to enhance the large-scale structure in the mixing layer whilst suppressing the fine-scale turbulence. Furthermore, the ratio of the intensity of the longitudinal to the transversal fluctuations as well as the magnitude of the Reynolds shear stresses were increased; the latter increase was associated with a more rapid growth in the width of the mixing layer in the polymer solution. These changes can be explained in terms of the rheological properties of the polymer additives.

Introduction

An interesting flow in which to study the influence of polymer additives on turbulence is a plane mixing layer, not least because of the detailed experimental and theoretical investigations available for Newtonian fluids [1 - 3]. Furthermore, since the development is governed by the momentum transfer processes, the results are simpler to interpret than for example for pipe flows, in which the presence of the wall has a strong influence on the flow. Studying the modification of the momentum transfer process in a mixing layer produced by the addition of polymer additives should also help to understand their influence on other types of turbulent flows and may in fact lead to an improved understanding of turbulent flows in general.

The Influence of Polymer Additives
on Velocity and Temperature Fields
IUTAM Symposium Essen 1984
Editor: B. Gampert

Since the polymer additives that produce drag reduction in pipe flows (so-called drag reducing additives) obviously exert a strong influence on the turbulence processes, it is just these additives in the same concentrations that are used when studying their influence on other turbulent flows, even when drag reduction as such is not present. In the current study the polymer additive was polyacrylamide (Separan AP30) added at a concentration of 50 ppm to water. Its effectiveness is indicated by the fact that in a 32 mm diameter pipe at a Reynolds number of 30 000 it produced about 60% drag reduction.

Experimental Equipment

The results were obtained from studies on plane mixing layers produced by allowing two parallel streams to join at the entrance to a rectangular channel about one metre in length. Some experiments were carried out in a channel with a cross-section of 40 x 20 mm, initial flow velocities (U_1 and U_2) of 2 and 1 m/s and free stream turbulence levels of about 0.5%, and others in a 40 x 40 mm channel with initial velocities of 1 and 0.5 m/s and turbulence levels of about 3%. Figure 1 shows a schematic representation of the formation and development of the mixing layer. The dashed lines f = 0.9 and 0.1 represent the positions at which the main stream velocity has the values $U_2 + 0.9\ \Delta U$ and $U_2 + 0.1\ \Delta U$ respectively, where $\Delta U = U_1 - U_2$. These lines

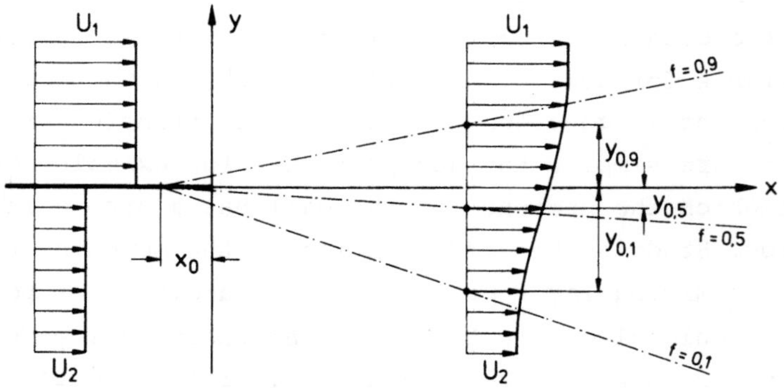

Fig.1 Schematic representation of the plane mixing layer

can be taken to define the edge of the mixing layer, although other criteria are also often used.

The flow circuit was a closed system with a total volume of 60 litres. In order to minimize mechanical degradation of the polymer solutions the liquid was pumped through the system by special low shear pumps. Several precautions were taken to suppress vibrations in the system, and before entering the channel the fluid passed through a series of sieves and a contraction to ensure low turbulence levels. Further details of the apparatus can be found in [4,5].

The structure of the mixing layer was investigated using flow visualization and laser-Doppler anemometry (LDA). To minimize disturbances to the flow, schlieren photography was used for vizualizing the development of the mixing layer; a small temperature difference of about 1 ^{o}C between the two streams provided suitable gradients in the refractive index field. Although the schlieren system integrates along lines through the mixing layer, the analysis of the resulting photographs was straightforward because the flow is principally two-dimensional. A more quantitative analysis was carried out using two LDA systems that enabled not only velocity and autocorrelation measurements to be made, but also space-time correlations and measurements of the Reynolds shear stresses [4,5].

Results and Discussion

Ideally the mixing layer should start with zero thickness and grow linearly with increasing x. However, the presence of boundary layers on the splitter plate leads to a displacement of the virtual origin of the mixing layer by a distance x_o (see fig.1). Moreover, the initial disturbances take some time to die away so that undisturbed development of the mixing layer first starts at x = 100 mm. Because of the limited cross-sectional area of the channel this development only proceeds up to about 300 mm, after which point the walls strongly disturb the flow.

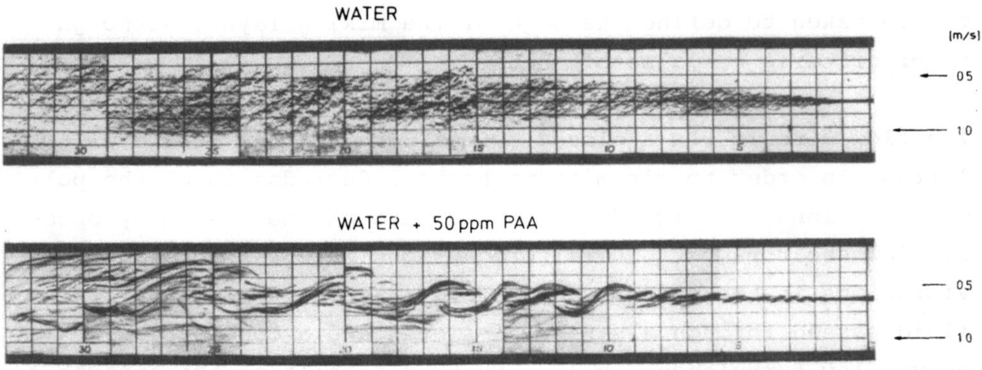

Fig.2 Schlieren photographs of the initial 330 mm of a mixing layer in a 40 X 40 mm channel for water and a drag reducing fluid

In figure 2 it can clearly be seen that the structure present in water is drastically altered by the addition of a small amount of polymer. In particular, the fine-scale turbulence is suppressed and the large-scale structures appear much more distinct. From the photographs alone it is not possible to decide whether these structures are in fact stronger or just more distinct because of the suppression of the fine-scale turbulence, but correlation measurements show that they are indeed stronger.

The discontinuities in the structure shown in the figure are due to the limited size of the lenses in the schlieren system, which necessitated a series of photographs to be taken to cover the whole field of view shown. The surprisingly good matching indicates the regularity of the structure. In interpreting the lower photograph it is necessary to recognize that the lines are not streamlines but rather lines of constant temperature gradient normal to the orientation of the knife-edge in the schlieren optics (in this case horizontal).

Profiles of the main stream velocity and the longitudinal and transversal fluctuations at various positions along the mixing layer are shown in figure 3. The main stream velocity profiles show that the mixing layer in the polymer solution is somewhat narrower at x = 100 and 200 mm but not thereafter. By 900 mm almost fully developed pipe flow is established. If the width of the mixing layer is defined as

$$b = \Delta U(\partial U/\partial y)^{-1}_{max}$$

then the growth rate db/dx is found to be up to 50% higher in the polymer solution and x_o is reduced, i.e. the virtual origin is moved closer to the end of the splitter plate.

Figures 3b and c show that fluctuations in the flow direction are enhanced whereas those across the mixing layer tend to be damped by the addition of polymer. An indication of their relative importances is given by the ratio V_{RMS}/U_{RMS}. Measurements of the intensity of the turbulent fluctuations in the neutral direction, W_{RMS}, show that W_{RMS}/U_{RMS} is also reduced by the addition of polymer.

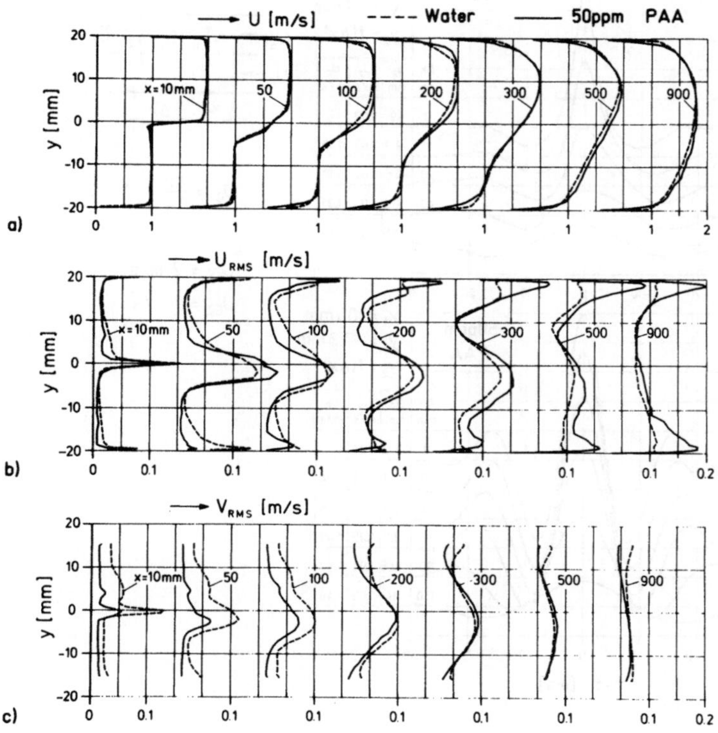

Fig.3 Profiles of the main stream velocity (U) as well as the longitudinal (U_{RMS}) and transversal (V_{RMS}) fluctuation levels at various positions along a mixing layer [4].

Space-time correlations such as those in figure 4, which can be interpreted as a measure of the similarity between structures at different times and positions, indicate that the large-scale structures preserve their identity for much longer in the polymer solution

In this case the correlations were made at the edge of the mixing layer with one point fixed at x = 100 mm and the other at various positions up to 160 mm downstream. The value of y_2 was varied for each x position to remain on the edge of the mixing layer. All the curves are plotted with a time shift so that their maxima lie directly under one another. This time shift, τ_m, is the time required for a structure to travel from point 1 to point 2. Using the separation Δx enables the transport velocity of the large-scale structures, U_c, to be calculated. It is seen to be considerably higher in the polymer solution.

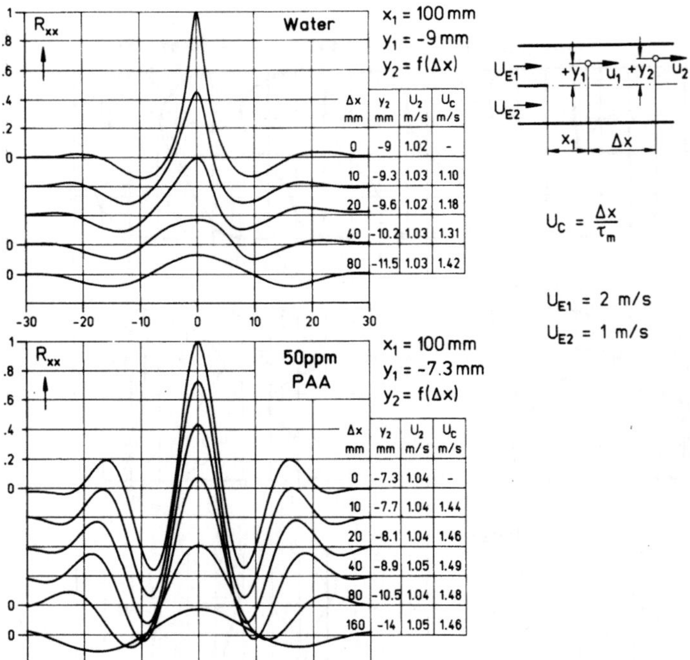

Fig.4 Longitudinal space-time correlations at the edge of a mixing layer. The delay time τ_m is the time required for the structure to travel from the point 1 to the point 2 [5].

Fourier transformations of the correlation curves provide an equivalent description of the mixing layer in terms of power spectra. Figure 5 shows a strong suppression of the higher frequencies at all positions. At lower frequencies the polymer solution shows more distinct peaks, sometimes with higher amplitudes

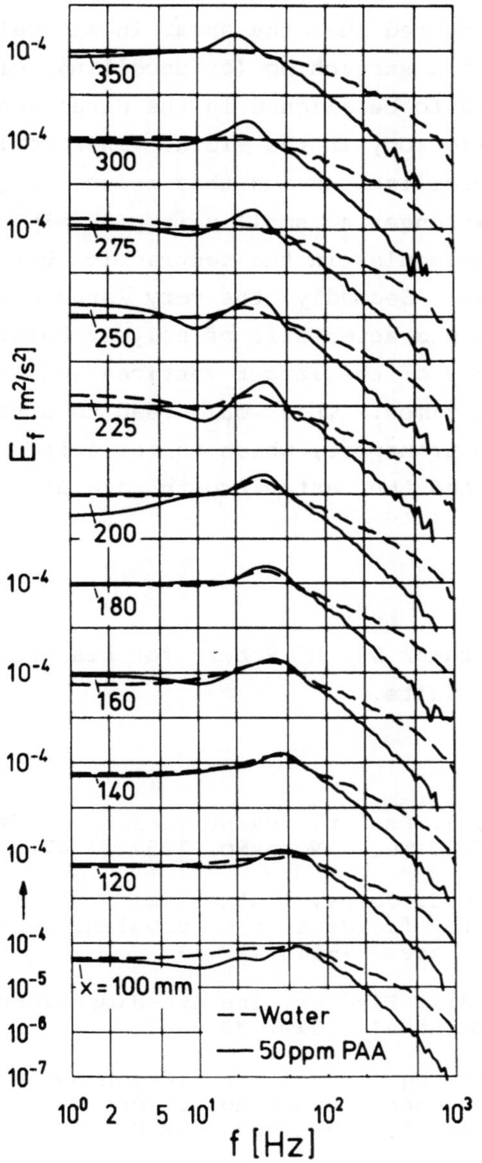

Fig.5 Power spectra of the transversal fluctuations at various positions along the middle axis of a mixing layer [4].

than for water. These correspond to the large-scale structures. At some positions two peaks were observed indicating locations at which vortex pairing occurs; the two frequencies being those of vortices just before and just after pairing.

It is appropriate to ask how these results can be understood in terms of the rheological properties of the polymer solutions. First, it is to be expected that the shear in the main flow direction will lead to a stretching (or uncoiling) of the molecules so that they tend to be aligned in the direction of flow, thus producing an anisotropy in the fluid. This would tend to inhibit transversal oscillations and thus reduce V_{RMS}/U_{RMS}. That this effect is stronger at small x is probably due to a prealignment of the molecules in the contraction before they enter the mixing layer. Secondly, the very large elongational viscosity known to be characteristic of polymer solutions probably prevents the decay of the larger vortices and leads to an increase in their lifetimes. This would lead to a higher rate of turbulent momentum transport, which is reflected in the larger Reynolds stresses and the faster growth rate of the mixing layer.

Acknowledgement

The author wishes to thank Dr. R. Scharf for his providing many of the experimental results.

References

1. Liepmann, H.W. and J. Laufer, Investigations of free turbulent mixing. N.A.C.A. Tech. Note No. 1257 (1947).

2. Spencer, B.W. and B.C. Jones, Statistical investigation of pressure and velocity fields in the turbulent two-stream mixing layer. AIAA Paper 71-613 (1971).

3. Wygnanski, I. and H.E. Fiedler, The two-dimensional mixing region. J. Fluid Mech. 41 (1970) 327-361.

4. Scharf, R., Die Wirkung von Polymerzusätzen auf die Turbulenzstruktur in der ebenen Mischungsschicht zweier Ströme. Dissertation, Universität Dortmund (F.R.G.) 1984.

5. Kwade, M., Beeinflussung der Turbulenzstruktur in der ebenen Mischungsschicht zweier Ströme durch Polymerzusätze. Rheol. Acta 21 (1982) 120-149.

The Turbulent Mixing of Viscoelastic Fluids in Pipe Flow

P.V. Bartels, A. Markus, and John M. Smith

Laboratory for Physical Technology,
Delft University of Technology, The Netherlands

Summary

The homogenisation of dilute polymer solutions in turbulent pipe flow has been measured by an iodine - thiosulphate reaction between parallel entering flows, following the method of Hiby, and by laser doppler anemometry. A qualitive comparison has been made between the results of the mixing, made visible by the redox reaction and the velocity fluctuations in the three directions.

Introduction

Much is known about the mechanism of drag reduction caused by polymers like polyethyleneoxide or polyacrylamide (PAA). In several pipe flow studies, for instance those of Mizushina and Usui (1), Thielen (2) and of McComb and Rabie (3), it has been shown that in pipe flow the boundary layer is thickend and the maximum turbulence intensity is located futher from the wall with increasing concentrations of polymer. Whilst the absolute maximum of intensity is reduced, the intensity in the centre is not affected. There is also a relative shift towards lower turbulence frequencies. As a result, it can be expected that the modification in turbulence structure resulting from the presence of polymer in the solution will not only decrease the friction at the wall but also the rate of homogenisation in the fluid bulk. For the logical development of in-line mixers for polymer reactors it is necessary to understand more about the energy requirement, turbulence and the mixing times in tube flow in relation to the visco elasticity of the fluids.

In this study the effect of the concentration of polymer on turbulent mixing has been investigated using a fast chemical reaction and by laser doppler anemometry. Because of the low concentrations used, the elastic effects produced by the polymer molecules will be local, whilst the fluid as a whole is virtually Newtonian.

The Influence of Polymer Additives
on Velocity and Temperature Fields
IUTAM Symposium Essen 1984
Editor: B. Gampert

Experimental

The apparatus for the reaction experiments consists basicly of a transparent tube 9 m long with an inner diameter of 16.4 mm. In two tanks of 250 l aqueous solutions of PAA with a special starch are prepared (0.2g/l). One solution is coloured blue by addition of iodine (0.13 mmol/l). The second solution contains a greater concentration of thiosulphate (0.52 mmol/l). These solutions were pumped through the pipe in equal volumetric flows, controlled by maintaining zero pressure difference between the identical entrances. Initially the two flows are separated by a partition in the tube, 0.4 m long. It is presumed that the iodine is oxidized in a rapid diffusion controlled reaction with the thiosulphate. The terminal point at which the last colouration disappears defines the mixing length. The method is based on the papers of Hiby (4) and Käppel (5).

The pressure drops have been measured over .5 and 5 m length of the 16.4 mm tube with a Depex DP15 pressure transducer. The average velocity was determined by an inductive flow meter (figure 1).

Turbulence profiles across the tube were measured by laser doppler anemometry in all three directions separately at 5 m from the tube entrance. A short thin wall section of plastic tube surrounded with a rectangular chamber filled with the experimental fluid served to minimize the optical distortion. The apparatus is similar to that used by Mizushina and Usui (1).

A reference beam with forward scatter method has been used with a TPD rotating grating preshift and tracking system (6) for the 16.4 mm tube. An average volt meter and a correlator HP 3720 with a spectrum analyser were used to interpret the tracker output. Complementary velocity field measurements were carried out in a 44 mm diameter tube. Simultane measurements in the radial and axial direction were possible, which were analysed by computer. Bragg cells were used for the preshift.

The solutions were made from Polyacrylamide of Dow Chemicals (Separan 30). A 2.5 % bulk solution was diluted to the appropiate concentrations. Measurements were done at least 20 hours after dilution to allow to stabilize before the measurements.

Reactive mixing

The effect of dissolved polymers on the turbulent diffusion was found to be significant. Figure 2 shows an increase in mixing length for a 300 ppm PAA solution to about 8 times that in water in the 16.4 mm pipe. This solution , which has a viscosity of 3.0 Pa s gives the maximal drag reduction of 73% in this tube at a Reynolds number of 20000. Despite the longer mixing lengthes, it is possible to get the same rate of homogenisation with less energy dissipation, at lower concentrations of polymer, when the maximal drag reduction does not occur (figure 3).

This results, obtained by the redox reaction, show that the influence of the polymers is very complicated. There must be a effect on the turbulence in the core of the tube, because of the considerably increased mixing length, however the influence of the polymers on the turbulence in the boundary layer is of greater importance. It is not yet possible to relate the measured mixing length directly to the energy dissipation or drag reduction, though there may be some relationship with the theory of Virk (7), relating to maximal drag reduction.

Turbulent flowfield data

Figure 4 shows the flow profiles for the solutions used at a Reynolds number of 20000. The 300 ppm and 500 ppm solution are close to the ultimate profile of Virk (7),however the three layer model is not appropiate to such a small tube. There is no evident core profile parallel to the the Newtonian velocity profile. This implies that the effects normally associated with the buffer layer are essential for mixing at this small scale. In the 44 mm tube the core region is more distinct (figure 5, the data are shown in table 1). In both figures the drag reduction is related to slopes of the velocity profiles. A relative reduction is found with higher concentrations in the normalised axial velocity fluctuations for the 16.4 mm pipe (figure 6). Thielen (2) has found a similar effect. Figure 7 shows the increase in the maximal turbulence fluctuations, normalised by the superficial velocity, in the axial direction in polymer fluids with a maximal drag reduction. This might explain the results of figure 3.

Similar results are shown in figure 8, this time for the 44 mm tube. Because of the larger scale and better instrumentation it was possible to approach the wall more closely and to get results that are in close agreement with Laufer (8), and Mishuzina and Usui (1). In this 2.7 times wider tube the more concentrated polymers do not show a decrease in the maximal velocity fluctuations, but the relative maximum increases futher as we move from the wall. This local effect will make the scale up of mixing devices for visco elastic fluids complicated.

Dispersion

For the mixing with parallel entrances the radial and especially the tangential velocity fluctuations are important for the dispersion. Figure 9 shows the radial velocity fluctuations. This results are in general agreement with those known in the litterature (2). In figure 10 the 200 ppm solution has smaller fluctuations than those we found for the lower and higher concentrations of PAA. This is consistent with the results found for the axial fluctuations.

Although these results for the velocity fluctuations are in agreement with the mixing length experiment, it is not directly possible to relate the mixing performance to the turbulence profiles found.

Conclusions

The velocity fluctuations measured give a qualitive explanation of the mixing length found for the redox reaction. The asymptote for drag reduction from the theory Virk seems to be significant for the maximal value of the velocity fluctuations and therefore for the homogenisation in the pipe. This scale of reactive mixing gives no fully developed turbulent core. This fact makes the up-scale complex.

Acknowledgement

Acknowledgement is made to J. Kunen (lab. for Aero and Hydrodynamica of D.U.) and G.L. van Lookeren for the experimental work done.

Symbols

f	Friction factor	-
R	Radius	m
Re	Reynolds number	-
U	Time average axial velocity	m/s
u	axial velocity component	m/s
$\bar{V}$	Superficial velocity	m/s
v	radial velocity component	m/s
w	tangential velocity component	m/s
ε	Energy dissipation needed for a specific rate of homogenisation	J/kg
ρ	Density	kg/m^3
τ_w	Wall shear stress	Pa

Subscripts

+	Normalised by $\sqrt{\tau_w/\rho}$
x	Normalised by $\bar{V}$

Superscript

'	Fluctuating component

References

1 Mizushina, T. and Usui, H., Phys. of Fluids, 20, (1977), S100

2 Thielen, W. , Thesis, TH Aachen 1977

3 McComb, W.R. and Rabie, L.H., A.I.Ch.E. J., 28, (1982), 547

4 Hiby, J.W., Verfahrenstechnik, 4, (1970), 538

5 Käppel, M., V.D.I. Forschungsheft 573, München 1976

6 Molen, K. van der and Maanen, H.R.E. van, Ch. Engng. Sci., 33, (1978), 1161

7 Virk, P.S., Mickley, H.S. and Smith, K.A., J. of Appl. Mech., (june 1970), 488

8 Laufer, J., NACA report 1174, (1954)

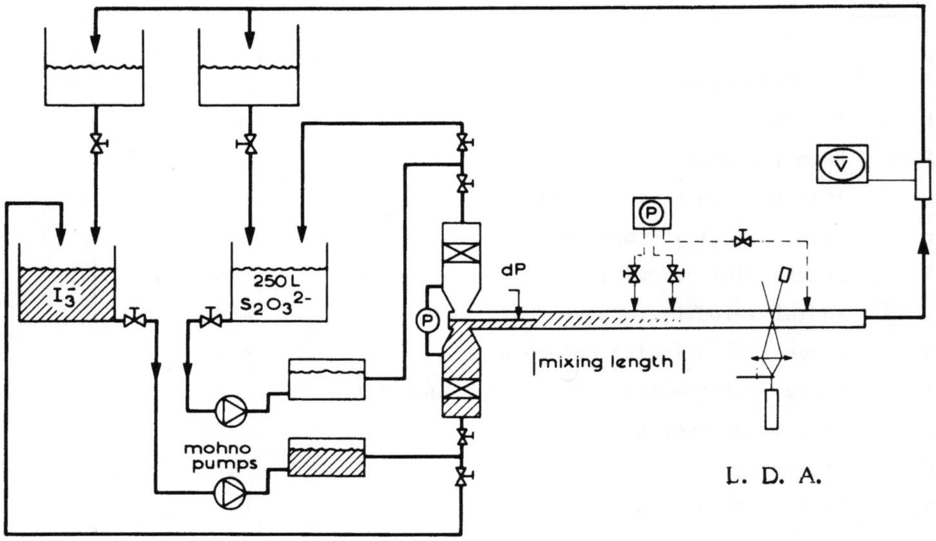

Figure 1. Experimental set up for measuring the mixing length by a redox reaction and the flow field by Laser Doppler Anemometry (LDA) in the 16.4 mm tube

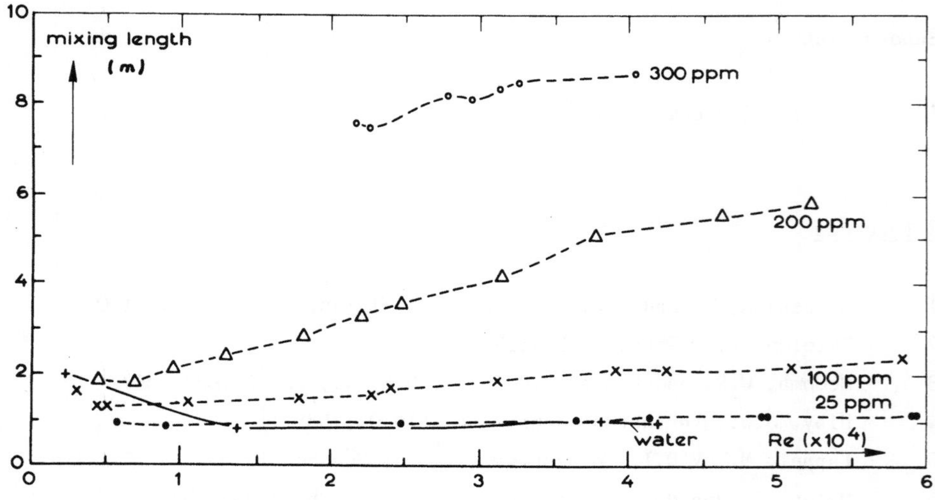

Figure 2. Mixing length versus Reynolds number for polyacrylamide solutions (PAA)

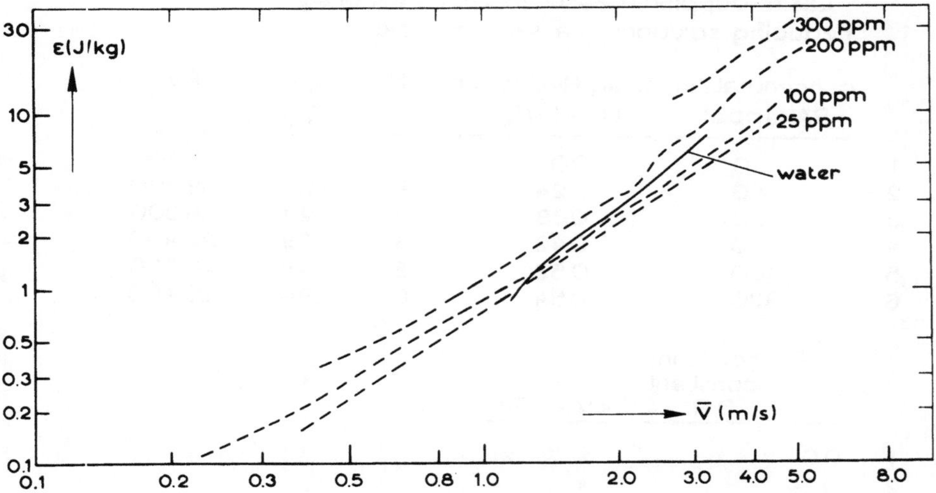

Figure 3. The energy dissipation ε needed to obtain a specific rate of homogenisation versus the superficial velocity $\bar{V}$ for diluted polyacrylamide solutions

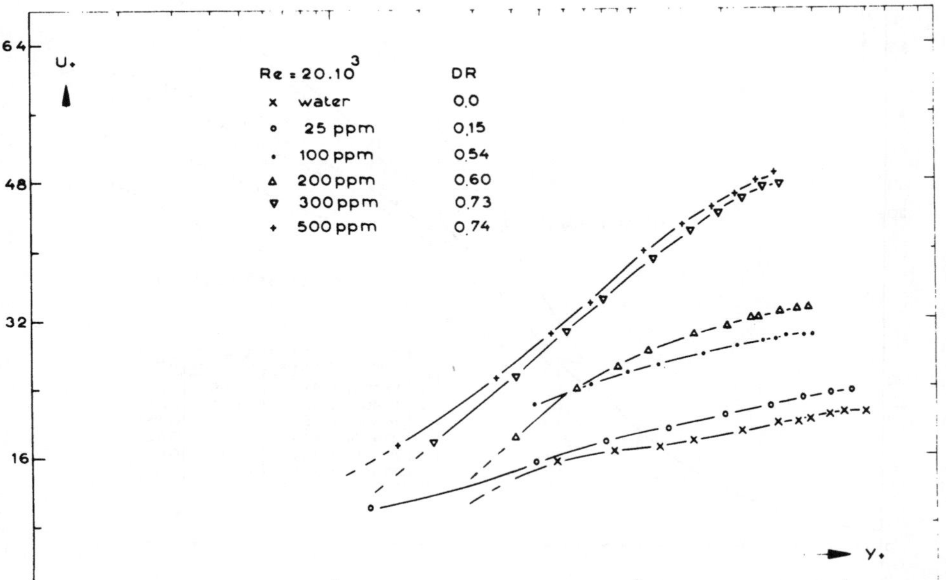

Figure 4. The velocity fields in the 16.4 mm tube for the drag reducing polyacrylamide solutions

Laser doppler measurements with drag reducing solutions in a 44 mm pipe.

	concentration PAA(ppb)	Drag Reduction $(f_s - f)/f_s$	pH	η_{rel} η_s/η_{water}	Re
1	0	0.0	8	1	19 700
2	40	0.24	8	1.11	18 800
3	80	0.38	8	1.29	16 300
4	80	0.42	8	1.29	20 400
5	160	0.52	8	1.48	20 700
6	320	0.54	6	1.94	20 400

	D.R.	relation: constant D.R.	<v>	Re
1	0.0		x	x
2	0.24		x	
3	0.38		x	
4	0.42			x
5	0.52	x		x
6	0.54	x		x

Table 1.
Data for figure 5 and figure 8

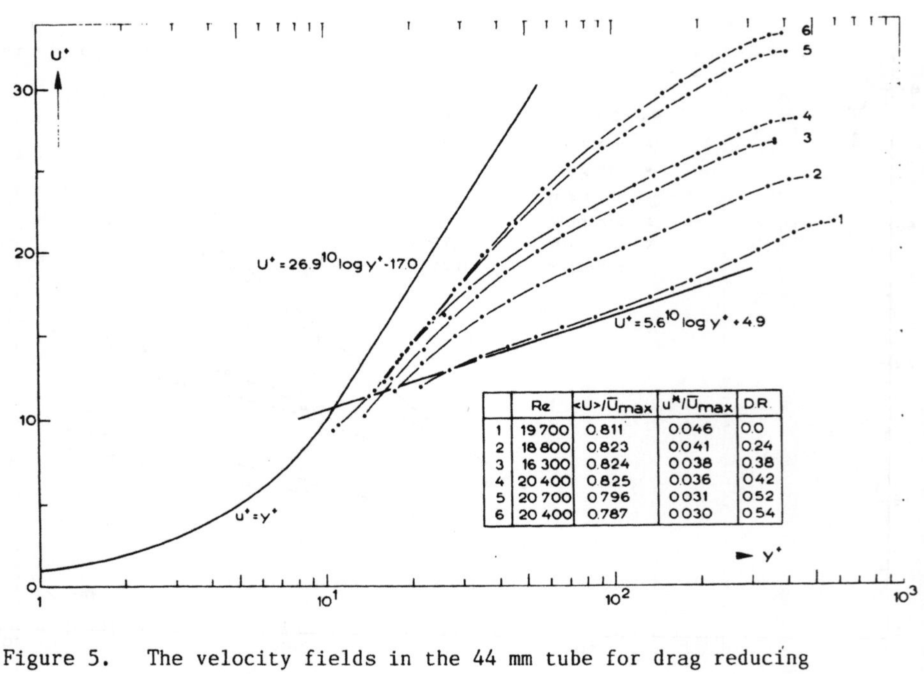

Figure 5. The velocity fields in the 44 mm tube for drag reducing polyacrylamide solutions

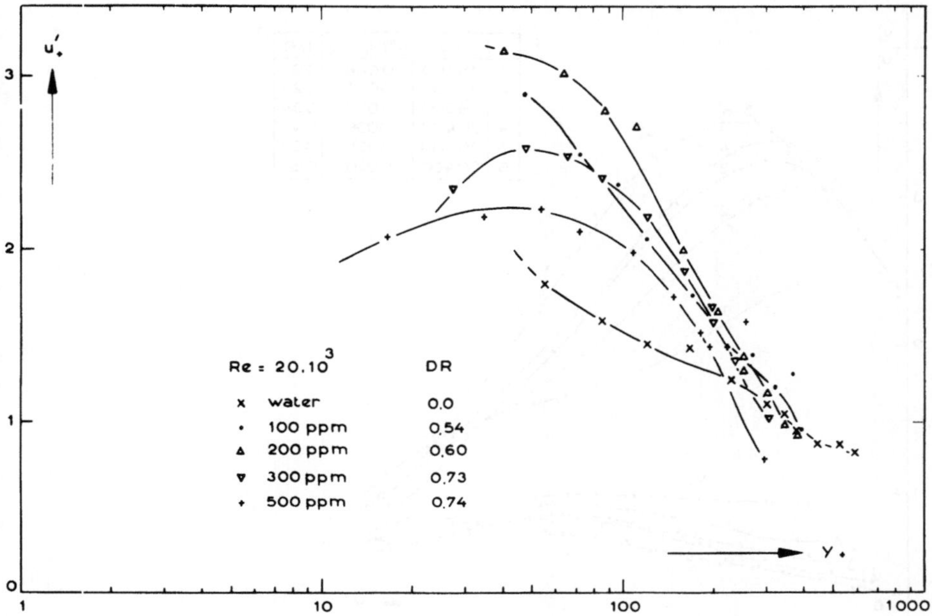

Figure 6. The axial velocity fluctuations u'_+ in the 16.4 mm tube

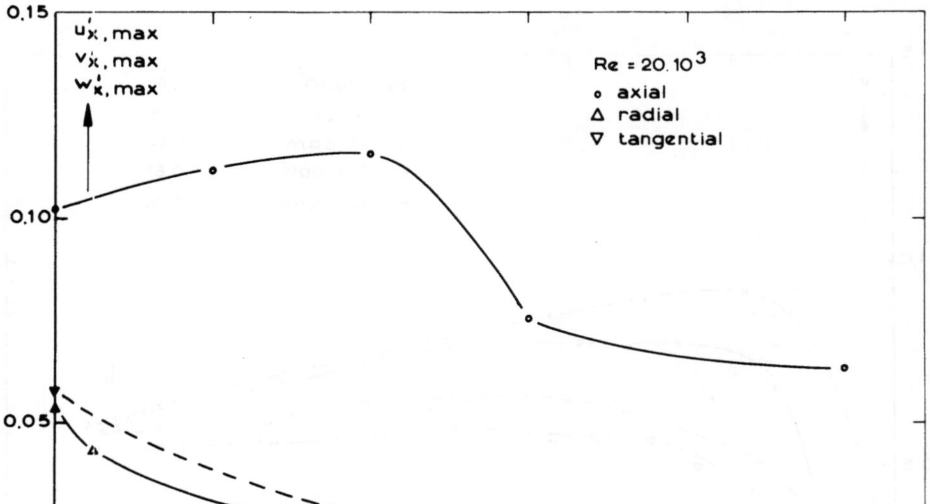

Figure 7. The maximal velocity fluctuations u'_x found in three directions versus the concentrations of the polyacrylamide solutions used

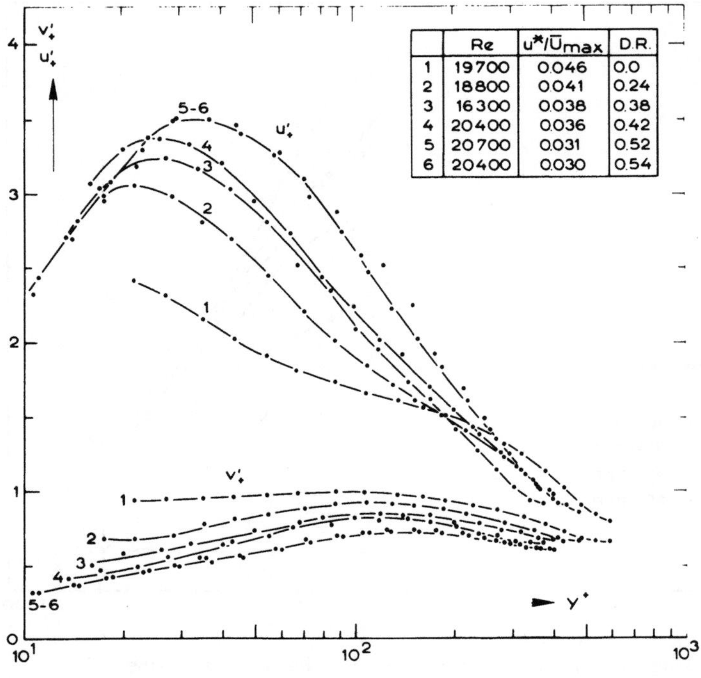

Figure 8. The axial and radial velocity fluctuations in the 44 mm tube

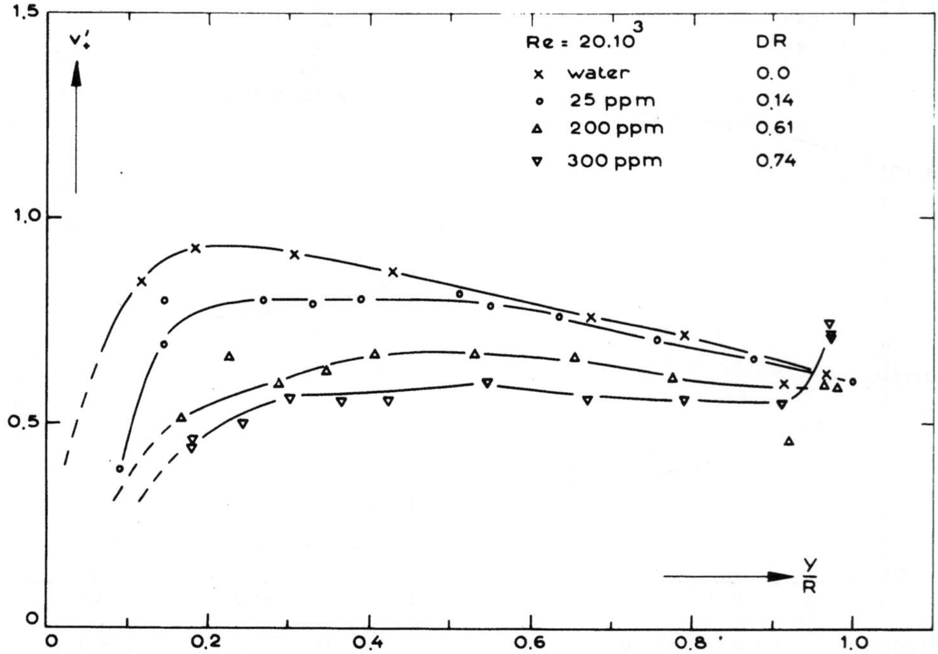

Figure 9. The radial velocity fluctuations in the 16.4 mm tube

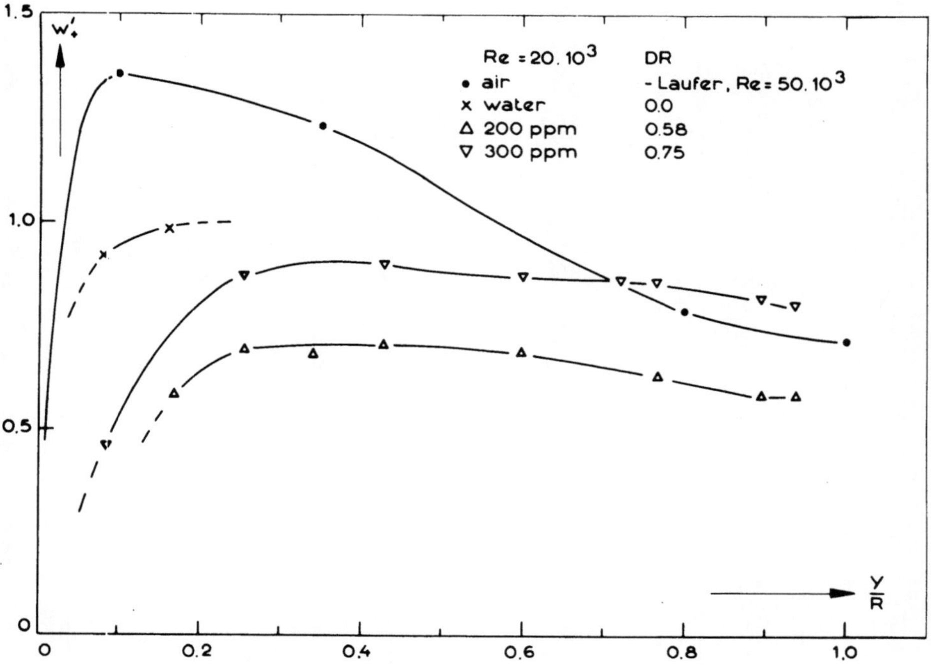

Figure 1Ɔ. The tangential velocity fluctuations in relation to the dimensionless radial distance in the 16.4 mm tube.

Session 9
Theories II

A Qualitative Understanding of Drag Reduction by Polymers

NEIL S. BERMAN

Department of Chemical and Bio Engineering
Arizona State University
Tempe, Arizona

Summary

The interaction of turbulent flow containing regions of high strain rate and high shear rate with polymer molecules can involve both individual molecules and polymer structures. The results of this interaction lead to suppressions of the small scale motions and to production of some large scale motions that do not contribute to turbulent diffusion. Experimental results are interpreted in terms of the interaction of molecules near the wall and movement of the polymeric large scale structures in the center of pipe flows.

Introduction

From a model of turbulent flow dynamics near a solid boundary it can be shown that drag reduction means that the scale of the dissipation is increased relative to that for the non-drag reducing fluid. The increase in the scale of the small eddies near the wall can arise in several ways. Among these are the following:

1) An increase in viscosity in the inertial sublayer while the viscous sublayer remains at a lower viscosity.
2) The presence of molecular aggregates in the flow which are larger than the microscale.
3) A change in the large scale structure.

All of these mechanisms probably can occur in turbulent flow of polymer solutions and in this review we are concerned with the experimental evidence for these mechanisms. We will examine evidence for the existence of drag reduction which can be extrapolated to the influence of a single molecule, intrinsic drag reduction. Then we will look at evidence for molecular interactions and finally we consider the interaction of polymer solutions with the turbulence.

The Influence of Polymer Additives
on Velocity and Temperature Fields
IUTAM Symposium Essen 1984
Editor: B. Gampert

Drag Reduction Due to Individual Molecules

Let us first look at very dilute solutions of linear flexible molecules which exist as random coils when the fluid is at rest. If no molecular interations are present, we expect to find that drag reduction is a linear function of concentration. In order to test this intrinsic drag reduction hypothesis, we require a monodisperse polymer, careful solution preparation, and precise measurements for small amounts of drag reduction. For polydisperse polymers increases in drag reduction can be the result of changes in the effective part of the distribution. Unless we can be sure that the entire distribution is equally effective, a polydisperse sample will not give the correct relationship between drag reduction and concentration.

Solution preparation is a complex problem and not very well understood. Recent experiments by Chang and Darby [1] and Oliver and Bakhtiyarov [2] indicate that the drag reduction effectiveness of partially hydrolyzed polyacrylamides increases with mixing time. At Arizona State University we have obtained similar results especially when the solvent is deionized water. On the other hand freshly mixed poly (ethylene oxide) (PEO) solutions give the most drag reduction. Impurities and ions present in water also affect the polymers. In addition agglomerates of molecules may be present in the solution or can be formed by the flow. These solution difficulties are present in both aqueous and nonaqueous solutions.

There is evidence that the flow polymer interaction occurs in a thin ring near the wall of the pipe. For a test of the concentration dependence, the size of this ring must be kept constant or changed in a known way. This can only be accomplished if the drag reduction remains low, probably less than 20%.

There are no reported experiments that fit all of the criteria and also consist of a range of Reynolds numbers leading to 20% drag reduction. Most articles report only data at one or two Reynolds numbers. The experiments of Berman and Yuen [3] used samples of poly (ethylene oxide) obtained from fractionation runs in a gel chromatography column. Seventeen different fractions with molecular weights that varied from 0.55×10^6 to 11.4×10^6 were obtained and diluted to final concentrations from 0.07 to 3.5 ppm. Complete drag reduction trajectories for each solution were measured from onset until the drag reduction reached a constant level. The final correlation of slope increment, δ, divided by intrinsic viscosity vs. concentration represented several hundred data points. It is remarkable that all of the data fit on the same curve, but the concentrations were not low enough to show any intrinsic drag reduction. Also, for the higher molecular weights the agarose columns were not very efficient in separation. Therefore the lowest concentration points were the most in doubt.

If the drag reduction is small, we can argue, based upon the separation results of Berman and Yuen [3], that a very high molecular weight sample can be used to test for intrinsic drag reduction even if polydisperse. Then most of the molecules are effective throughout the Reynolds number range of the test. Only the true intrinsic viscosity is unknown, but we can plot slope increment vs. concentration. Sufficient information is given by Oliver and Bakhtiyarov [2] to make such a plot when we assume constant onset independent of concentration. In Figure 1 the results are plotted along with those of Berman and Yuen [3] adjusted for the differences in intrinsic viscosity, and we arbitrarily match the curves at a concentration of 0.25 ppm. Below this apparent concentration we find a linear relationship between slope increment and drag reduction. The real concentration of effective molecules may be less, so 0.05 ppm may be a more realistic value for these molecules with molecular weights approximately 10^7.

Part of the problem in determining the effective concentration concerns the location in the pipe or boundary layer where the molecules interact with the turbulence. McComb and Rabie [4] showed that the polymer molecules need to be in a thin annular ring near the wall to be effective in drag reduction. To proceed with a further discussion of the interaction of polymer molecules with the flow we must have some sort of model. Lumley's [5] hypothesis is that the molecules become stretched in the region of high fluctuating strain rate near the wall and these elongated molecules increase the viscosity of the fluid in the pipe except in the viscous sublayer. The fluctuating strain rate has a peak near $y_+ = 10$ and the inverse molecular time scale is exceeded only in a ring near the wall. The ratio of polymer time scale to flow time scale is usually called the Deborah number, De.

As shown by Durst et al. [6] we can plot the average fluctuating strain rate vs. y_+ and then determine the effect of increasing De on the fraction of the fluid that is interacting with the polymers at any cross-section. For a 6.35 mm tube we find that when the Deborah number is approximately four times the onset Deborah number, De_o, that fraction is 33%. There is roughly a linear relationship between the fraction and the Deborah number when $De > De_o$. However, only 4% of the flow area is effective for a 57 mm tube when the Deborah number is four times the onset value. Therefore, when the concentration of polymer is low there can be differences between large and small pipes which are not explained by the difference in onset time scale. It is reasonable to assume that in the model the increase in viscosity in the center of the pipe is related to the fraction of molecules that are elongated by the flow.

Other evidence for the model of molecular stretching comes from the qualitative relationship between onset and molecular time scale [7,8]. The solution viscosity, temperature, and molecular weight all change onset in the way predicted by the time scale hypothesis. The magnitude of the change is not exactly as predicted, but the experiments did not attempt to control the variables previously listed. Concentrations were much larger than the individual molecule limit which also means that the time scale would depend upon concentration. Since the direction of the change is correct, the time scale relationship to onset must be correct also.

The Expanded Viscous Sublayer

The time scale interaction leads to elongation of polymer molecules and an increase in the extensional viscosity of the solution. This change alone can be shown to predict drag reduction and an increase in the size of the viscous sublayer. When the viscosity is increased in the inertial sublayer, the dimensionless large eddies, l, scale with $y_+(p)$ where $y_+(p) = u_* y/\nu_p$ and ν_p is the increased viscosity due to the stretched polymers. In the viscous sublayer the smallest possible dimensionless eddies, η_+, scale with $y_+^{1/4}$, where the unchanged viscosity is used to calculate y_+. When $l_+ > \eta_+$, turbulence can exist although large fluctuations may still be present when $\eta_+ > l_+$. This analysis as first proposed by Lumley [9] shows that the viscous sublayer is increased in size when the stretched polymers are present. The smallest turbulent eddies are larger in the polymer solution flow than the corresponding smallest turbulent eddies in pure solvent flow. Laser Doppler velocity fluctuation measurements confirm the disappearance of these smaller eddies in drag reducing flows but the viscous sublayer is apparently unchanged. The problem concerns the details of the fluctuations that exist near the wall. We will return to some other aspects of this problem later, but here let us look at the types of fluctuations that are observed in such an expanded viscous sublayer. Only the motions which average to non-zero Reynolds stress should count as turbulence, but most one dimensional velocity measurements have no way to distinguish these correlated fluctuations from the uncorrelated ones.

Tan and Berman [10] have developed a Lagrangian technique to measure velocity fluctuations which distinguish random from periodic motions. Although we have not as yet applied this method to boundary layer flows, we do have some data for a free shear layer (submerged jet). When both Eulerian and Lagrangian measurements are made in a submerged jet, the Lagrangian mean square velocity fluctuations are less than the Eulerian fluctuations. The Lagrangian method correlates with Reynolds stress measurements, but the Eulerian velocity fluctuations for single components can be larger or smaller than in the corresponding pure solvent jet [11]. Therefore, even though the viscous sublayer may be expanded, the experimental measurements include large fluctuations and the separation of the viscous sublayer from the buffer layer is not clear.

Molecular Interactions

Lumley's model also predicts a maximum drag reduction when the smallest turbulent eddy becomes equal to the pipe diameter. If the viscosity is increased beyond this point, the drag reduction will decrease. However, it is doubtful that this mechanism can occur except for small capillaries. The reason is that higher concentrations of polymers are required so that molecular interactions cannot be neglected. If no weak bonds between molecules are formed, a solution of overlapping long thin rods will lead to a large increase in extensional viscosity with a concentration dependence of $c^{1/2}$. This is what is observed.

There are many examples of experimental data in the literature which do not have an apparent time scale relationship. Fibers, micelles and extended polyelectrolytes in deionized water are drag reducing but have no onset. Freshly mixed poly (ethylene oxide) does not have an onset of pitot tube error [12], and there is no onset of excess pressure drop in some constrictions. After similar poly (ethylene oxide) solutions are circulated through a pipe in turbulent flow or passed through a porous plug, the behavior is different. The freshly mixed solutions are more viscoelastic and appear to be composed of associations or networks of polymer molecules [14]. Thus the effective molecular size is much larger than suggested by the size of individual molecules. Since the drag reduction increases with an increase in flexibility of the molecule or the flexibility of the structure, there is still a time scale coupling with the flow but the time scale is different from that of a single molecule.

Several experiments in my laboratory have been designed to test some aspects of the molecular interaction and flexibility. Intermolecular bonds between poly (ethylene oxide) and red dyes are readily formed in aqueous solutions [15]. At the lowest dye concentrations the dye molecules bond at two points within a single poly (ethylene oxide) molecule and drag reduction is reduced because the molecule becomes more rigid. When the dye concentration is high, bonds are formed between two polymer molecules and the resulting solution of agglomerates has increased drag reduction.

Similarly xanthan gum molecules are prevented from expanding in acid solution because internal hydrogen bonds are formed with the random coil. Figure 2 shows some other interesting behavior of xanthan gum solutions as measured by Suaysompol [16]. In neutral salt solution or in basic solution the xanthan gum molecule is a partially extended flexible rod, and in deionized water the molecule is more extended and more rigid. When the molecule is extended as in deionized water, xanthan gum solutions are drag reducing and have no onset. The trajectory for deionized water shows a drop in drag reduction when $Re\sqrt{f}$ exceeds 3000 (Re = 7000) and a subsequent increase beyond $Re\sqrt{f}$ of 8000 (Re = 15,000). Similar trajectories are found with other polyelectrolytes dissolved in deionized water [17,18]. In all these cases the higher Reynolds number increase in drag reduction corresponds to the trajectory after onset of the polymer in salt solution where the molecules would exist as random coils in a stagnant solution. The implication is that the expanded molecules are collapsed by the flow. Possible mechanisms for this are

(a) The molecules are flexible but interact to form agglomerates or networks and these are broken by high shear. Then when the strain rate is increased, the molecules can be stretched by the flow and drag reduction increases.

(b) The molecules have side groups that collapse along the backbone due to self association. The stiffened backbone gives an extended rigid rod and solutions of these rods are drag reducing when the rods align in the flow to give large increases in extensional viscosity. High shear causes the side groups to flip up away from the backbone and the molecule collapses to a random coil. An increase in strain rate enables the molecules to stretch out again.

Polyacrylamides seem to follow the first mechanism and xanthan gum the second.

At low Reynolds numbers larger pipes result in greater drag reduction for these extended polyelectrolytes but this pipe diameter effect is reversed at higher Reynolds numbers when the stretching of molecules in high strain becomes the dominant mechanism for drag reduction. In order to have drag reduction at the lower Reynolds numbers, the molecules must be aligned or the aggregates or local networks must be larger than the microscale. Although it is possible that isolated aggregates exist that are larger than the microscale, most experiments involve solutions which are more concentrated and the aggregates fill the entire volume. In either case interations between aggregates, individual polymer molecules, trace impurities and the turbulent flow all seem to be important.

Interactions with Large Eddies

The turbulent flow contains locally high strain near the wall possibly in the ejections of slow moving fluid from the wall in the bursting process. The polymer molecules would behave as they do in a submerged jet where there are adjacent layers of high speed and low speed fluid. This analogy has been developed extensively by Landahl [19]. Experiments illustrating drag reduction with polymer threads in the middle of a pipe flow suggest that the production of such threads in the bursting process leads to another drag reduction mechanism. We have found that when a one mm. polymer jet is injected in the center of a turbulent pipe flow containing no polymers, drag reduction is obtained if the jet is unstable and wavy or if the jet mixes completely. A stable thread that remains fixed in the center of the pipe is not drag reducing. Figure 3 shows the results of Sinha [20] for three different concentrations of injected solution of poly (ethylene oxide) where the final concentration was the same as the premixed polymer solution (also shown in the figure). The three concentrations illustrate the different flow patterns that were observed when dye was added to the injected fluid. The 4000 ppm thread becomes unstable at a low

Reynolds number, and drag reduction onset occurs when the instability begins. Premix solution has an onset at a higher Reynolds number, and the 8000 ppm thread is stable until an even higher Reynolds number. The 500 ppm jet mixes and does not act like a thread so the drag reduction is close to the premixed case. Additional experiments with polyacrylamide (Separan AP-273) show that the 4000 ppm poly (ethylene oxide) injection curve is duplicated for 500 ppm of PAM when the flow visualization looks the same [21]. However, premix PAM solutions after 14 hours of mixing that result in a homogeneous solution have much higher drag reduction.

Additional evidence for a different mechanism of the drag reduction by viscoelastic threads comes from laser Doppler velocity fluctuations. Since most local velocity measurements in premixed polymer solutions are for concentrations high enough to include the possible formation of polymer threads, it is appropriate to compare the premix results with the nonhomogeneous polymer injection results. Figure 4 shows some typical velocity profiles from the literature [22-26] and from Sinha [20]. The lower concentration premix solutions show a displacement of the velocity profile in the inertial sublayer but no change of slope on the semi-log plot. The largest change is in the buffer layer where the buffer layer increases in size as drag reduction increases. Although a variety of Reynolds numbers, concentrations and polymer solvent types are represented, the curves are functions of drag reduction amount only for the premix solutions. At very high drag reductions, there is also a dependence on Reynolds number. One reason for this is that total suppression of turbulence for low Reynolds numbers represents less drag reduction than at high Reynolds numbers. The asbestos fiber curve comes close to this limit, but large scale fluctuations were still present in the experimental measurements. The slope of the velocity profile in the inertial sublayer is changed for the non-homogeneous polymer injection. This suggests that the large scale structure of the turbulence is changed compared to the premix solutions with comparable drag reduction.

In contrast to the behavior of the velocity profile data, axial turbulent intensities near the wall for the same amounts of drag reduction do not lie on the same curve. Typical differences for three solutions which give the same velocity profile and drag reduction are shown in Figure 5. When drag reduction is increased there is a tendency toward a reduction in the turbulence intensity and a shift of the peak value of turbulence intensity to higher y_+. This shows up well only at the highest values of drag reduction. The premix and nonhomogeneous injection experiments of Sinha show some difference, but not any more difference than the premix experiments with the same drag reduction. Unfortunately there were no Reynolds stress measurements for these experiments and we do not know what part of the fluctuations are turbulent. Radial turbulent intensities give the same scatter, so we find both increases and decreases compared to the pure solvent.

Similar variations in turbulent intensity were found in submerged jets [11]. When flow visualization was combined with laser Doppler velocimetry, we found that increases in turbulent intensity compared to pure solvent flow were always accompanied by the appearance of wavy strings in the jet. Some movements of the strings in the axial and radial directions were uncorrelated and did not give an increased Reynolds stress. An example of a poly (ethylene oxide) jet is shown in Figure 6. Even when the turbulent intensities in the jet were reduced at higher polymer concentrations or for higher molecular weights, the reductions in Reynolds stress exceeded the reductions in turbulent intensity. Turbulence intensity measurements alone do not tell us much about the dynamics.

Frequency spectra or moments of the velocity distribution do give some additional information. A sketch of Sinha's results for water, premix polymer solution and nonhomogeneous injection is shown in Figure 7. The comparison is for equal Reynolds numbers and equal drag reduction for the polymer flows. Premix solutions show a lower skewness of axial fluctuations than pure solvent in the center of the pipe and no change near the wall.

The injection experiments showed that the positive skewness persisted much farther from the wall than for the premix solution. These observations mean that the slow moving fluid that is carried from the wall into the pipe center persists longer in the center in the case of the premix solution compared to the other two. The unsteady polymer thread pushes fast moving fluid from the center to the wall thus increasing the average size of the eddies near the wall in the nonhomogeneous case. At higher amounts of premix drag reduction Berner [27] found that the skewness curve crosses zero farther out from the wall and becomes more negative in the center of the pipe compared to the premix curve for 35% drag reduction shown in Figure 7. If the bursting process in the boundary layer is the source of large interactions between the flow and polymers, there should be changes in the character of the bursts. From flow visualization experiments in polymer solutions the average dimensionless spanwise distance between streaks and the average dimensionless length of streaks are increased compared to pure solvent [28]. (The scales in this discussion are assumed to be the inner scales.) However, data on the time between bursts do not clearly show differences between polymer solutions and solvent and we will explore the reasons for this. Landahl and Bark [29] found that the increased extensional viscosity of a non-dilute suspension of long thin rods had a large stabilizing effect on high frequencies or small scale motions. These small scale motions eventually lead to the bursts so a reduced frequency of bursts should be detected in polymer solutions. The bursts themselves are accompanied by a large scale instability which persists in the polymer solutions but is rapidly eaten up by the small scales in pure solvent. Thus velocity observations at a point in the flow for solvent alone will show that the largest fluctuations are related to the bursts [30]. However, both bursts and the instability of the ejected fluid will give large velocity excursions for polymer solutions. Different kinds of conditioned sampling may be required to distinguish these two separate processes.

We have looked at the short time sampling of laser Doppler

velocity signals near the wall to verify these differences. Random sampling leads to increased average times between bursts in polymer solutions compared to the solvent alone. If we trigger on the largest fluctuations and record the time to the next peak, this time is less than the bursting time obtained from random sampling. For a 1.27 cm. pipe at Re = 9000, the dimensionless conditionally sampled time between peaks for 100 ppm PEO is about equal to the bursting time for the pure solvent. The result is similar to measurements of the preferred mode in submerged jets [31]. Conditioned sampling shows that the frequency of the preferred mode in a polymer solution jet is nearly the same as that for the solvent jet.

Conclusions

The mechanism of drag reduction that involves elongation of molecules and orientation of the elongated molecules appears to occur by itself only for ultra low concentrations. At higher concentrations or for extended polyelectrolytes, the presence of entanglements or molecular associations gives interactions with the flow at lower Reynolds numbers. One form of the interaction is the suppression of small scale motions in the regions of high shear. That is the aligned molecules or molecular structures are reluctant to undergo rapid changes. If long strings of elongated polymers are formed with superimposed large scale instabilities, drag reduction can also occur by modification of the large eddies. Although the suppression of small scales can apparently occur by different mechanisms, it may be difficult to separate them in turbulent flow and all will occur together.

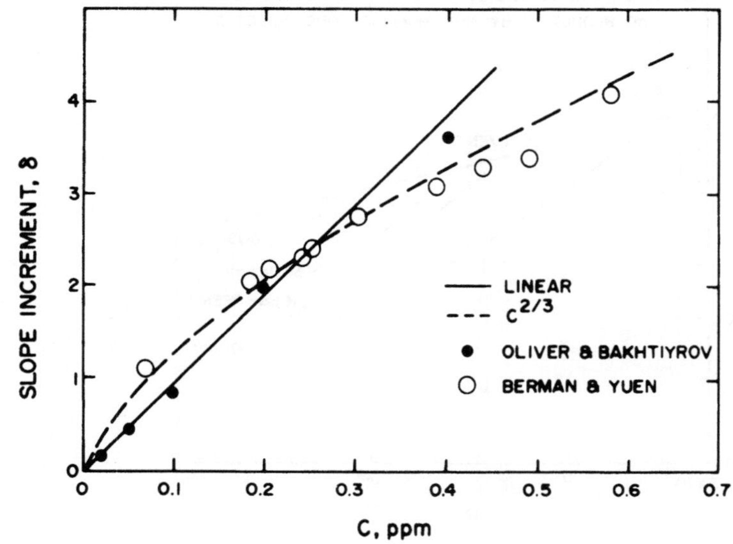

Fig. 1. Drag reduction slope increment (slope of polymer $1/\sqrt{f}$ vs. Re $\sqrt{f}$ curve less that for solvent) at very low polymer concentrations.

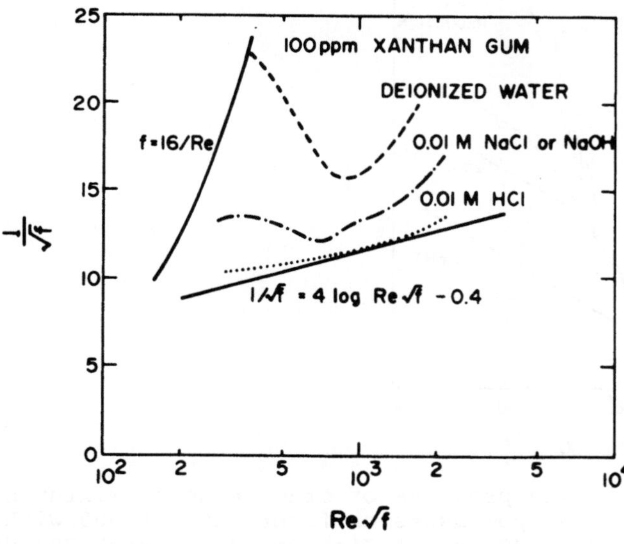

Fig. 2. Drag reduction trajectories for xanthan gum in a 0.554 cm. tube in three different solutions from Suaysompol [16].

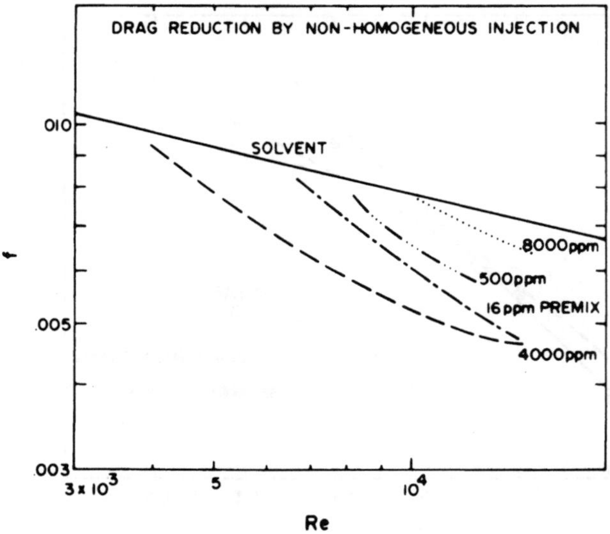

Fig. 3. Comparison of injection of a 0.1 cm. thread of PEO at the center of a 1.27 cm. tube with premix for equal well mixed concentrations of 16 ppm. Data are from Sinha [20].

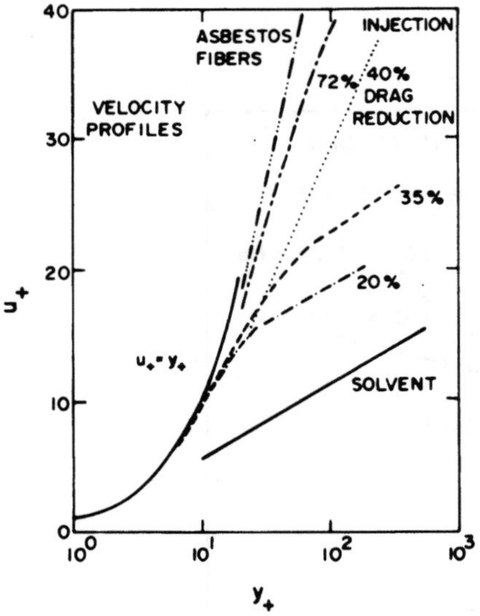

Fig. 4. Typical velocity profiles of drag reducing mixtures. The curves represent 300 ppm asbestos fibers at Re=9000 with 63% drag reduction [26], 300 ppm PEO at Re=14,400 with 72% drag reduction [25], 4000 ppm PEO injection [20], 2000 ppm polyisobutylene in oil at Re=7500 with 20% drag reduction [22], and three 35% drag reduction experiments: (1) 100 ppm polyacrylamide at RE=20,000 [24], (2) 800 ppm polyisobutylene in oil at Re= 22,500 [23], (3) 100 ppm PEO at Re=9000 [20].

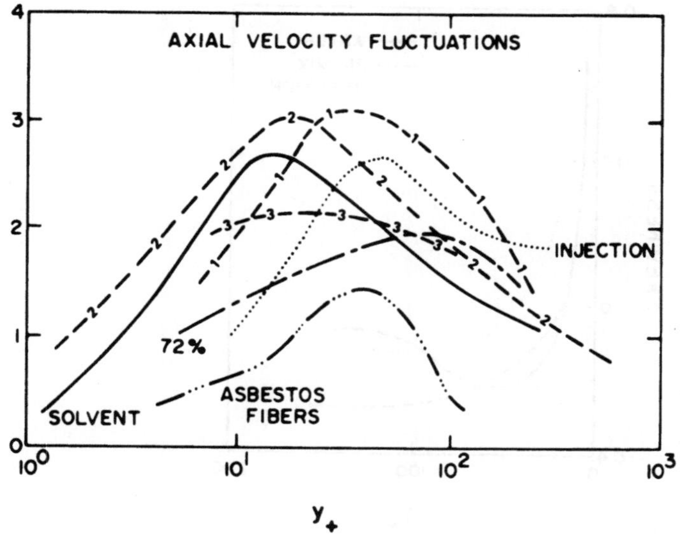

Fig. 5. Root mean square axial velocity fluctuations near the wall where the experiments correspond to those cited for Fig. 4. The results of the curves labeled -1-, -2-, and -3- gave the same 35% drag reduction curve in Fig. 4.

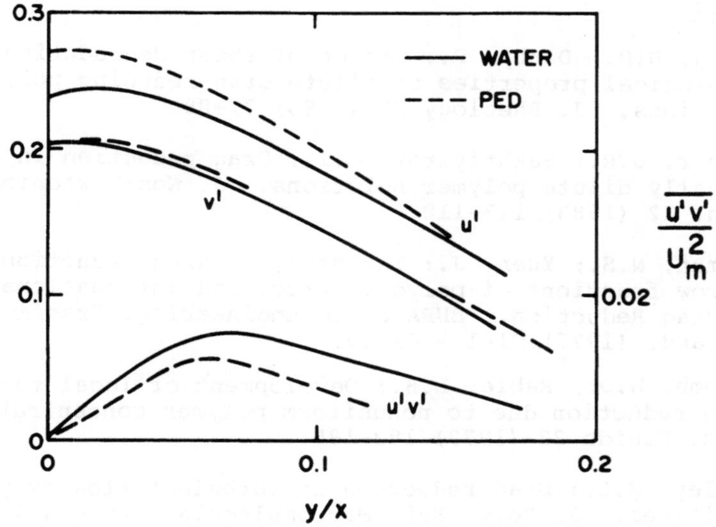

Fig. 6. Turbulent intensities and Reynolds stress in a 2 mm submerged jet 60 diameters from the source with Re=14,000 for water and 100 ppm PEO from Berman and Tan [11].

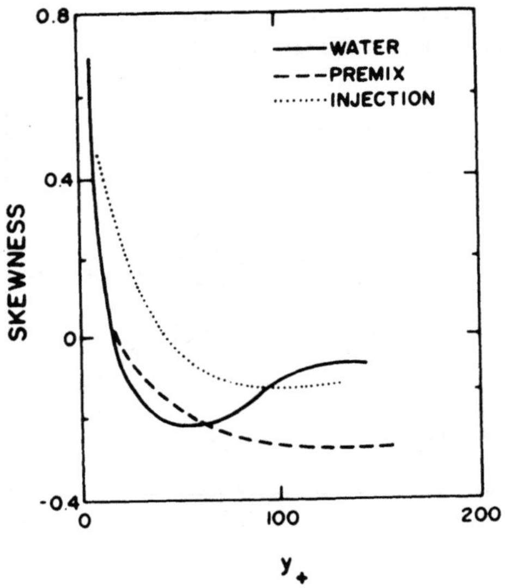

Fig. 7. Skewness of the axial velocity fluctuation distribution near the wall in pipe flow for water, 100 ppm PEO premix, and 4000 ppm injection from Sinha [20].

References

1. Chang, H.D.; Darby, R.: Effect of shear degradation on the rheological properties of dilute drag-reducing polymer solutions. J. Rheology 27 (1983) 77-88.

2. Oliver, D.R.; Bakhtiyarov, S.I.: Drag reduction in exceptionally dilute polymer solutions. J. Non-Newtonian Fluid Mech. 12 (1983) 113-118.

3. Berman, N.S.; Yuen, J.: The study of drag reduction using narrow fractions of polyox. Proc. 2nd International Conf. on Drag Reduction. BHRA Fluid Engineering. Cranfield, England. (1977) C1-1 - C1-10.

4. McComb, W.D.; Rabie, L.H.: Development of local turbulent drag reduction due to nonuniform polymer concentration. Phys. Fluids 22 (1979) 183-185.

5. Lumley, J.L.: Drag reduction in turbulent flow by polymer additives. J. Poly. Sci. Macromolecular Reviews 7 (1973) 283-290.

6. Durst, F.; Haas, R.; Interthal, W.; Keck, T.: Polymerwirkung in stromungen-mechanismen und praktische anwendungen. Chem. Ing. Tech. 54 (1982) 213-221.

7. Berman, N.S.: Drag reduction by polymers. Ann. Rev. Fluid Mech. 10 (1978) 47-64.

8. Hunston, D.L.; Zakin, J.L.: Effect of molecular parameters on the flow rate dependence of drag reduction and similar phenomena. In Hough, G.R. ed.; Viscous Flow Drag Reduction (1980) 373-385.

9. Lumley, J.L.: Drag reduction by additives. Ann. Rev. Fluid Mech. 1 (1969) 367-384.

10. Tan, H.; Berman, N.S.: A laser method to determine turbulent intensity using probability after effects. J. Phys. E. Sci. Instrum. 15 (1982) 906-910.

11. Berman, N.S.; Tan, H.: Two component laser Doppler velocimeter studies of submerged jets of dilute polymer solutions. AIChE Journal (1985) to be published.

12. Berman, N.S.; George, W.K.: Time scale and molecular weight distribution contributions of dilute polymer solution fluid mechanics. Proc. 1974 Heat Trans. Fluid Mech. Inst. Davis, L.R.; Wilson, R.E., eds. Stanford Univ. Press (1974) 348-364.

13. Ou, J.J.: The behavior of dilute polymer solutions: flow through constrictions. M.S. Thesis Arizona State University (1981).

14. Hinch, E.J.; Elata, C.: Heterogeneity of dilute polymer solutions. J. Non-Newtonian Fluid Mech. 5 (1979) 411-425.

15. Berman, N.S.; Berger, R.B.; Leis, J.R.: Drag reduction of well mixed solutions of poly (ethylene oxide) and organic dyes in water. J. Rheology 24 (1980) 571-587.

16. Suaysompol, K.: Effects of ionic species on turbulent drag reduction of xanthan gum solutions. M.S. Thesis Arizona State University (1983).

17. Virk, P.S.: Drag reduction by collapsed and extended polyelectrolytes. Nature 253 (1975) 109-110.

18. Berman, N.S.: Flow time scales and drag reduction. Phys. Fluids 20 (1977) S168-S174.

19. Landhal, M.: Drag reduction by polymer addition. in Proc. Thirteenth Int. Cong. Theor. Appl. Mech. (Moscow) Becker, E.; Mikhaelov, G.K. eds. Springer-Verlag, Berlin (1973) 177-199.

20. Sinha, P.K.: Drag reduction by centrally injected polymer jet. M.S. Thesis Arizona State University (1984).

21. Berman, N.S.; Sinha, P.K.: Drag reduction in pipe flow for non-homogeneous injection of polymer additives. in Proc. International Conf. Fund. and Appl. of Poly. Drag Red. in Liquid Flows. Bristol, England (1984).

22. Patterson, G.K.; Chosnek, J.; Zakin, J.L.: Turbulent structure in drag reducing polymer solutions. Phys. Fluids 20 (1977) S89-S99.

23. Durst, F.; Keck, T.; Kleine, R.: Turbulence quantities and Reynolds stress in pipe flow of polymer solutions measured by two channel laser-Doppler anemometry. in Turbulence in Liquids, Patterson, G.K.; Zakin, J.L. eds. (1981) 55-65.

24. Reischman, M.M.; Tiederman, W.G.: Laser-Doppler anemometer measurements in drag-reducing channel flows. J. Fluid Mech. 70 (1975) 369-392.

25. Mizushina, T.; Usui, H.: Reduction of eddy diffusion for momentum and heat in viscoelastic fluid flow in a circular tube. Phys. Fluids 20 (1977) S100-S108.

26. McComb, W.D.; Chan, K.T.J.: Drag reduction in fibre suspensions: transitional behavior due to fibre degradation. Nature 280 (1979) 45-46.

27. Berner, C.: Etude statistique de la structure de turbulence dans les fluides non-Newtoniens. Thesis Louis Pasteur University of Strasbourg (1980).

28. Tiederman, W.G.; Smith, A.J.: Structure of the viscous sublayer in drag-reducing channel flows. in Turbulence in Liquids, Zakin, J.L.; Patterson, G.K. eds. (1977) 312-322.

29. Landahl, M.T.; Bark, F.H.: Application of a two-scale boundary layer turbulence model to drag reduction. in Colloques Internationaux du CNRS No. 233, Polymers et Lubrification (1975) 249-258.

30. Bogard, D.G.; Tiederman, W.G.: Evaluation of burst detectors with single point velocity measurements. Bull. Am. Phys. Soc. 28 (1983) 1396-1397.

31. Berman, N.S.; Berman, J.C.; Perkins, M.; Lindquist, W.; Hwang, J.J.: Initial development of submerged circular jets of dilute polymer solutions. in Turbulence in Liquids (Proc. Eighth Symposium on Turbulence, Rolla, MO, 1983) to be published 1984.

Predictions of Turbulent Drag Reduction for a Linear Viscoelastic Fluid

S. LYONS AND C. A. PETTY

Department of Chemical Engineering
Michigan State University
East Lansing, Michigan 48840

SUMMARY

A link between the decoupling of the Reynolds stress and the temporal relaxation of turbulent fluctuations, controlled by a finite propagation of momentum at small spatial scales, has been developed for turbulent flow of a Maxwell fluid. A characteristic time between turbulent *bursts* from the wall region, a relaxation time for the velocity, space-time correlation, and the intensity of turbulent fluctuations normal to the mean flow govern the behavior of the Reynolds stress in the near wall region. The theory predicts quantitative extents of drag reduction in agreement with experiments and, more importantly, contains the familiar *saturation* and *onset* features of turbulent drag reduction.

INTRODUCTION

Because dilute polymer solutions have molecular relaxation times much larger than Newtonian fluids, momentum transfer between radial and axial velocity fluctuations in the near wall region of a fully developed turbulent pipe flow may be delayed enough to explain the decoupling of the Reynolds stress observed indirectly by Seyer and Metzner (1) and, more directly, by Logan (2). This phenomenon necessarily occurs in the constant stress region near the wall where molecular transport processes become important and where axial velocity fluctuations are coupled to the mean field by fluctuations in the radial velocity. The physical model of drag reduction developed here stems from an earlier application by Ruckenstein (3) of the Einstein and Li (4) renewal theory for a Maxwell fluid. Ruckenstein correctly emphasized the fundamental fact that momentum is transported at finite speeds. Here this idea is used to estimate the extent of drag reduction in terms of the von Karman Number and a characteristic

The Influence of Polymer Additives
on Velocity and Temperature Fields
IUTAM Symposium Essen 1984
Editor: B. Gampert

Deborah Number. The theory shows the familiar *onset* and *saturation* features of drag reduction and offers an explanation for both within a specific mathematical framework.

BACKGROUND

The turbulent flow analyzed is statistically stationary and statistically homogeneous in the axial and transverse directions; however, because of no slip at rigid interfaces, the mean velocity and the Reynolds stress are strong functions of the distance from the wall (see Fig. 1). Near the wall, the total mean stress is constant (see 5, 6). Therefore,

$$-\rho\langle u_1' u_3'\rangle(x_1) + \langle T_{13}\rangle(x_1) = T_w \,. \tag{1}$$

For statistically stationary flows of very dilute polymer solutions, the mean molecular stress and the mean strain rate are proportional,

$$\langle T_{13}\rangle = \mu \frac{d\langle u_3\rangle}{dx_1} \,. \tag{2}$$

Eq. (2), follows from the constitutive model for a Maxwell fluid, viz.,

$$\underline{\underline{T}} + \lambda \frac{\delta}{\delta t}\underline{\underline{T}} = 2\mu \underline{\underline{S}} \tag{3}$$

λ and μ are phenomenological coefficients characteristic of the fluid, not the turbulence. The time derivative in Eq. (3) is the upper convected derivative of Oldroyd (7). In what follows, λ and μ are assumed to be independent of the kinematics (cf. Lumley (8); and, Armstrong and Jhon (9)). The parameter $(\nu/\lambda)^{1/2}$ physically represents a finite propagation speed for the molecular transport of momentum. Using the Rouse-Zimm theory, Hershey and Zakin (10) estimated that $\lambda \simeq 10^{-3} - 10^{-6}$ sec. for dilute concentrations of high molecular weight polymers, (also see Patterson and Zakin (11)). Therefore, with $\nu = 10^{-2}\,cm^2/s$ and $\lambda = 10^{-4}$s, the propagation speed (10 cm/s) for dilute polymer solutions is about three orders of magnitude smaller than that of the solvent (10^4 cm/s).

For $\lambda = 0$, Eq. (3) reduces to the constitutive model for a Newtonian fluid and, for statistically stationary turbulence, the average of Eq. (3) gives

$$\langle \underline{\underline{T}} \rangle + \lambda \langle \underline{u} \cdot \nabla \underline{\underline{T}} - \nabla \underline{u} \cdot \underline{\underline{T}} - \underline{\underline{T}} \cdot (\nabla \underline{u})^T \rangle = 2\mu \langle \underline{\underline{S}} \rangle \tag{4}$$

Large scale eddies act to stretch fluid elements in the near wall region on a time scale comparable to ℓ_1/u^*, where the spatial scale ℓ_1, is defined in Fig. 1. The magnitude of the non-Newtonian term in Eq. (4) is $\lambda u^*/\ell_1$, so if λ is small relative to the Newtonian time scale $\overset{o}{\ell}_1/\overset{o}{u}{}^*$, then Eq. (4) reduces to

$$\langle \underline{\underline{T}} \rangle = 2\mu \langle \underline{\underline{S}} \rangle \tag{5}$$

inasmuch as $u^*/\overset{o}{u}{}^* \ll 1$ and $\overset{o}{\ell}_1/\ell_1 \ll 1$, a *posteriori*. (The superscript 'o' denotes a property of the solvent.) Therefore, a feature of the theory developed herein is that the viscoelastic properties of a dilute polymer solution affect the mean field only indirectly through the Reynolds stress.

PRELIMINARIES

Eq. (1) can be integrated across the constant stress region by introducing an *eddy* viscosity, which formally relates the Reynolds stress and the mean gradient. The result is

$$\langle u_3 \rangle(\infty) = u^* \int_0^{200} \frac{dx_1^+}{\left(1 + \frac{\nu_e}{\nu}\right)} . \tag{6}$$

If R denotes the ratio of the *eddy* viscosity with and without polymer and if R is spatially homogeneous (see Fig. 17 in Virk (12)), then Eq. (6) can be integrated for a specific representation of $\overset{o}{\nu}_e(x_1)$ The *eddy* viscosity in the near wall region for a Newtonian fluid is (5, 6, 12)

$$\overset{o}{\nu}_e/\nu = \begin{cases} .00087\, x_1^{+3}, & 0 \leq x_1^+ \leq 20 \\ .4\, x_1^+, & 20 \leq x_1^+ \leq 200. \end{cases} \tag{7}$$

Early experiments of Seyer and Metzner (1) and others show that the maximum velocity (i.e., $\langle u_3\rangle(\infty)$) scales with the bulk average velocity. Therefore, an additional assumption in the development includes

$$\langle u_3\rangle(\infty)/u_b = \langle \mathring{u}_3\rangle(\infty)/\mathring{u}_b \; . \tag{8}$$

Now with $f/2 = (u^*/u_b)^2$, Eqs. (6) and (8) imply that

$$(\mathring{f}/f)^{1/2} = \int_0^{200} \frac{dx_1^+}{1+R\mathring{\nu}_e/\nu} \Bigg/ \int_0^{200} \frac{dx_1^+}{1+\mathring{\nu}_e/\nu} \; . \tag{9}$$

Eq. (9) provides a link between the reduction in the *eddy* viscosity due to the polymer and the friction factor. The goal of this communication is to develop a theory for R .

THEORY

The set of equations for turbulent flows of an incompressible Maxwell fluid includes the continuity equation, the equation of motion, and Eq. (3). For short times and within the constant stress region, we neglect the convective contributions to Eq. (3) and assume that fluctuations in the molecular stress are governed by

$$\underline{\underline{T}}' + \lambda \frac{\partial}{\partial t} \underline{\underline{T}}' = 2\mu \underline{\underline{S}}' \; . \tag{10}$$

Thus, the temporal behavior of dilute polymer solutions for short times will determine the extent of drag reduction rather than the large resistance to extensional flows for long times.

A model for axial fluctuations stems directly from the equation of motion and can be developed in the usual way by introducing Reynolds' decomposition (5). Physically, we visualize a process in which interfacial fluid undergoes a renewal process. Following a *burst*, a relatively quiescent period of *weak* turbulence occurs. The time between *bursts*, τ_M, is controlled by whatever (nonlinear) processes cause the ejection of fluid from the near wall region into the core flow. We assume that most of the momentum lost to the wall occurs during the quiescent stage and that during this period the nonlinear mixing of axial momentum is unimportant. This hypothesis is used only for the axial velocity fluctuations; the other two components of the velocity fluctuation need not be analyzed directly. Eq. (20) and the linearized equation of motion for the fluctuating velocity yield an inhomogeneous, hyperbolic equation for u_3':

$$-\mathcal{L} u_3' = f \tag{11}$$

where
$$f = \left(u_1' + \lambda \frac{\partial u_1'}{\partial t}\right) \frac{d\langle u_3 \rangle}{d x_1} \tag{12}$$

and
$$\mathcal{L} = \left(1 + \lambda \frac{\partial}{\partial t}\right)\left(\frac{\partial}{\partial t} + \langle u_3 \rangle \frac{\partial}{\partial x_3}\right) - \nu \nabla^2 . \tag{13}$$

The form of Eq. (11) shows that axial momentum fluctuations between *bursts* are generated locally throughout the constant stress domain by two physical processes which couple directly with the mean gradient. The acceleration coupling is a new feature related to the elastic nature of the fluid. Furthermore, spatial mixing of axial momentum occurs by mean convection and viscous transport, but for $\lambda > 0$ a finite propagation velocity slows down the mixing process. Non-linear mixing effects could be included by using an approach similar to Snellenberger and Petty (13) and Hill and Petty (14).

ANALYSIS

The solution to Eq. (11) is determined by three factors: (1) initial conditions; (2) boundary conditions; and, (3) the inhomogeneous source term. If $u_3'(\underline{x},t)$ is statistically stationary and if fluctuations on the boundary are zero (or are far from the constant stress region), then the solution to Eq. (11) depends only on the inhomogeneous source of axial momentum:

$$u_3'(\underline{x},t) = -\int_{t-\tau_M}^{t} d\hat{t} \int d\hat{\Omega}\, G(\underline{x},t|\hat{\underline{x}},\hat{t}) f(\hat{\underline{x}},\hat{t}) \tag{14}$$

$G(\underline{x},t|\hat{\underline{x}},\hat{t})$ is a Green's function associated with the convective-viscoelastic operator $\mathcal{L}$ on the domain $\hat{\Omega}$. For small values of $\hat{\tau} = t-\hat{t}$, the Green's function is sharply peaked for small values of $|\underline{x}-\hat{\underline{x}}|$. If τ_M represents the average time between *bursts*, then the nonzero domain of the Green's function is bounded above by $\tau_M(\nu/\lambda)^{1/2}$.

A model for $\langle u_1' u_3' \rangle$ follows directly from Eq. (14) by multiplying the equation by $u_1'(\underline{x},t)$ and averaging the result. The two space-time correlations and the mean gradient within the resulting integral vary over distances on the order of $10\nu/u^*$ or larger. It is noteworthy that these correlations are multiplied by the Green's function which has an "effective" spatial domain of influence in a frame of reference moving with the local mean velocity of only $(\tau_M \nu)^{1/2}$. For small values of τ_M (i.e., $\tau_M^+ \ll 100$), a spatial smoothing approximation can be used to estimate $\langle u_1' u_3' \rangle$. This leads to

$$\langle u_1' u_3' \rangle = -\langle u_1'^2 \rangle (x_1) \tau_c(x_1) \frac{d\langle u_3 \rangle}{dx_1} \tag{15}$$

where

$$\tau_c(x_1) = \left(1 + \frac{\lambda}{\tau_H}\right) \int_0^{\tau_M} d\hat{\tau}\, K(x_1,\hat{\tau})\, e^{-\hat{\tau}/\tau_H} \tag{16}$$

and

$$K(x_1,\hat{\tau}) = \int d\hat{\Omega}\, \frac{G(\underline{x},t|\hat{\underline{x}},\hat{t})}{4\pi\nu} . \tag{17}$$

Eq. (15) follows by using the exponential model for the space-time correlation defined in Fig. 1. Three spatial scales and an Eulerian relaxation time τ_E characterize the space-time structure of $\langle u_1'(\underline{x},t) u_1'(\hat{\underline{x}},\hat{t}) \rangle$. Because the Green's function for the wall eddy is locally convected by the mean axial velocity, the effect of the smoothing approximation employed above is to replace the coordinates $(\hat{x}_1, \hat{x}_2, \hat{x}_3)$ in $\langle u_1' \hat{u}_1' \rangle$ with $(x_1, x_2, x_3 + u_G \hat{\tau})$ where u_G represents an average convection velocity for the wall eddy. Thus, τ_H in Eq. (16) is

$$\frac{1}{\tau_H} = \frac{1}{\tau_E} + \frac{u_G}{\ell_3} . \tag{18}$$

With $u_G^+ \simeq$ 5-10 and $\ell_3^+ \simeq$ 1000, the hydrodynamic relaxation time τ_H can be interpreted as an Eulerian integral time scale for velocity fluctuations normal to the rigid interface.

$\tau_c(x_1)$ represents a characteristic time for transporting axial momentum fluctuations by velocity fluctuations normal to the mean flow. It depends upon the molecular parameter λ as well as the two turbulent time scales τ_H and τ_M. Because the Green's function is zero on the boundary, $\tau_c(x_1)$ is also zero for $x_1=0$; however, it rapidly approaches its limiting value $\tau_c(\infty)$ for $x_1 >> (\tau_M \nu)^{1/2}$. Because $\nu_e << \nu$ deep in the viscous sublayer, $\tau_c(x_1)$ in Eq. (15) can be replaced by $\tau_c(\infty)$; the turbulent intensity still makes ν_e zero at the wall. Thus,

$$\nu_e(x_1) = \langle u_1'^2 \rangle (x_1) \tau_c(\infty) \tag{19}$$

where $\tau_c(\infty)$ is given by Eq. (16) with the memory kernel replaced by

$$K(\infty,\hat{\tau}) = 1 - e^{-\hat{\tau}/\lambda}. \tag{20}$$

Eq. (20) follows from Eq. (17) by using a Green's function associated with the hyperbolic operator $\mathcal{L}$ and integrating over an infinite spatial domain (13, 14, 15). For $\lambda=0$, the Green's function is viscous and $K(\infty,\hat{\tau})=1$ for all values of $\hat{\tau}$. With $\lambda>0$, the elastic behavior of the dilute polymer solution retards the transport process and effectively reduces the eddy viscosity coefficient. Note, however, that for $\hat{\tau}>3\lambda$ the memory function approaches the Newtonian limit.

Inserting Eq. (20) into Eq. (16) and carrying out the integration yields the major result of this communication, viz.,

$$\frac{\tau_c(\infty)}{\tau_H} = 1 - e^{-\alpha_M} - \alpha_\lambda e^{-\alpha_M}(1 - e^{-\alpha_M/\alpha_\lambda}) \tag{21}$$

where

$$\alpha_M = \tau_M / \tau_H \tag{22}$$

and

$$\alpha_\lambda = \lambda / \tau_H . \tag{23}$$

Fig. 2 illustrates the behavior of $\tau_c(\infty)/\tau_H$. An interesting feature of Eq. (21) is that for fixed α_M a positive lower bound occurs. This phenomenon, which leads to a maximum drag reduction asymptote, physically stems from a balance between an increase production of axial momentum due to acceleration coupling with the gradient of the mean field and the retarded mixing of this momentum due to a finite propagation speed.

SCALING ASSUMPTIONS AND DRAG REDUCTION

Experimental data in the near wall region partially support the additional assumption that (12, 16, 17)

$$\langle u_1'^2\rangle^{1/2}/u^* = \langle \mathring{u}_1'^2\rangle/\mathring{u}^*. \tag{24}$$

With this scaling hypothesis the reduction factor R can be estimated without developing a specific theory for the normal fluctuations. Furthermore, if we also assume that τ_M and τ_H both scale with wall parameters, then

$$\langle u_1'^2\rangle\tau_H = \langle \mathring{u}_1'^2\rangle\mathring{\tau}_H$$
$$\alpha_M = \mathring{\alpha}_M$$
$$\tau_H^+ = \mathring{\tau}_H^+ .$$

It now follows from Eqs. (19) and (21) that

$$R = 1 - \alpha_\lambda e^{-\alpha_M}\,\frac{1 - e^{-\alpha_M/\alpha_\lambda}}{1 - e^{-\alpha_M}} . \tag{25}$$

Fig. 3 is a contour plot of R and shows what combinations of $(\alpha_M, \alpha_\lambda)$ lead to significant reductions in the *eddy* viscosity.

The dimensionless group α_λ, which compares the molecular relaxation time to the fluid relaxation time, can be written as

$$\alpha_\lambda = De\,\Lambda^2/2\tau_H^+ \tag{26}$$

where D_e and Λ are, resp., a characteristic Deborah number, $\lambda\nu/v^2$, and von Karman number, $Re\sqrt{f}$.

The extent of drag reduction can be estimated for a specific D_e and Λ once the *universal* parameters α_M and τ_H^+ have been specified. Fig. 4 shows the behavior of $1/\sqrt{f}$ predicted by the theory for $\alpha_M = 0.3$, and $\tau_H^+ = 150$. As already anticipated, a maximum drag reduction asymptote obtains for large D_e at fixed Λ. Because of the exponential character of the memory function $K(\infty,\hat{\tau})$, the reduction factor R changes very rapidly over a decade variation in α_λ (see Fig. 3) and this leads to an apparent onset phenomenon as illustrated in Fig. 4.

CONCLUSIONS

A theory for drag reduction of a linear viscoelastic fluid has been developed which is qualitatively and quantitatively consistent with the drag reduction phenomena. The following conclusions follow from this analysis:

1. A *linear* theory of viscoelasticity is sufficient for obtaining a significant drag reduction effect; qualitative arguments which assume that either λ or μ depend on the kinematics of turbulent flows appear unnecessary (see, esp., Lumley (8)).

2. The constitutive model was restricted to small time scales (i.e., $\tau_M^+ \ll 100$) and, thereby, does not retain the *extensional* viscosity effect. Although resistance to extensional flows may govern some aspects of drag-reducing flows, this feature also seems unnecessary in view of the quantitative predictions presented here.

3. The production of axial momentum fluctuations by acceleration coupling with the mean field is an essential aspect of the saturation phenomenon. Without this effect, $R \to 0$ as $\alpha_\lambda = \infty$ for all α_M. Figs. 3 and 4 show that a non-zero asymtote for R exists which depends on α_M.

4. For $\alpha_M = 0.3$ and $\alpha_\lambda = \infty$., $R = 0.14$. Although this agrees qualitatively with the *eddy* viscosity data summarized in Fig. 17 of Virk's review article, the maximum reduction in drag based on the model is higher than Virk's asymptote. The temporal time scales used here were suggested by Fortuna and Hanratty (19) and Achia and Thompson (20). Obviously, lowering α_M will decrease the extent of drag reduction but the results of Fig. 4 may be high because the Green's function used in the analysis does not contain hydrodynamic mixing effects.

5. Different drag reduction asymptotes have been reported in the literature (12, 18). This would be possible for different values of α_M. Achia and Thompson (20) have noted that local pressure variations characteristic of a specific experimental set-up could influence τ_M^+. The main theoretical conclusion here, however, is that the same drag-reducing system responds differently to coherent structures having different time scales.

6. Fig. 3 illustrates that for a fixed $\alpha_M < 1$, a significant reduction in the eddy viscosity begins to occur when $\lambda \sim \tau_M$, the mean time between *bursts*.

7. The parameter De in Fig. 4 contains the familiar diameter effect associated with the drag reduction phenomenon. For fixed Λ and for a given polymer system, $De \propto 1/D^2$. For $\lambda = 10^{-2}$ sec and for $\nu = 10^{-2}$ cm^2/s, the calculations portrayed in Fig. 4 correspond to pipe diameters ranging from 0.1 cm to 10 cm.

Acknowledgement

The methodology used here was developed earlier by Dr. H. T. Yao (20) to study turbulent mass transfer near interfaces; his research was supported by the National Science Foundation (Grant ENG 79-15256). S. Lyons received summer support from the Olin Corporation for Senior Research in Chemical Engineering. The final manuscript was prepared by C. A. Petty while on sabbatical leave at the Dorr-Oliver Company in Stamford, CT. The many encouragements of Dr. C. A. Willus and his colleagues are sincerely appreciated. Yvonne Grimm typed the final draft of the manuscript for which we are very thankful.

References

1. Seyer, F. A. and A. B. Metzner; Turbulence Phenomena in Drag-Reducing Systems. AIChE J. 15 (1969) 426.

2. Logan, S. E.; Laser Velocimeter Measurement of Reynold's Stress and Turbulence in Dilute Polymer Solutions. AIAA J. 10 (1972) 962.

3. Ruckenstein, E; On the Mechanism of Drag Reduction in Turbulent Flow of Viscoelastic Liquids. Chem. Eng. Sci. 26 (1971) 1075.

4. Einstein, H. A. and H. Li; The Viscous Sublayer Along a Smooth Boundary, Proc. Am. Soc. Civil Engrs., J. Eng. Mech. Div., 82,293 (1956).

5. Monin, A. S. and A. M. Yaglom; Statistical Fluid Mechanics: Mechanics of Turbulence, Vol. 1. Cambridge: The MIT Press 1971.

6. Tennekes, H. and J. L. Lumley; A First Course in Turbulence. Cambridge: The MIT Press 1972.

7. Astarita, G. and G. Marrucci; Principles of Non-Newtonian Fluid Mechanics. New York: McGraw-Hill 1974.

8. Lumley, J. L.; Drag Reduction by Additives. Ann. Rev. Fluid Mech. 1 (1969) 367.

9. Armstrong, R. and M. S. Jhon; A Self-Consistent Theoretical Approach to Polymer Induced Turbulent Drag Reduction. Dept. Chem. Eng., Carnegie-Mellon University (personal communication) 1984.

10. Hershey, H. C. and J. L. Zakin; Existence of Two Types of Drag Reduction in Pipe Flow of Dilute Polymer Solutions. Ind. Eng. Chem. Fund. 6 (1967) 381.

11. Patterson, G. K. and J. L. Zakin; Prediction of Drag Reduction with a Viscoelastic Model. AIChE J. 14 (1968) 434.

12. Virk, P. S.; Drag Reduction Fundamentals. AIChE J. 21 (1975) 625.

13. Snellenberger, R. W. and C. A. Petty; Estimates of Average Mass Transfer Rates Using An Approximate Hydrodynamic Green's Function. Chem. Eng. Commun. 20 (1983) 311.

14. Hill, J. C. and C. A. Petty; A Statistical Theory of Turbulent Mass Transfer Induced By a Mean Concentration Gradient. PACHEC III, Seoul, Korea, May 8-11, 1983.

15. Morse, P. M. and H. Feshbach; Methods of Theoretical Physics, Part I. New York: McGraw-Hill 1953.

16. Berman, N. S.; Drag Reduction by Polymers. Ann. Rev. Fluid Mech. 10 (1978) 47.

17. Schummer, P. and W. Thielen; Structure of Turbulence in Viscoelastic Fluids. Chem. Eng. Commun. 4 (1980) 593.

18. Lee, W. K., R. C. Vaseleski, and A. B. Metzner; Turbulent Drag Reduction in Polymeric Solutions Containing Suspended Fibers. AIChE J. 20 (1974) 128.

19. Fortuna, G. and T. J. Hanratty; The Influence of Drag-Reducing Polymers on Turbulence in the Viscous Sublayer. J. Fluid Mech. 53 (1972) 575.

20. Achia, B. V., and D. W. Thompson; Structure of the Turbulent Boundary in Drag-Reducing Pipe Flow. J. Fluid Mech. 81 (1977) 439.

21. Yao, H. T.; The Application of Turbulent Relaxation Models To Mass Transfer Near Interfaces At High Molecular Schmidt Numbers. Ph.D. Dissertation, Michigan State University 1982.

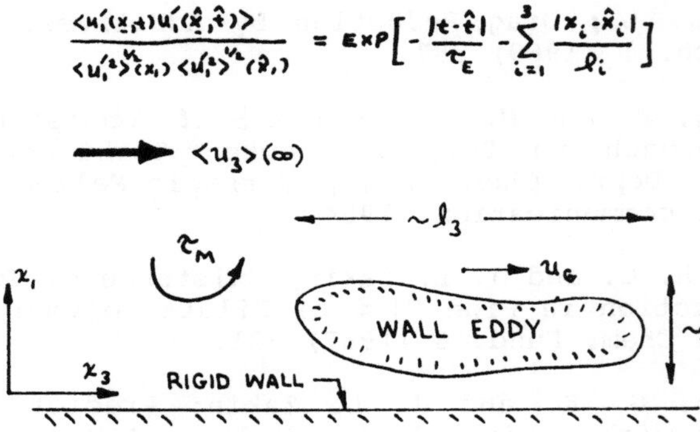

Fig. 1 Schematic of the transport eddy model; definition of the space-time velocity correlation for fluctuations normal to the interface

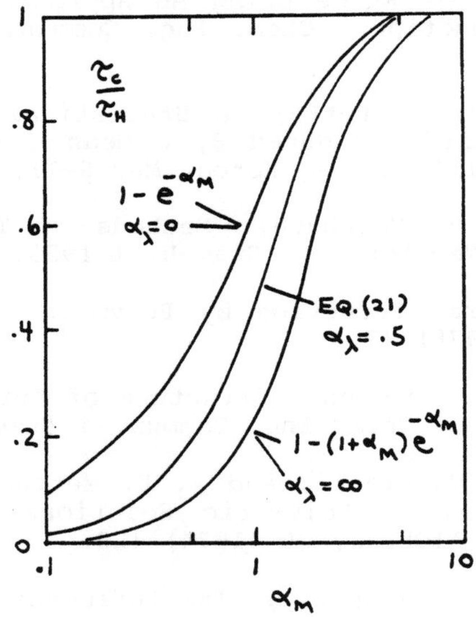

Fig. 2 Dependence of the transport relaxation time on the time between bursts and the molecular relaxation time

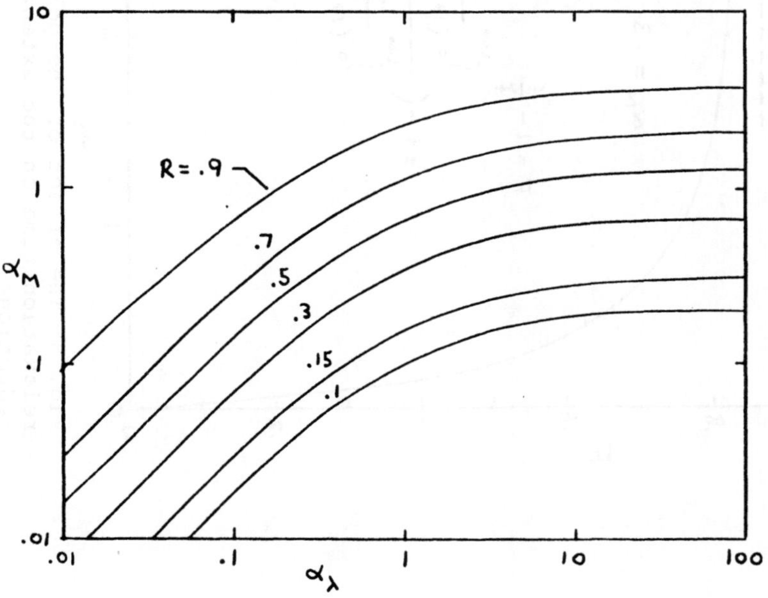

Fig. 3 The effect of the time between bursts and the molecular relaxation time on the eddy viscosity coefficient

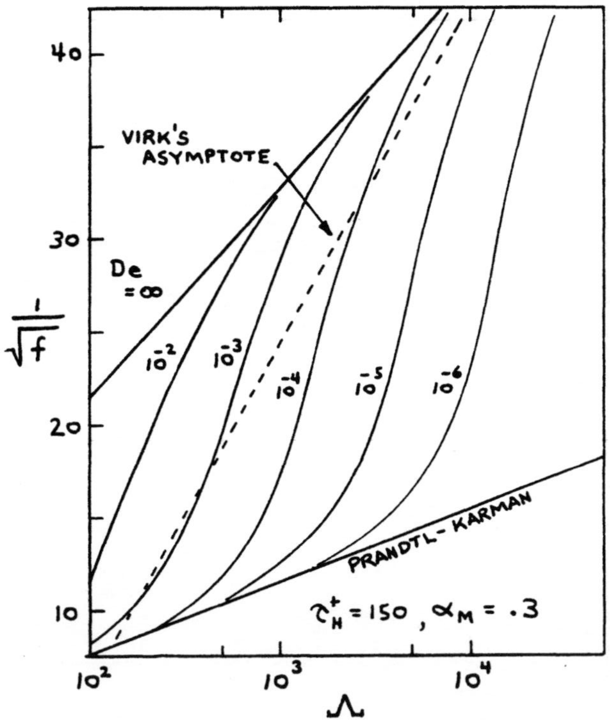

Fig. 4 Prediction of friction factor for a linear viscoelastic model

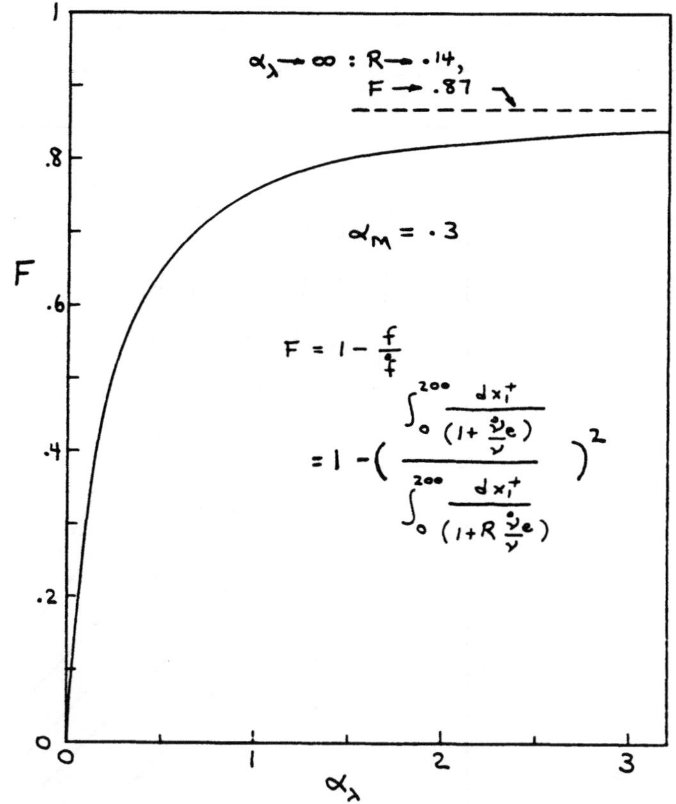

Fig. 5 The effect of the polymer relaxation time on the extent of drag reduction

An Engineering Approach to Modeling Complex Flow Behaviour of Polymer Solutions

R. Darby

Chemical Engineering Department
Texas A&M University
College Station, TX 77843

Summary

The concept of energy dissipation and storage in a linear viscoelastic fluid in oscillatory deformation is used as a basis for generalizing the definition of the Fanning friction factor. First order characteristics of dissipative eddies are used to apply the model to turbulent flow, and the definition of the fluid relaxation time is extended to account for nonlinear properties. The resulting model is used as a basis for correlation of drag reduction data for both fresh and degraded polyacrylamide solutions of various concentrations in a range of tube sizes on a common line with the Newtonian solvent. The model predictions are consistent with other observations on the relation between drag reduction and polymer concentration, degradation, tube diameter, and boundary layer properties.

Introduction

Solutions of high polymers may exhibit very striking behavior in complex flow systems as a consequence of their nonlinear viscoelastic properties. A common example is the drag or energy dissipation experienced in the turbulent flow of very dilute solutions of certain high polymers, which may be as much as 80% less than that of the solvent alone at the same flow rate. The rheological characteristics of these solutions are not only quite complex, but they are generally difficult to evaluate quantitatively because of the low concentrations of interest. These properties, coupled with the complex character of turbulent flow, result in an exceedingly complex system which is virtually impossible to analyze precisely. Consequently, various approximations and simplifying assumptions are necessary in order to obtain a relationship between observable quantities.

The Influence of Polymer Additives
on Velocity and Temperature Fields
IUTAM Symposium Essen 1984
Editor: B. Gampert

The approach which is taken here, and applied to the analysis of turbulent drag reduction in dilute polymer solutions, is typical of the classical engineering approach to system analysis and prediction. We start with fundamental principles to develop a concept or idea of the basic mechanism which is responsible for, or at least consistent with, the phenomenon to be described. The next step is to formulate a quantitative description, or "model", incorporating the essence of the basic concept. Rather than expressing this model in the most general terms including all possible details, the engineering approach involves starting with first order concepts and simplifying assumptions to describe the basic elements of the model in simple terms, but which may not involve a complete representation of the minute details of the system. The resulting model is a first order approximation,and may include parameters representative of an ideal system rather than actual properties of the real system. The process must be complemented, or completed, therefore, by development of some relation (by either theoretical or experimental means) connecting the model parameters with real system properties. Examples of this approach include the classical boundary layer and friction factor models for turbulent flows, film models and transfer coefficients for interphase transport of heat and mass, etc. The following analysis of turbulent drag reduction may be considered representative of this approach.

Energy Dissipation Model

Friction loss or drag is synonymous with the dissipation of energy. For flow in a pipe, the rate of energy dissipation per unit mass of fluid can be expressed in terms of the mean flow velocity and either the wall stress, pressure drop, or Fanning friction factor, as follows:

$$\dot{e} = \frac{4\tau_w v}{\rho D} = -\frac{\Delta P v}{\rho L} = \frac{2fv^3}{D} \qquad (1)$$

Now for purely viscous materials, all the energy associated with deformation is ultimately dissipated, contributing to friction loss or drag. However, for viscoelastic materials, only a portion of this energy is dissipated, the rest being stored by elastic deformations, and is therefore conserved rather than dissipated. This is easily illustrated by considering a linear viscoelastic fluid (e.g. a Maxwell fluid), with a viscosity of η_o and a relaxation time of λ subjected to a local oscillatory velocity of frequency ω and amplitude v_o. The solution of the momentum equation for this situation yields the following expression for the mean rate of energy dissipation per unit mass [1]:

$$\dot{e} = \frac{v_0^{2}\omega}{2(1+\omega^{2}\lambda^{2})^{1/2}} \qquad (2)$$

For a purely viscous Newtonian fluid (which can be considered a Maxwell fluid with $\lambda = o$) this confirms that all of the kinetic energy of the motion is dissipated. However, the presence of elasticity ($\lambda \neq 0$) results in a reduction in the dissipated energy. Since polymer solutions are viscoelastic, and since turbulence constitutes an unsteady deformation which may be described by a Fourier spectrum of oscillating eddies, this simple model provides a basic explanation for drag reduction by polymer solutions in turbulent flow. From (1) and (2), the friction factor for the Newtonian solvent (f_s) and polymer solution (f_p) under comparable flow conditions may be related by

$$f_p = \frac{f_s}{(1+N_{De}^{2})^{1/2}} \quad \text{where } N_{De} = \omega_p\lambda \qquad (3)$$

Here, ω_p represents a characteristic frequency of the most dissipative turbulent eddies in the polymer solution, which may be expected to differ somewhat from that for the solvent under comparable flow conditions. This quantity must be determined for the solution, as well as an appropriate equivalent value of λ, inasmuch as the solutions of interest are not linear Maxwell fluids but instead exhibit marked nonlinear viscoelastic properties.

Dissipation in Turbulent Eddies

The power density associated with a turbulent eddy for a Newtonian fluid can be represented as follows:

$$\dot{e}_s = \tau^+\dot{\gamma}^+/\rho \doteq (\bar{v}')^2\,\bar{v}'/l_e \doteq (\bar{v}')^2\omega \qquad (4)$$

where l_e is the characteristic eddy length, related to the corresponding eddy frequency by $l_e = \bar{v}'/\omega$. According to Davies [2], the maximum energy dissipation rate occurs when the equivalent eddy Reynolds number has a value of about 9, i.e.

$$N_{Re}' = \frac{l_e\bar{v}'}{\nu} \cong 9, \text{ or } (\bar{v}')^2 \cong 9\omega\nu \qquad (5)$$

Eliminating $\bar{v}'$ from (4) and (5), and expressing $\dot{e}_s$ in terms of f_s from (1) leads to:

$$\omega \cong \frac{V}{D}\left(\frac{2f_s N_{Res}}{9}\right)^{1/2} \qquad (6)$$

If the Blasius equation (e.g. $f_s = 0.0791/N_{Re}^{1/4}$) is used to eliminate f_s, this becomes:

$$\omega \cong 0.0166\left(\frac{8V}{D}\right) N_{Res}^{3/8} \qquad (7)$$

which shows that ω is of the order of $8V/D$ for typical turbulent Reynolds numbers (e.g. ~ 5×10^4).

For polymer solutions, for the lack of more specific information, we assume that an expression analogous to (6) for the dissipative eddy frequency holds as well, i.e.

$$\omega_p \cong \frac{V}{D}\left(\frac{2f_p N_{Rep}}{9}\right)^{1/2} \qquad (8)$$

When this is combined with (6) and (3) to eliminate f_p and f_s, we see that

$$\omega_p = \frac{\omega}{(1+\omega_p^2\lambda^2)^{1/4}}\,(\mu_s/\eta_o)^{1/2} \qquad (9)$$

where η_o is the Newtonian (limiting) viscosity of the solution. Equation (9) shows that the dissipative eddy spectrum for the polymer solution is shifted to slightly lower frequencies relative to that of the solvent, which is consistent with experimental observations. [3,4]. Combining (7) and (9), and replacing ω_p in the denominator by $8V/D$, which is of the same order, we find

$$N_{De} = \frac{0.0166\, N_\lambda\, N_{Res}^{3/8} (\mu_s/\eta_o)^{1/2}}{(1+N_\lambda^2)^{1/4}} \tag{10}$$

where $N_\lambda = 8v\lambda/D$

Rheological Properties

Equation (10) cannot be applied until λ, the equivalent relaxation time for the fluid, is known. However, λ is defined by the linear Maxwell model, and actual polymer solutions exhibit marked nonlinear viscoelastic properties [5,6,7]. For both fresh and shear degraded solutions of Separan AP-30 polyacrylamide, which exhibit pronounced drag reduction in turbulent flow, Chang and Darby [8] have measured the nonlinear viscosity and first normal stress functions, and Tsai and Darby [6] have reported transient elastic properties of similar solutions. A nonlinear hereditary integral function containing six parameters has been proposed to represent the measured properties [8]. The apparent viscosity function predicted by this model is:

$$\eta = \eta_\infty + \frac{\eta_o - \eta_\infty}{(1+\zeta^2\dot{\gamma}^2)^\Omega} \tag{11}$$

This represents the solid lines in Fig. 1, which shows data for three concentrations (500, 250 and 100 wppm) each of the fresh and shear degraded solutions. The parameters that govern the viscosity function in (11) also strongly influence other predicted response functions such as normal stress behavior, elongational viscosity, transient response, etc.

The only parameter in (11) having dimensions of time is ζ. Although this is not a relaxation time per se', it can be associated with a "Maxwell-type" relaxation time, as follows. Although the linear Maxwell model predicts a constant (Newtonian) viscosity, it may be generalized by utilizing a co-rotational reference frame which follows the local rotation and translation of each fluid element [9]. When a term is added to account for the high shear limiting behavior, the result is the co-rotational form of the Jeffreys model:

$$\tau_{ij} + \lambda \frac{D\tau_{ij}}{Dt} = \eta_o \Delta_{ij} + \lambda\eta_o \frac{D\Delta_{ij}}{Dt} \quad (12)$$

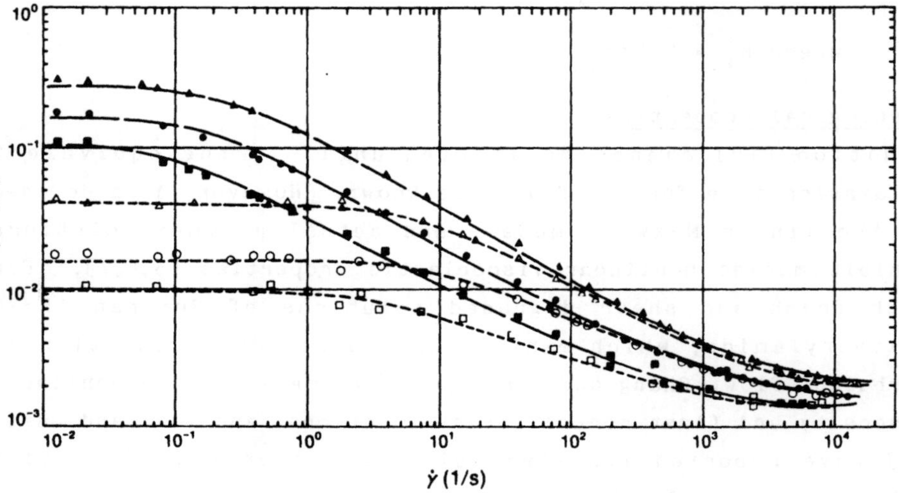

Fig. 1. Apparent Viscosity of Fresh and Degraded Polyacrylamide Solutions

which is taken to be the defining equation for λ. This model predicts the following viscosity function:

$$\eta = \eta_o + \frac{\eta_o - \eta_\infty}{(1+\lambda^2\dot{\gamma}^2)} \quad (13)$$

Here, λ is still the relaxation time of the fluid, but now we see that it is also responsible for the shear thinning character of the viscosity. It is a consequence of the finite time required for the polymer molecules to recover to their equilibrium configuration relative to the rate at which they are deformed. Because of the general similarity between (11) and (13), a quantitative expression for λ can be obtained by equating these expressions and solving for λ:

$$N_\lambda = \frac{8v\lambda}{D} = [(1+N_\zeta^2)^\Omega - 1]^{1/2} \quad (14)$$

$$\text{where } N_\zeta = \zeta\dot{\gamma} \doteq 8v\zeta/D \quad (15)$$

where we have used the result that the shear rate is equivalent to the eddy frequency in turbulent flow, which is of the order 8v/D. Equation (14) is a dimensionless "relaxation time function", which for a nonlinear fluid is analogous to the apparent viscosity function.

<u>Drag Reduction Correlation</u>

Our approximate model is now complete, in that it involves only parameters which can be measured directly. A complete viscosity determination establishes values for ζ, Ω and η_o, where ζ is the reciprocal of the shear rate at which transition from Newtonian to shear thinning viscosity occurs, and $\Omega = (1-n)/2$ where n is the flow index in the power law region. For given values of v and D, N_ζ is obtained from (15), which is used in (14) to get N_λ. N_{De} is then obtained from (10), and is used with (3) to relate the polymer solution friction factor, f_p, to that of the Newtonian solvent, f_s, at the known solvent Reynolds number, N_{Res}.

Turbulent tube flow data for the six solutions shown in Fig. 1 were obtained in seven different tubes ranging from 0.18 to 1.02 cm diameter [1,10]. A representative plot of f versus N_{Res} for the 0.46 cm tube is shown in Fig. 2. Because of the

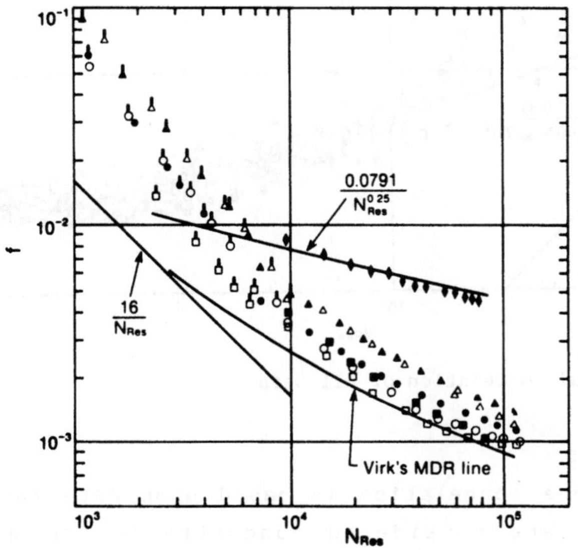

Fig. 2. Friction Factor vs. Solvent Reynolds Number
All Six Solutions in 0.460 cm. Tube

approximate nature of to equation (10) for N_{De}, it does not provide a precise representation of all of the data in its original form.. However, if the numerical exponents are allowed to vary somewhat, a regression of all data, including all concentrations of fresh and degraded solutions in all tubes, yields the following expression, with a correlation coefficient of 0.956:

$$N_{De} = \frac{0.0867\ N_\lambda\ N_{Res}^{\ 0.34}\ (\mu_s/\mu_o)^{0.237}}{(1+N_\lambda^{\ 2})^{0.33}} \tag{16}$$

with N_λ given by (14). This correlation is shown in Fig. 3 with representative data points covering the entire range of the data for all conditions. The solid line in this plot is the Colebrook equation for Newtonian fluids in smooth tubes which is similar to the Blasius equation but slightly more accurate.

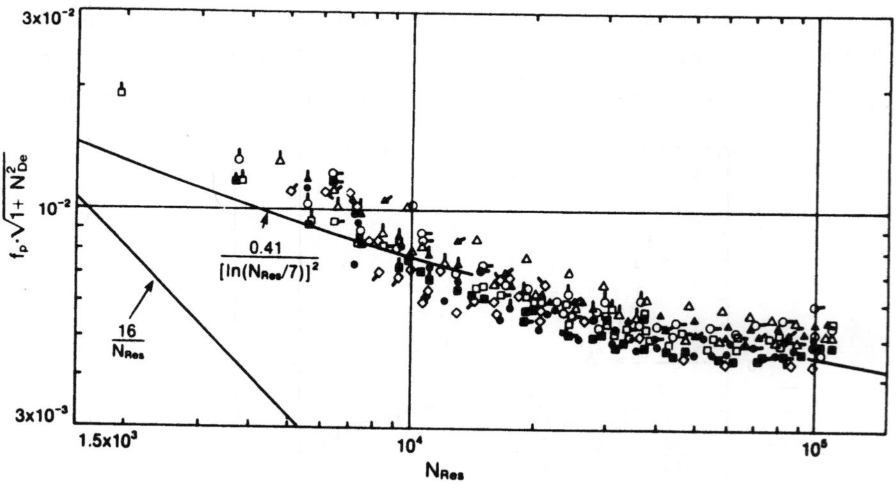

Fig. 3. Final Correlation of All Data

Model Predictions

Although the above correlation is based upon data for polymer solutions which are considered concentrated for most drag reduction applications, it has been shown to be consistent with observed dependence of drag reduction upon both concentration and degradation in more dilute solutions, as well as observed dependence upon pipe diameter [1,10]. For example, Fig. 4 shows predicted friction factor characteristics for both fresh and degraded 250 wppm solutions, for four larger pipe sizes. It is noted that drag reduction decreases with increasing diameter, as is well known. It also shows that although the degraded solution is a better drag reducer in small pipes, it is less effective than the fresh solution in large pipes. Also

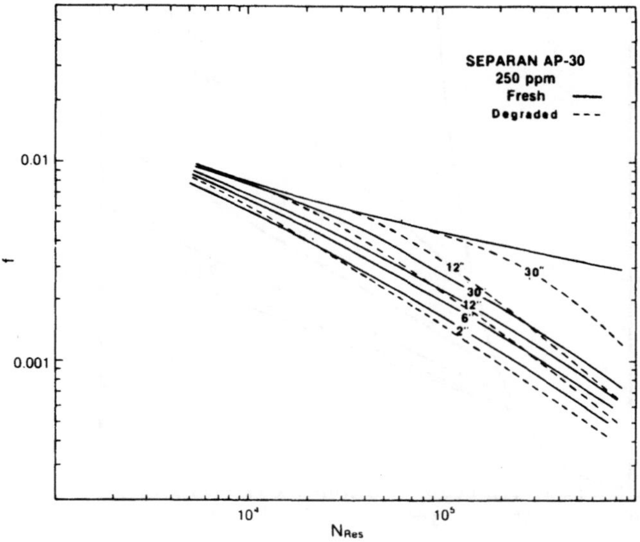

Fig. 4. Predicted Friction Loss for 200 ppm Fresh and Degraded Solutions in Various Size Pipes

evident is the trend of the curves to follow the Newtonian curve to higher Reynolds numbers before deviating as the diameter increases or elasticity decreases (i.e. lower values of N_{De}). This is also consistent with observations. Many interpretations of drag reduction have centered about the effect of the additive on boundary layer characteristics of the flow. The boundary layer implications of the present model are easily shown simply by replacing the friction factor in the von Karman "law of the wall" equation by equation (3) for the polymer solution. This leads to the following equation for the velocity profile:

$$v^+ = \frac{(1+N_{De}^2)^{1/4}}{\kappa} \ln y^+ + \frac{(1+N_{De}^2)^{1/4} \ln(1+N_{De}^2)^{1/4}}{4\kappa} + A \tag{17}$$

where $\kappa = 0.4$, $A = 5.5$ are the usual Newtonian boundary layer parameters. This expression is plotted in Fig. 5, which clearly shows that an increase in boundary layer thickness results from an increase in N_{De}, which governs the degree of drag reduction.

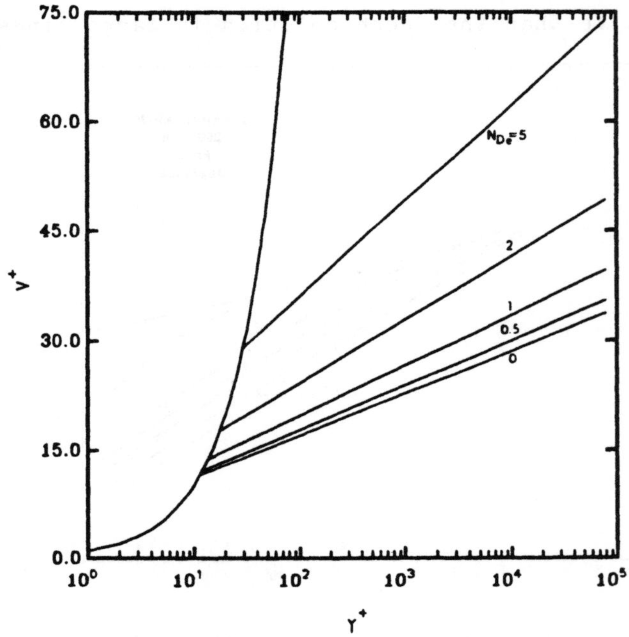

Fig. 5. Predicted Boundary Layer Velocity Profiles

References

1. Darby, R., "A Generalized Correlation For Friction Loss in Drag Reducing Polymer Solutions", Encyclopedia of Fluid Mechanics, N.P. Cheremisinoff (ed.), Gulf Publishing Co. (in press)
2. Davies, J.T., "Turbulence Phenomena", pp 51-62, Academic Press, N.Y. (1972)
3. Baker, S.J., J. Fluid Mech., 60, 721 (1973)
4. Fortuna, G. and T.J. Hanratty, CEP Symp. Ser. No. 111, 67, J.G. Savins and P.S. Virk (ed.), p. 90 (1971)
5. Darby, R., Trans. Soc. Rheol., 14, 185 (1970)
6. Tsai, C.F. and R. Darby, J. Rheol., 22, 219 (1978)
7. Argumedo, A., T.T. Tung, and K.I. Chang, J. Rheol., 22, 449, (1978)
8. Chang, H.D. and R. Darby, J. Rheol., 27, 77 (1983)
9. Bird, R.B., R.C. Armstrong, O. Hassager, "Dynamics of Polymeric Liquids", Vol. 1, Wiley & Sons, N.Y. (1977)
10. Darby, R. and H.D. Chang, AIChE J., 30, 274 (1984)

Session 10
Injection Experiments

Heterogeneous Drag Reduction in Turbulent Pipe Flow

Hans-Werner Bewersdorff

Lehrstuhl für Strömungsmechanik, University of Dortmund, F.R.G.

ABSTRACT

Experiments are reported in which a concentrated polymer solution is injected into the centre of a turbulent pipe flow. Drag reduction is obtained even if the polymer forms a liquid thread which is conveyed in the core region of the flow, i.e. no significant part of the injected polymer is present in the near-wall region. This type of drag reduction differs from that found in homogeneous solutions and seems to be due to an interaction between the polymer thread and the large-scale structure.

INTRODUCTION

Drag reduction has been found to be accompanied by a modification of the structure of turbulence. For dilute polymer solutions in pipe flow the mean velocity profile and the turbulence intensities are modified. In the velocity profile the transition layer between the viscous sublayer and the turbulent core, called the buffer layer or elastic sublayer, increases with the level of drag reduction. The turbulent core shows a parallel displacement [1,2]. The maximum of the axial turbulence intensity (normalized with the friction velocity) is shifted more and more towards the centre of the pipe with increasing drag reduction, while the turbulence intensity in the core region seems to be unchanged. The radial turbulence intensity is also damped in the buffer layer [3,4]. Because the turbulence production and dissipation takes place mainly in the buffer layer and the turbulent burst rate is reduced in the case of drag reduction [5,6], it is normally assumed that the addition of polymers only changes the structure of turbulence in the near-wall region.

To investigate this assumption experiments were carried out in which a dilute polymer solution was injected into the centre of a turbulent pipe

The Influence of Polymer Additives
on Velocity and Temperature Fields
IUTAM Symposium Essen 1984
Editor: B. Gampert

flow [7] and the pressure drop was measured downstream of the injection point. The results demonstrated that drag reduction only occurred when the polymer had dispersed into the near-wall region. If concentrated polymer solutions are injected in the same way, an elastic polymer thread is observed in the centre of the pipe and surprising drag reduction also occurs [8-10]. This type of drag reduction was called h e t e r o g e - n e o u s drag reduction by VLEGGAAR [8,9]. Because in this case the reduction in heat transfer can be correlated with the drag reduction [11], it seems that the polymer thread damps the radial momentum transfer. STENBERG [12,13] also reports experiments with a heterogeneous polymer distribution in which a polyox solution (WSR 301; c_p = 0.2% and 1%) was injected into a turbulent pipe flow. Before entering the measuring section the fluid flowed through a rotational mixer. The onset point of drag reduction was shifted to higher Reynolds numbers as the mixer speed is increased. Flow visualization showed the existence of polymer threads in the measuring section.

All these experimental results demonstrate that drag reduction can also be obtained for a heterogeneous polymer distribution. It may be concluded that the mechanism of this type of drag reduction is different from that of a dilute polymer solution. The aim of the present work is to obtain more information about the characteristic features of this type of drag reduction by central injection of concentrated polymer solutions into the centre of a turbulent pipe flow.

EXPERIMENTAL SET-UP

A description of the experimental set-up can be found in [10] and [14]. For determing the drag reduction from pressure drop measurements the test pipe of acrylic glass (l = 16 m, d = 50 mm) was provided with 14 pressure taps at 1 m intervals. A sufficient entrance pipe length was provided to ensure fully developed flow in the measuring section. The flow rate was estimated by a flowmeter. The polymer solution was injected with a needle into the centre of the pipe and a dosing pump was used to provide the desired flow rate. The diameter of the needle was chosen so that the bulk velocities of the water and the polymer solution were nearly equal. Colouring the polymer solution provided the possibility of looking for turbulent diffusion. The polymer, Separan AP30, was suspended in isopropanol. This suspension was added to the desired amount of water, which was stirred at low speed for 48 hours. At a distance of 12 m from the injection point the concentration profile measurements were carried out.

Rheology of the injected polymer solutions

The rheological behaviour of the injected polymer solutions was investigated by the Mechanical Spectrometer which has an open cone and plate arrangement. The measurements were restricted at lower shear rates by the measuring range of the force gauges and at higher shear rates by the deformations of the free gap surface. In the whole measuring range and for all polymer concentrations the first normal-stress difference was higher than the shear stress at the same shear rate (fig. 1). This material

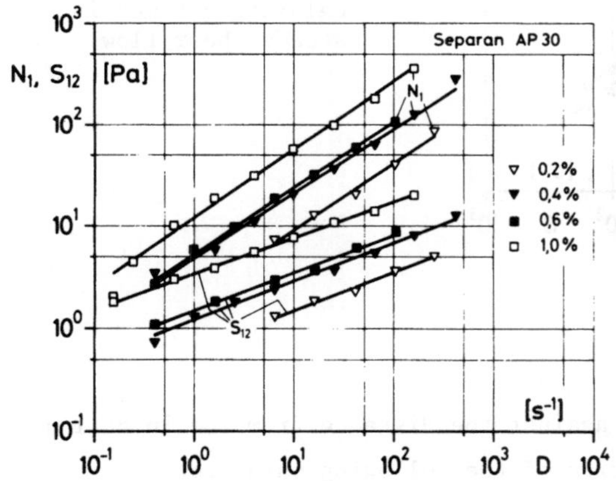

Figure 1: Shear stress and first normal stress difference for polyacrylamide solutions.

behaviour is typical for viscoelastic fluids. Furthermore the slope of the first normal-stress difference was nearly twice the slope of the shear stress. This behaviour was found in the whole concentration range used.

An important quantity for the interaction between polymer solutions and the turbulent flow is the relaxation-time behaviour of the polymer solution. The experimental results of MICHELE [15] demonstrate that in the upper (Ostwald-de-Waele-)power-law region the measured relaxation times from start up tests are identical with those calculated from the measurements of the first normal-stress difference and the shear stress with a Maxwell model. Therefore "effective" relaxation times λ were calculated by

$$\lambda = \frac{N_1}{2\,S_{12}\,D} \quad , \tag{1}$$

where N_1 is the first normal-stress difference, S_{12} the shear stress, and D designates the shear rate. In a double-logarithmic plot the relaxation times decreased linearly with the shear rate (fig. 2). Furthermore it can be seen that the concentration has only little influence on the effective relaxation time.

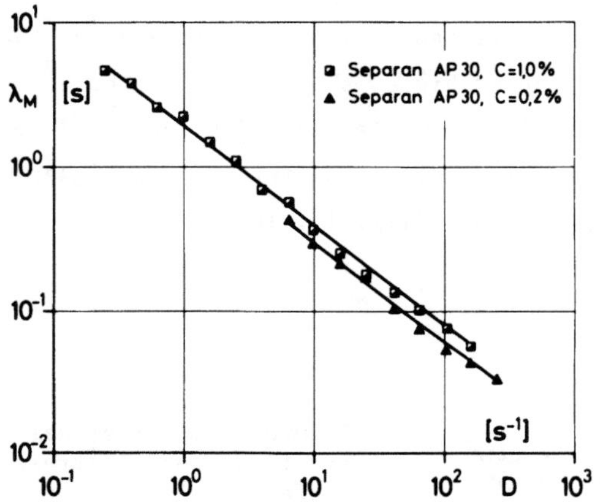

Figure 2: Effective relaxation times in steady shear flow.

LOCAL DRAG REDUCTION

If drag reduction DR is plotted against the distance from the injection point, it is found to be a function of the following parameters

$$DR = f\,(Re,\ c_R,\ c_p,\ d,\ P) \tag{2}$$

where Re is the Reynolds number, c_R the polymer concentration averaged over the cross-section of the pipe, c_p the concentration of the injected polymer solution, d the pipe diameter and P designates the polymer type. By flow visualization it was found that the injected polymer solution was not dispersed by the turbulent flow provided a certain value of c_p (0.25% for SEPARAN AP30) was exceeded. The injected polymer solution formed a long thread, which remained intact down the entire length of the pipe, although it was deformed by the turbulent flow, sometimes being stretched or forming loops, as shown in [14] and [16].

Although there is no important mixing of the injected polymer solution, a remarkable drag reduction is found. Drag reduction as a function of the distance from the injection point shows an increase at the beginning and a nearly constant value at the end of the measuring section (fig. 3).

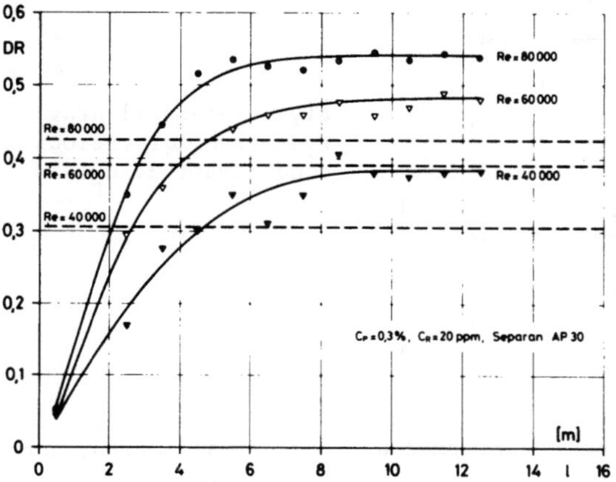

Figure 3: Drag reduction versus distance from the injection point. The dashed line shows the drag reduction of a homogeneous polymer solution.

By flow visualization of a coloured polymer solution it was observed that in the first third of the measuring section the amplitude of the radial oscillations of the polymer thread was increased, while the frequency was decreased. Comparing these ultimate values of drag reduction found in the injection experiments with those of a homogeneous polymer solution (c_R = const) shows large differences in the measured friction factor (fig. 3). These differences increase with the Reynolds number. For c_p = 0.25% and 0.3% the polymer thread disintegrated into small polymer strings that were similar to those obtained by schlieren photography in [13]. These polymer strings were distributed over the whole cross-section of the pipe. However, for these two values of c_R the differences from a homogeneous polymer solution are the largest.

If the concentration of the injected polymer solution is decreased, drag reduction increases much more quickly from the injection point. This may be explained by the fact that no polymer threads or strings are observed. Therefore the polymer is transported very quickly to the wall region. The ultimate values of drag reduction are nearly identical with those of homogeneous polymer solutions. The results for $c_p \leq 0.2\%$ are in agreement with those published by Mc COMB [17, 18]. From fig. 4 it may be concluded that for c_p = 0.2% the polymer distribution is homogeneous and for c_p = 0.25% it is remarkably heterogeneous. This can be recognized by the different kinds of increase of the local drag reduction. For c_p = 0.25% a higher ultimate value of drag reduction is also found.

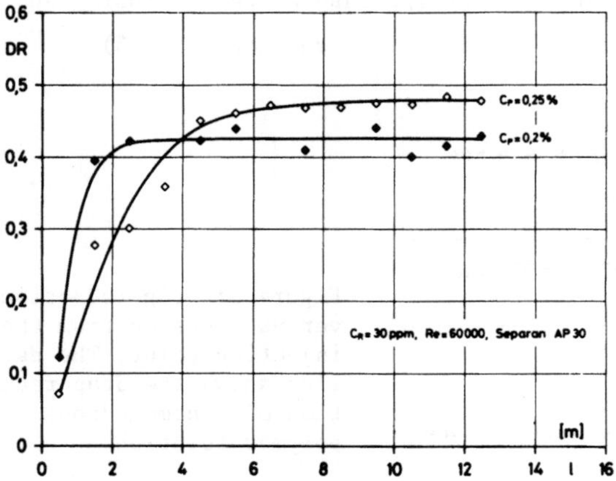

Figure 4: Local drag reduction for various concentrations c_p.

To check these results, which were confirmed by flow visualization, concentration profiles were measured 12 m downstream of the injection point. A needle was used to take samples over the cross-section of the pipe. Because the samples were colourless reagents were added to achieve a turbidity which was necessary for photometric measurements. The concentration of the sample was estimated from a master curve. The results (fig. 5) are in agreement with those obtained by flow visualization.

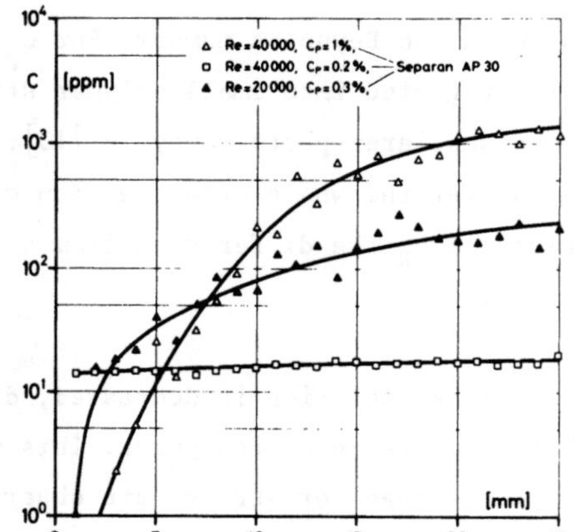

Figure 5: Concentration profiles.

For c_p = 0.2% the polymer distribution is nearly homogeneous, whereas for c_p = 0.3% the polymer concentration is only 1 ppm at a wall distance of 1 mm. The concentration profile for c_p = 1% demonstrates the existence of a near-wall region in which almost no polymer can be detected. Only at a wall distance of 2 mm does the polymer concentration reach 2 ppm.

If Simpson's rule is used to calculate an averaged polymer concentration

$$\bar{c} = \frac{2}{R^2} \int_0^R c(r)\ r\ dr\ , \qquad (3)$$

where R is the pipe radius, a good agreement with the polymer concentration regulated by the dosing pump is obtained for c_p = 0.2%. In contrast, for c_p = 0.3% and 1% much higher averaged concentrations are found. The reasons for this behaviour is that the polymer threads wind around the needle and so are sucked in with the sample. This assumption was confirmed by flow visualization with a coloured polymer solution. Therefore the concentration profile measurements are only valid in the near-wall region for higher values of c_p where no polymer thread can be sucked in.

If the ultimate values of drag reduction from the injection experiments are used to calculate a friction factor, the friction behaviour as a function of Reynolds number can be compared for the various injection experiments (fig. 6). All values measured at one concentration c_p lie on a straight line. The onset-point for drag reduction depends on the concentration of the injected polymer solution. Especially at high Reynolds numbers, the differences between the injection experiments and the homogeneous polymer solution are remarkable. For all concentrations c_p a higher level of drag reduction was obtained than for a homogeneous polymer solution.

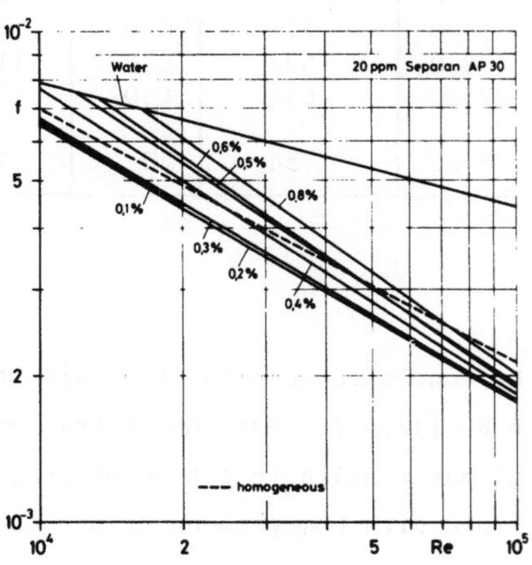

Figure 6: Friction behaviour for homogeneous and heterogeneous drag reduction.

Table 1 gives a review of the data of the friction behaviour in the injection experiments. It includes the constants A and B of the friction law

$$\frac{1}{\sqrt{f}} = A \log (Re \sqrt{f}) + B \quad . \tag{4}$$

In addition according to BERMAN [19] the product of the velocity gradient at the wall u_τ^2/ν for the onset-point and the relaxation time at the same shear rate (s. fig. 2) was calculated. The product does not have a characteristic value like that found by BERMAN for homogeneous polymer solutions. The Reynolds number in all experiments is based on the solvent kinematic viscosity ν.

Table 1:

c_p [%]	c_R [ppm]	A	B	Onset Re	u_τ^2/ν [s^{-1}]	λ [s]	$\lambda u_\tau^2/\nu$
0.1	20	10.6	-18.5	6000	65	0.08	5.2
0.2	20	10.1	-17.1	5400	54	0.09	4.8
0.3	20	10.1	-17.1	5800	61	0.1	6.1
0.4	20	13.1	-27.1	9400	142	0.055	7.8
0.5	20	15.3	-34.8	12000	218	0.04	8.7
0.6	20	17.3	-41.2	13400	264	0.035	9.2
0.8	20	18.2	-45.1	16200	368	0.03	11
1	20	25.7	-72.6	27600	934	0.018	16.8
homogeneous	20	8.6	-13,2	6000	65	?	?
0.2	50	18.3	-40.9	6100	67	0.07	4.7
0.3	50	17.1	-37	5600	57	0.1	5.7
0.4	50	22.7	-57.6	11800	211	0.04	8.4
0.5	50	22.7	-58.3	12400	230	0.035	8.1
0.6	50	22.5	-57.7	13600	271	0.03	8.1
0.8	50	21.2	-55.8	19800	523	0.022	11.5
1	50	21.0	-55.9	22500	654	0.02	13.1
homogeneous	50	14.1	-27.1	5400	54	?	?

Conclusions

From the experimental results it may be concluded in connection with the results of VLEGGAAR [8,9,11] and STENBERG [12,13], that the centre-line injection of concentrated polymer solutions results in a type of drag reduction, that differs from that obtained with homogeneous dilute polymer solutions. Therefore, it seems that the mechanism of these two types of drag reduction are different. For homogeneous polymer solutions it is normally assumed that the increased extensional viscosity influences the burst phenoma in the near-wall region, which results in a reduced burst rate and so reduces the energy dissipation. If the injection of concentrated polymer solutions results in a larger degree of drag reduction than in the homogeneous case it must be concluded that the small amount of polymer which is in the near-wall region is not dissolved in molecular form but rather is present as polymer threads or associates. This was also found by wall injection experiments of GOREN [20] and RAMU and TULLIS [21,22]. A theory for "heterogeneous dilute" polymer solutions was developed by HINCH [23]. Using the assumption of a minimum length/diameter ratio of the polymer strands, polymer networks can be built. They have only little influence on the flow behaviour at low extensional rates but at high extensional rates strain hardening dominates, which influences the flow more than for homogeneous polymer solutions. With this explanation the results of the injection experiments with c_p = 0.1% and 0.2% and the results of STENBERG [12, 13] may be explained. In these experiments the polymer distribution was nearly homogeneous, and the drag reduction was only a little higher than in the homogeneous case.

However, this explanation is not able to describe the results for $c_p >$ 0.3%. In these experiments the interaction between the polymer thread and large-scale structures should be the dominating effect. This may be concluded from the data of Table 1. The characteristic times of the flow in the wall region ν/u_τ^2 and of the polymer solution are not identical. Furthermore velocity profile measurements [14,16] showed an increased slope in the core region in comparison to a Newtonian fluid or a homogeneous polymer solution, i.e. Prandtl's mixing length and therefore the turbulent momentum transfer is decreased. A characteristic time of the large-scale structure in a turbulent pipe flow can be represented by the ratio of the pipe diameter to the mean velocity. For the onset conditions this time lies in the range of 0.1 to 0.5 s. The estimated effective relaxation times in the shear rate range from 5 to 50 s^{-1} are of the same order. This also indicates an interaction between the big eddies and the polymer thread. On the other hand a direct interaction between the polymer thread and the dissipating little eddies hardly seems possible. A disturbance of the movement of the big eddies can result in a decreased energy dissipation, because the big eddies transport the fluctuation energy to the dissipating small eddies. To obtain more information about the proposed mechanism LDA-measurements of the large-scale structure are planned.

REFERENCES

[1] VIRK, P.S., Drag reduction fundamentals, AIChE J., 21 (1975), 625-656.

[2] ELATA, C., J. LEHRER, A. KAHANOVITZ, Turbulent shear flow of polymer solutions, Israel J. Techn., 4 (1966), 87-95.

[3] LOGAN, S.E., Laser velocimeter measurement of Reynolds stress and turbulence in dilute polymer solutions, AIAA J., 10 (1972) 962-964.

[4] SCHÜMMER, P., W. THIELEN, Structure of turbulence in viscoelastic fluids, Chem. Eng. Commun., 4 (1980) 593-606.

[5] ACHIA, B.U., D.W. THOMPSON, Structure of the turbulent boundary layer in drag-reducing pipe flow, J. Fluid Mech., 81 (1977) 439-464.

[6] MIZUSHINA, T., H.USUI, Reduction of eddy diffusion for momentum and heat in viscoelastic fluid flow in a circular tube, Phys. Fluids, 20 (1977) S100-S108.

[7] WELLS, C.S., J.G. SPANGLER, Injection of a drag-reducing fluid into turbulent pipe flow of a Newtonian fluid, Phys. Fluids, 10 (1967) 1890-1894.

[8] VLEGGAAR, J., Studies on drag reduction, Thesis, University of Amsterdam (1973).

[9] VLEGGAAR, J., M. TELS, Drag reduction by polymer threads, Chem. Eng. Sci., 28 (1973) 965-968.

[10] BEWERSDORFF, H.W., K. STRAUSS, Turbulente Diffusion und Widerstandsverminderung in Rohrströmungen, Rheol. Acta, 18 (1979) 104-107.

[11] VLEGGAAR, J., M. TELS, Heat transfer in heterogeneous drag reduction system, Int. J. Heat and Mass Transfer, 16 (1973) 1629-1632.

[12] STENBERG, L.-G., T. LAGERSTEDT, O. SEHLEN, E.R. LINDGREN, Mechanical mixing of polymer additive in turbulent drag reduction, Phys. Fluids, 20 (1977), 858-859.

[13] STENBERG, L.G., T. LAGERSTEDT, E.R. LINDGREN, Polymer additive mixing and turbulent drag reduction, Phys. Fluids, 20 (1978) S276-S279.

[14] BEWERSDORFF, H.W., Heterogene Widerstandsverminderung bei turbulenten Rohrströmungen, Rheol. Acta, 23 (1984).

[15] MICHELE, J., Zur Rheometrie viskoelastischer Fluide mit der Kegel-Platte-Anordnung, Rheol. Acta, 17 (1978) 42-68.

[16] BEWERSDORFF, H.W., Effect of centrally injected polymer thread on drag in pipe flow, Rheol. Acta, 21 (1982) 587-589.

[17] Mc COMB, W.D., L.H. RABIE, Development of local drag reduction due to nonuniform polymer concentration, Phys. Fluids, 22 (1979) 183-185.

[18] Mc COMB, W.D., L.H. RABIE, Local drag reduction due to injection of polymer solutions into turbulent flow in a pipe, AIChE J., 28 (1982) 547-565.

[19] BERMAN, N.S., Flow time scales and drag reduction, Phys. Fluids, 20 (1977) S168-S174.

[20] GOREN, Y., J.F. NORBURY, Turbulent flow of dilute aqueous polymer solutions, Trans. ASME, J. Basic Eng., 89 (1967) 814-822.

[21] TULLIS, J.P., K.L.V. RAMU, Drag reduction in developing pipe flow with polymer injection, Int. Conf. on Drag Reduction, Cambridge (1974) Paper G3.

[22] RAMU, K.L.V., J.P. TULLIS, Drag reduction and velocity distribution in developing flow, J. Hydronautics, 10 (1976) 55-61.

[23] HINCH, E.J., C. ELATA, Heterogeneity of dilute polymer solutions, J. Non-Newtonian Fluid Mech., 5 (1979) 411-425.

Annular Injection of Concentrated Polymer Solutions into the Wall Region of a Turbulent Pipe Flow

Bodo Frings

Lehrstuhl für Strömungsmechanik, University of Dortmund, F.R.G.

ABSTRACT

The results of an experimental study of the injection of concentrated polymer solutions into the near-wall region of a turbulent pipe flow are presented. Local drag reduction and friction behaviour was obtained by measuring pressure differences over a test section 13 m in length. The results show a different type of behaviour than that obtained with homogeneous solutions as well as for centre-line or direct wall injection.

INTRODUCTION

A widely accepted hypothesis of the mechanism responsible for drag reduction is that the microscale eddies in the bufferzone of the turbulent boundary layer are suppressed. It is thought that the macromolecules (injected or pre-mixed) affect the sublayer instability which produces jet-like vortices, the so-called "bursts", which erupt into the boundary layer and lead to the production of much turbulent energy. Recent experiments have shown [1] that the polymer molecules seem to act as a "barrier" restricting communication across the boundary layer so that the fluid near the wall flows in a more laminar manner [2].

The aim of this paper is to analyse the effectiveness of injecting a circumferential ring of relatively concentrated viscoelastic fluid into a turbulent boundary layer in a pipe flow. In contrast to wall injection experiments, the polymer solution was injected into the buffer layer, in the region that shows the maximum energy dissipation rate. The presence of such a closed ring of polymer should have a significant influence on the larger eddies present in turbulent pipe flow and enable the processes responsible for drag reduction to be understood more fully.

The Influence of Polymer Additives
on Velocity and Temperature Fields
IUTAM Symposium Essen 1984
Editor: B. Gampert

EXPERIMENTAL SET-UP

The experimental system is shown schematically in figure 1. Experiments were performed in a straight, smooth-walled acrylic glass pipe 50 mm in diameter. In order to determine the drag reduction from pressure drop measurements the test pipe was provided with 14 pressure taps positioned at 1 m intervals down the pipe. A sufficient entrance pipe length was provided to ensure fully developed flow at the injection point. The water was pumped from a reservoir tank by a Mohno pump, the speed of which was controlled by a microcomputer to maintain a constant Reynolds (Re) number, during the experiments (between 10^4 and 10^5). The microcomputer continously measured the temperature of the water with a resistance thermometer and the discharge with a turbine wheel flow meter.

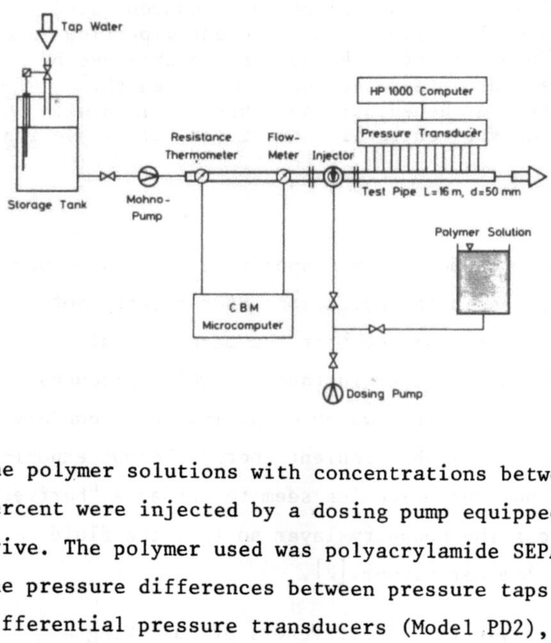

Fig. 1 Experimental set-up.

The polymer solutions with concentrations between 0.2 and 1.0 weight percent were injected by a dosing pump equipped with a variable speed drive. The polymer used was polyacrylamide SEPARAN AP45 (M_w = 4-5 x10^6). The pressure differences between pressure taps were measured with Philips differential pressure transducers (Model PD2), the outputs of which were connected to a computer (Hewlett Packard 1000) to enable direct analysis of the experimental data.

INJECTOR SYSTEM

The design criteria for the injector were that it should be able to inject an annular ring of polymer in the near-wall region at various distances from the wall with various thicknesses (i.e. gap widths). Furthermore, the degradation of the polymer in the injector and the disturbance of the flow by the injector were to be as small as possible. Figures 2 and 3 of the injector show that the polymer solution was supplied to the injector by an equalizing device 1, which split the polymer feed into four singular streams to ensure a nearly uniform polymer supply around the whole injector slit. In addition, after leaving the annular chamber 2, the polymer solution passed through a homogenizer 3. At the downstream end of the injector two exchangeable rings 4 were used to create the annular slit with exact dimensions. The gap width was adjustable between 0.1 and 0.5 mm. Varying the distance of the injector slit from wall required the inset consisting of the parts 2,3,4 to be exchanged.

Fig. 2 Polymer injector

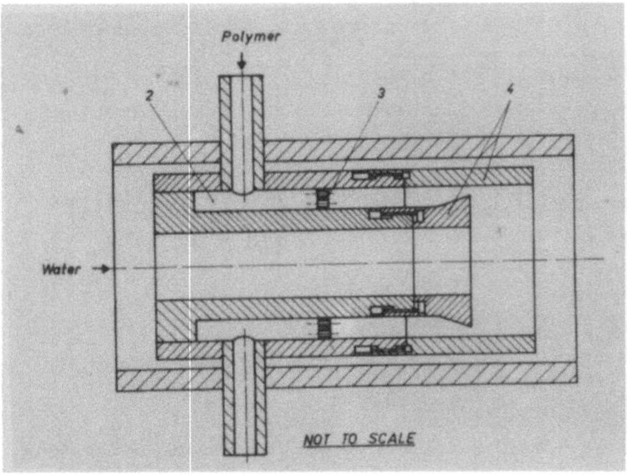

Fig. 3 Details of the polymer injector.

EXPERIMENTAL RESULTS

For investigating the behaviour of the injected polymer some of the polymer were dyed (before injection) with crystal violet, which has been shown not to influence the rheological properties of PAA-solutions.

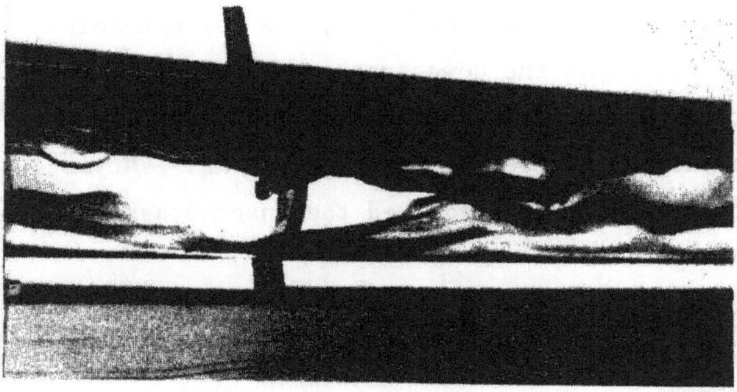

(a) Re = 5000, $\dot{V}$ = 22 [cm^3/s]

(b) Re = 30000, $\dot{V}$ = 8.7 [cm^3/s] x/d = 5

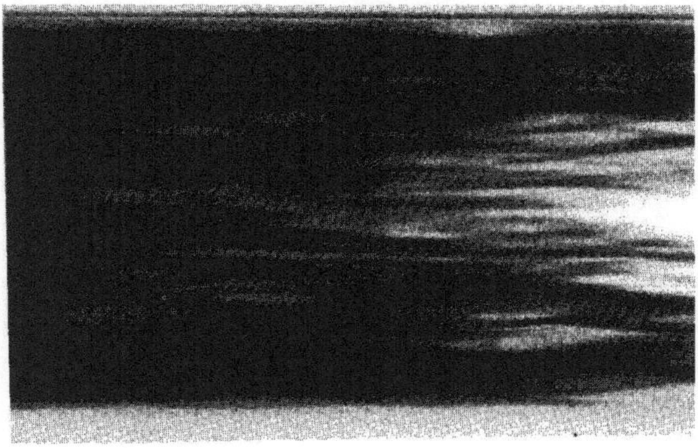

(c) Re = 30000, $\dot{V}$ = 44 [cm^3/s] x/d = 5

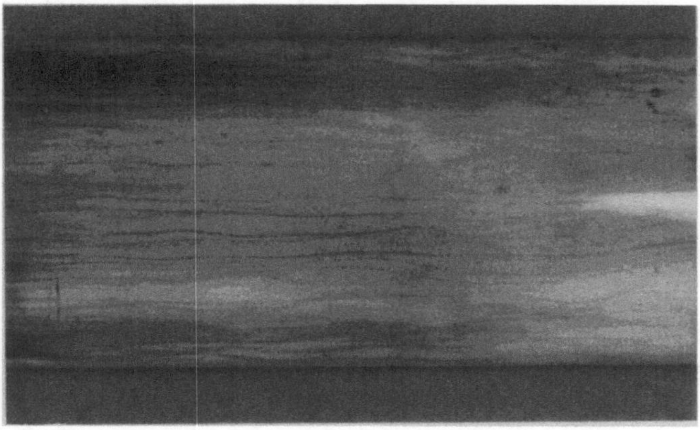

Re = 50000,
$\dot{V}$ = 8.7 [cm^3/s]
x/d = 5

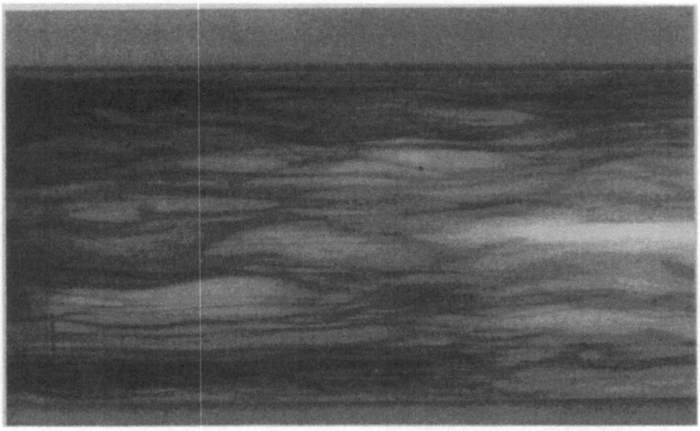

Re = 50000,
$\dot{V}$ = 44 [cm^3/s]
x/d = 5

Fig. 4 Injection of dyed polymer solution.

Figure 4 shows the behaviour of the injected solution for various Re-numbers and polymer injection rates $\dot{V}$. The distance of the injector from the wall was 2.5 mm, the gap width 0.4 mm. The polymer streaks are seen to become thinner as the Re-number increases and it is evident that there is no contact of the injected polymer with the wall. For Re = 3×10^4 and 5×10^4 a closed annulus of polymer was obtained.

In all the experiments shown in figure 4 the injected polymer seemed to remain intact over the entire length of the test section and it is believed that they formed an elastic network. This conclusion was confirmed by an experiment in which the discharge of water was abruptly stopped. For several meters from the injection point the polymer strands were seen to relax like a stretched rubber band.

Although no significant mixing of the injected polymer solution was noticed in the visual observations, a large amount of drag reduction was obtained. In the following, drag reduction is defined as

$$DR = 1 - \frac{\Delta P_P}{\Delta P_W}$$

where ΔP_P is the pressure loss with polymer present and ΔP_W is the pressure loss for normal water flow at the same flow rate. The drag reduction was found to depend on the Reynolds number (here based on the viscosity of water), the distance from the injection point x, the averaged polymer concentration c_R, the concentration of the injected polymer solution c_P, the pipe diameter d and the injector gap width s as well as the type of polymer used.

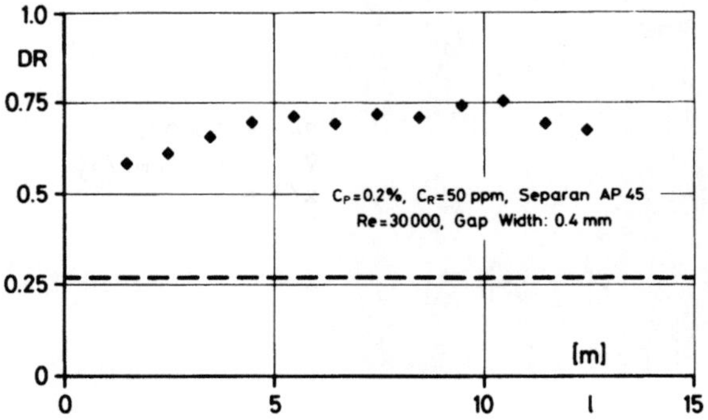

Fig. 5

Local drag reduction. The dashed line indicates the drag reduction of a homogeneous polymer solution.

It is evident from the plot of drag reduction versus distance from injection point in figure 5 that even at the beginning of the test section (x/d = 30) there is significantly more drag reduction for near-wall injection than for pre-mixed solutions with the same averaged concentration. The drag reduction reaches a quasi-stationary value at x/d = 70 and then remains nearly constant down the following eight meters of the test section. Qualitatively similar behaviour was obtained at all other experimental conditions described below.

Because the possibility of inlet and outlet flow effects at the beginning and the end of the test section could not be excluded the following friction factor versus Re-number plots were evaluated only from pressure loss measurements between x/d = 70 and x/d = 210.

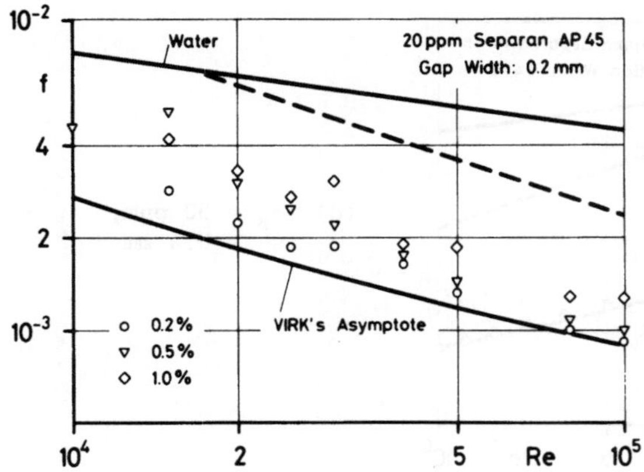

(a) c_R = 20 ppm, s = 0.2 mm.

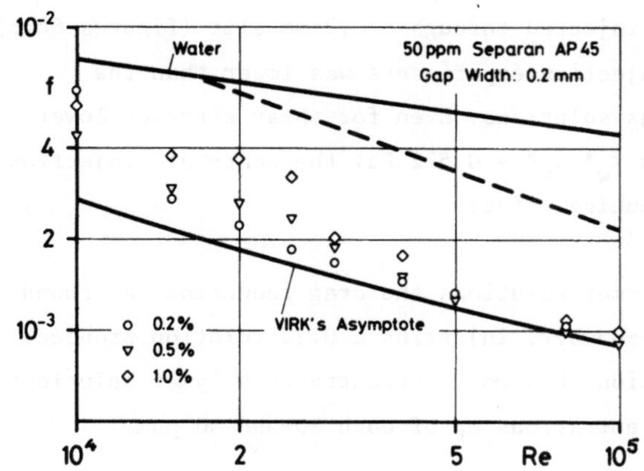

(b) c_R = 50 ppm, s = 0.2 mm.

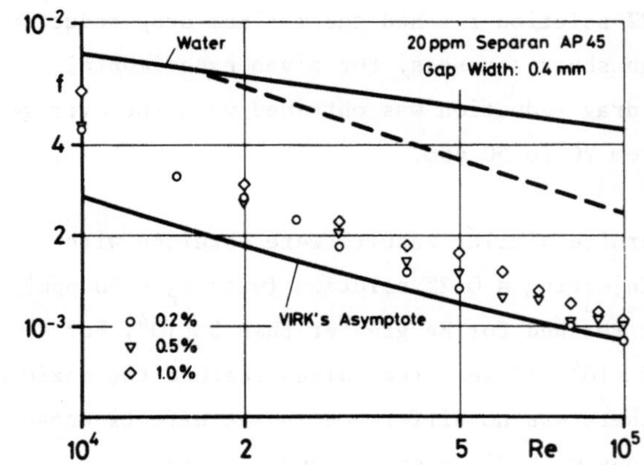

(c) c_R = 20 ppm, s = 0.4 mm.

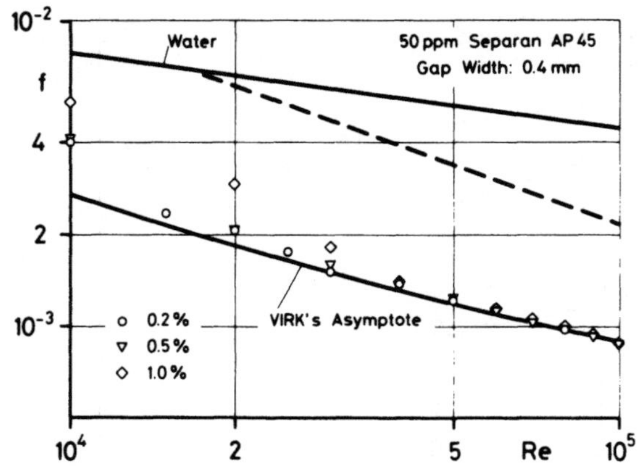

(d) c_R = 50 ppm, s = 0.4 mm.

Fig. 6 Friction behaviour for homogeneous (dashed lines) and heterogeneous drag reduction.

When the polymer solution was injected through a 0.2 mm slit (figures 6a,b) the friction factor for the injection experiments was lower than the values obtained for homogeneous solutions. Even for shear stresses lower then the critical shear stress τ_w^* (τ_w^* = 0.572 Pa) the near-wall injection produced a remarkable drag reducing effect.

In all investigations with polymer solutions the drag reduction was found to increase with increasing Re-number. Injecting a 0.2% solution produced a larger amount of drag reduction than more concentrated polymer solutions. This was true for average concentrations c_R of both 20 and 50 ppm.

With increasing Re-number all data approached Virk's asymptote and the differences in effectiveness between the three polymer solutions was reduced, although only the 0.2% solution reached the maximum drag reduction asymptote for sufficiently high shear stresses. For given experimental conditions a larger amount of drag reduction was obtained when the average concentration was increased from 20 to 50 ppm.

Larger drag reduction but otherwise similar results were obtained with a 0.4 mm gap (figures 6c,d). Injecting a 0.2% solution (with c_R = 50 ppm) enabled Virk's asymptote to be reached for Re greater than 3×10^4. For Reynolds numbers larger than 5×10^4 all measured values reached the maximum drag reduction asymptote and there was no difference in the effectiveness of the injected polymer solutions with different concentrations.

These results indicate that there is a limiting lower value for the thickness of the injected polymer annulus (gap width) to ensure effective drag reduction. It should be mentioned that injecting 0.5% solutions through a 0.4 mm gap reaching an average concentration of 50 ppm, Virk s value of maximum drag reduction was reproducible exceeded from x/d = 170 to 230 for Re = 10^5.

CONCLUSIONS

The annular injector described is able to place concentrated polymer solutions in the near-wall region. Visual observations showed that there was no contact of the injected polymer with the wall. In spite of this strong drag reduction was observed down the entire length of the pipe.

The drag reduction was found to vary with the amount of injected polymer solution, the Reynolds number and the gap width of the injector. There was only a little influence of the distance from the injection point. In contrast to wall injection experiments [3] all the experiments described here showed drag reduction that was significantly larger than the drag reduction obtained in homogeneous solutions with the same averaged concentration, even at large distances from the injection point. Whereas in the centre-line injection of concentrated polymer solutions [4] an asymptotic value of drag reduction is only reached a considerable distance downstream from the injection point, for near-wall injection a nearly constant value is reached very quickly. This may be due to the design of the injector which tends to stretch the polymer molecules before they are injected into the pipe [5]. The maximum drag reduction obtained was about 84% meaning that Virk's asymptote was exceeded by 4 percent for a distance of 3m.

More detailed investigations of the influence of the injection of polymer in the near-wall region of a pipe flow are still in progress.

ACKNOWLEDGEMENT

The author thanks Prof. H. Giesekus, H.W. Bewersdorff and D.H. Fruman for their invaluable advice and interest.

REFERENCES

[1] Kowalski, T.: "Turbulence suppression and viscous drag reduction by non-Newtonian additives". Trans. R.I.N.A., 110 (1968) 207 - 219.

[2] Kowalski, T., E. Brundrett,: "Macromolecular entanglement hypothesis in drag reduction flows". Int. Conf. on Drag Reduction, Cambridge, (1974) Paper C1.

[3] Tullis, I.P., K.L.V. Ramu : "Drag reduction in developing pipe flow with polymer injection". Int. Conf. on Drag Reduction, Cambridge, (1974) Paper G3.

[4] Bewersdorff, H.W.: "Effect of a centrally injected polymer thread on drag in pipe flow". Rheol. Acta, 21 (1982) 587 - 589.

[5] Hinch, E.J., C. Elata : "Heterogeneity of dilute polymer solutions". Journal of Non-Newtonian Fluid Mechanics, 5 (1979) 411 - 425.

Turbulence Structure of a Developing Duct Flow with Near-Wall Injection of Drag Reducing Polymers

ČEDO MAKSIMOVIĆ

Faculty of Civil Engineering
University of Belgrade
Belgrade, Yugoslavia

SUMMARY

The paper deals with results of an experimental study on drag reduction by adding polymer solution into a developing flow within a rectangular duct. The length of the duct was 2.5 m. The idea that considerable savings on polymers can be obtained by this method is tested.

Only the results of injection into top and bottom and side walls of the duct will be presented whereas the results on circumferential injection will be shown elsewhere.

The measurements of two velocity components were carried out by an LDV in five cross sections downstream from the point of injection.

The local values of the drag reduction, mean values of the velocities and turbulence intensities are presented along with the power spectrum density functions. It is shown that extremely high local values of drag reduction can be obtained by this method with very low average concentration of polymer solution.

Introduction

The effect of uniformly distributed polymer solution on drag reduction is well documented, Virk [1], Hoyt [2], as far as mean velocity and turbulence structure are concerned. An idea that, at least on short distances, significant drag reduction can be obtained by having the required concentration of polymers only in the near wall zone attracted many researchers:(Latto et al [3], Shieh et al [4], Tullis et al [5], for example), since it suggests much lower consumption of polymers.

The present experimental study follows the same idea, but it is aimed to provide more insight not only into drag reduction value, but also into structure of turbulence.

The water flow within a rectangular duct is used for investigations on the effect of near wall injected polymer solution on development and characteristics of boundary layer, comprising both bulk mean flow characteristics and turbulence structure. The author of the study had in mind possible

The Influence of Polymer Additives
on Velocity and Temperature Fields
IUTAM Symposium Essen 1984
Editor: B. Gampert

application of the method in engineering practice, Maksimović [6], in cases where relatively short ducts (length approximately equal to 100 or 200 diameters) are subject to peak discharges of short duration.

Experimental Rig and Procedure

A schema of the rig is shown in Fig. 1. The rectangular duct (width 29 mm hight 100mm) was supplied by constant discharge of ordinary laboratory water. The downstream level was kept constant and the upstream one was formed by itself depending on total energy losses in the duct. The details of the entrance portion of the duct are shown in Fig. 2, since it turned out that the geometry of the inlet had some effect on the behaviour of the flow within the duct.

Two working fluids were used:
- water with injection of constant discharge (q_p) of POLYOX WSR 301 solution (C_p= 1000 p.p.m. by weight) on top and bottom of the duct;
- water with top and bottom injection of dyed water (q_w=q_p) in order to have similar entrance conditions in both cases. In both cases RODAMIN B dye was used as a tracer.

The following quantities were measured:
a) total energy losses and drag reduction
b) two velocity components (separately) in a vertical plane of duct's symetry (plane x_1,x_2)
c) pressure fluctuations on the bottom wall
d) mean values of point concentration of polymers (by sample withdrowing)

The dynamic characteristics of velocity and pressure fluctuations were determined by coupling the analog output signal from the amplifier (for pressure) and from frequency tracker (for velocity) with a digital computer for final processing.

In this paper only the results of local values of polymer drag reduction, mean values of velocity and turbulence structure will be presented.

The results of pressure and concentration measurements are presented elsewhere, Maksimović [7] and [8].

Some preliminary results on turbulence structure are presented by Maksimović [9] and Maksimović and Philipson [10] and complete results are presented by, Maksimović [11].

Local Values of Drag Reduction

Local values are obtained by processing the mean values of piezometric levels. Distances between piezometric tappings were 10 cm in the zone of maximum drag reduction and 20 cm in the rest of the duct.

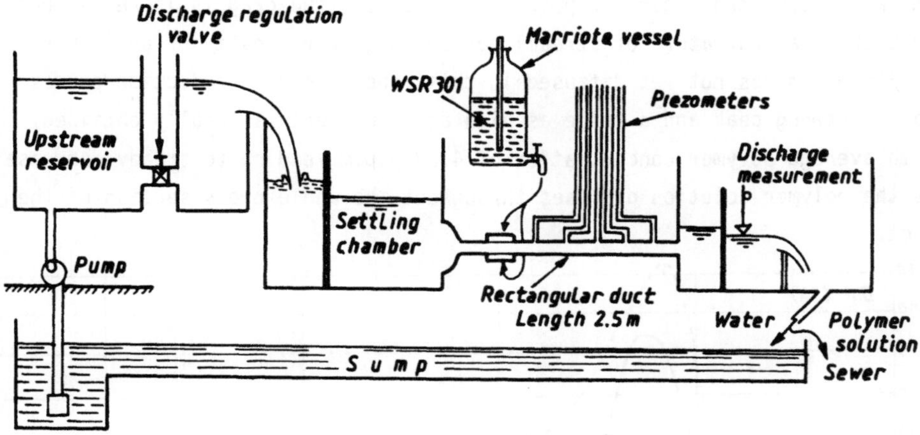

Fig. 1. Experimental rig

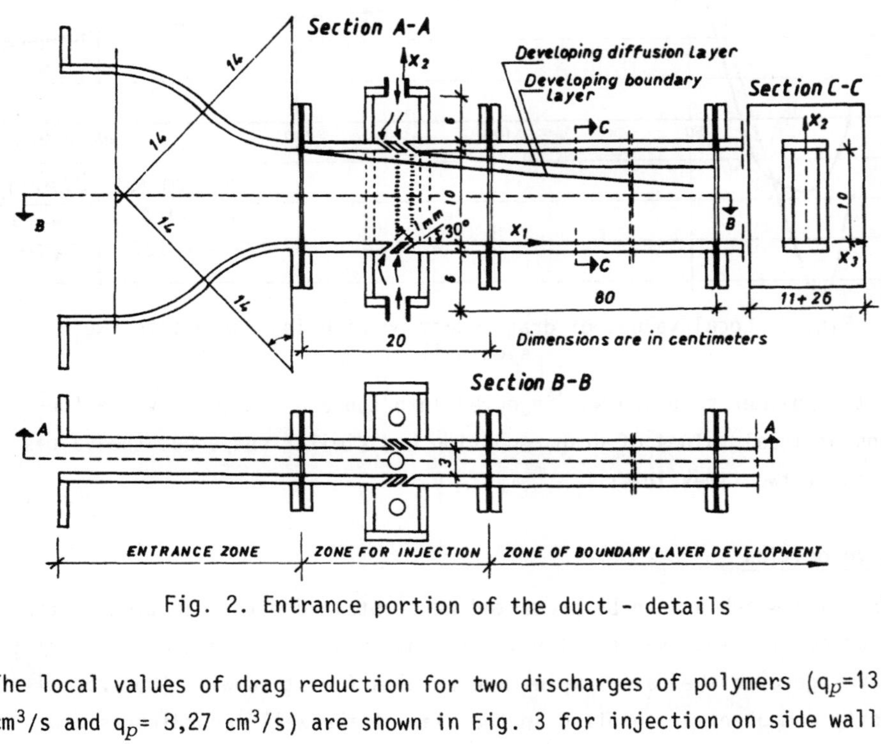

Fig. 2. Entrance portion of the duct - details

The local values of drag reduction for two discharges of polymers (q_p=13.8 cm^3/s and q_p= 3,27 cm^3/s) are shown in Fig. 3 for injection on side wall slots. The average concentration of polymers ($\overline{C}_p$) in these two cases was 4.23 p.p.m and 1.00 p.p.m respectively since the discharge of water was 3.26 l/s and the concentration of polymers in injected solution was 1000 p.p.m.

It can be seen that near the point of injection, the drag is increased due to both local curvature of streamlines and higher viscosity of polymer solution which has not yet diffused in this zone. The drag reduction curves have a strong peak and a value as high as 92 percent is locally obtained with average polymer concentration of 4.23 p.p.m leading to steady decrease as the polymer solution diffuses throughout the whole cross section of the duct.

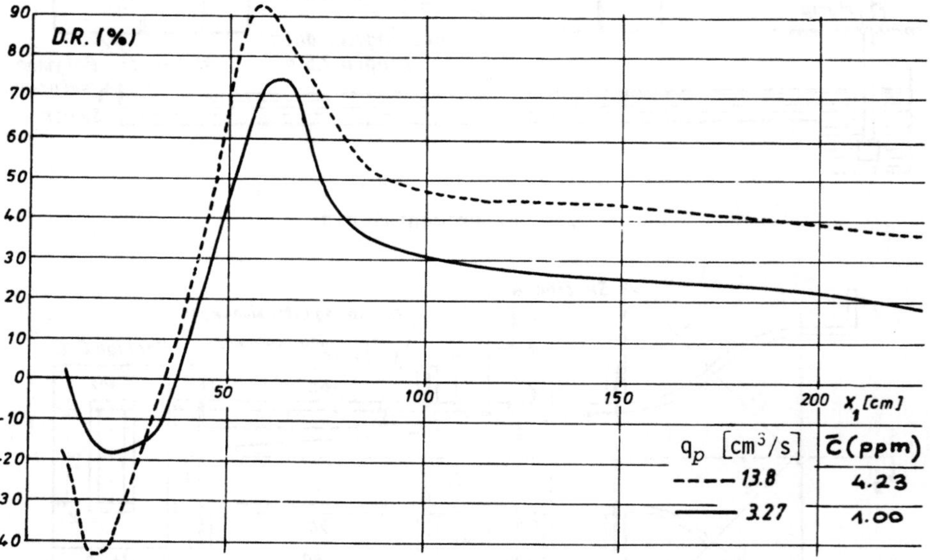

Fig. 3. Local values of drag reduction with injection into two side walls

When the polymer solution was injected into top and bottom only the functions of local values of drag reduction had similar shape but the values were about two times lower.

Mean Velocities

The results which are to be presented were obtained with injection in top and bottom. The mean values of velocity components were obtained by taking the average over the data sample. The results for streamwise (horizontal) velocity component are plotted in the "law of the wall" coordinates and presented in Fig. 4.

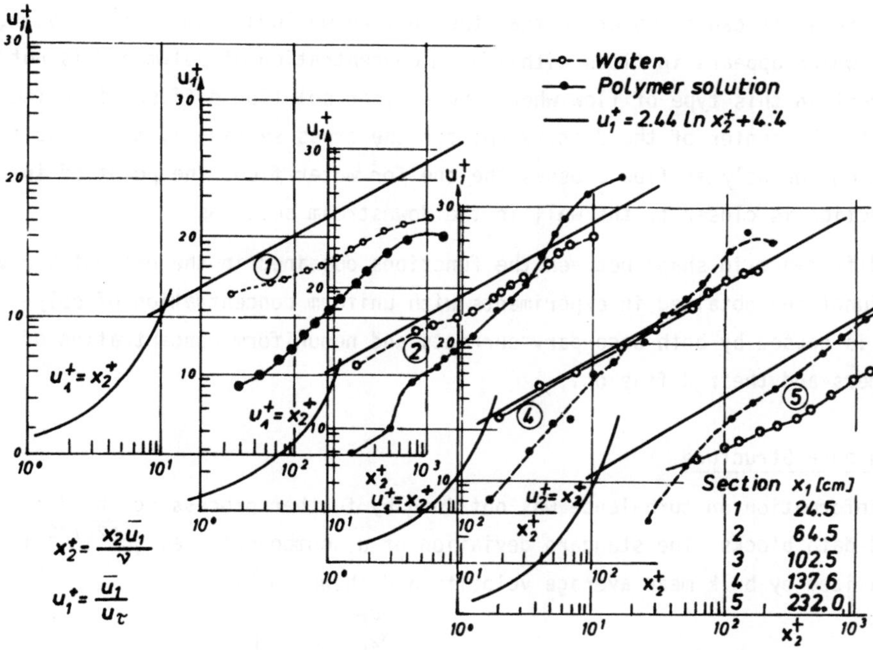

Fig. 4. Mean values of streamwise velocity component ($\overline{u}_1$) ploted as "wall functions"

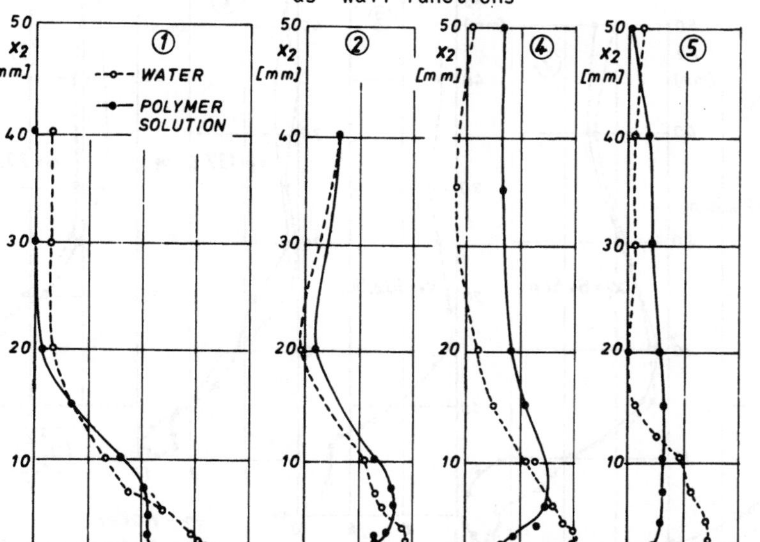

Fig. 5. Mean values of spanwise velocity component ($\overline{u}_2$)

Due to the well known corner effect (Nikuradse [12], Gessner [13]) the secondary current occur in noncircular ducts. In this study the values of u_2 component were measured only in (x_1, x_2) plane. The mean values $\overline{u}_2$ are shown in Fig. 5.

From Fig.4. it can be observed that the well known logarithmic shift (Virk, |14|) which appears in flows with uniform concentration of polymers, is not apparent in this type of flow where the polymer solution diffuses from the wall to the center of the duct. Except for the cross section 1, the velocity function for polymer flow crosses the one for water flow. The point of intersection is closer to the wall in the downstream sections.

The difference in shape between the functions obtained in the present study and functions obtained in experiments with uniform concentration of polymers is caused by both secondary currents and nonuniform concentration of polymers and their diffusion.

Turbulence Structure

The information on turbulence was obtained by further processing the digitized data blocks. The standard deviation of u_1 component i.e. $\overline{(u'^2)}^{1/2}$ is normalized by bulk mean average velocity and shown in Fig. 6.

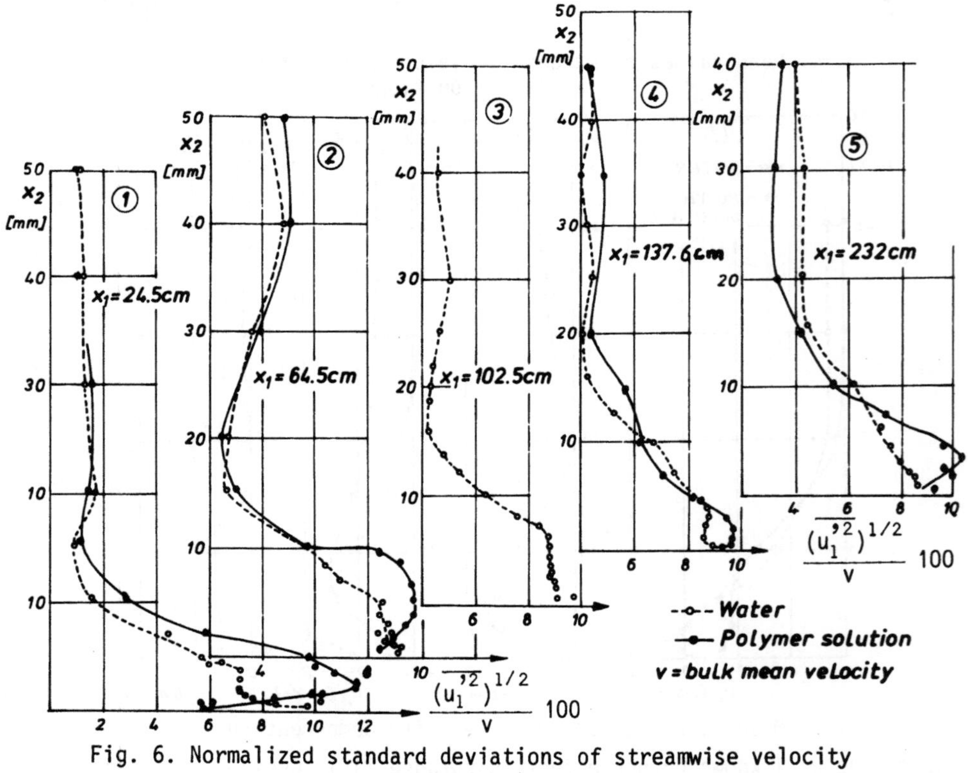

Fig. 6. Normalized standard deviations of streamwise velocity component

The standard deviations for spanwise velocity component are shown in Fig.7.

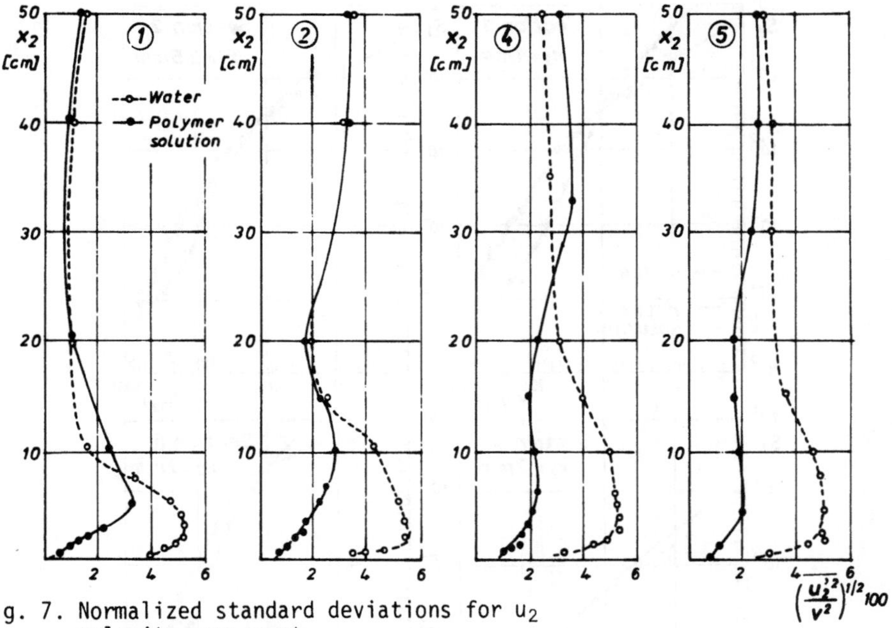

Fig. 7. Normalized standard deviations for u_2 velocity component

From the above two figures it is evident that the turbulence in streamwise direction is increased as confirmed by others (See Rudd [15], for example). In spanwise direction however it is significantly decreased leading to increased anisotropy of normal Reynolds stresses, Maksimović [9]. It is speculated that "decoupling" of fluctuations of these two velocity components is a direct consequence of different shape of vortices in the polymer flow. "Decoupled" velocity fluctuations decrease the turbulent energy production. (Khabakhpasheva et al [16]).

Power Spectrum Density Functions

The power spectrum density functions were obtained by applying the Fourier transform to the original set of data (time series of velocity). The Fast Fourier Transform algorithm described by Singleton [17] was applied.

For two cross sections and two points the functions are shown in Figure 8. for u_1 component and in Fig. 9. for u_2 component. The difference in shape of the two functions indicate that vorticity structure is changed for polymer flow. The power density function for u_1 is shifted towards low frequency portion and increased at part with lowest frequenccies.

For spanwise (u_2) component the energy content in all frequency domains is decreased. This is in agreement with data obtained by measurement of standard deviations.

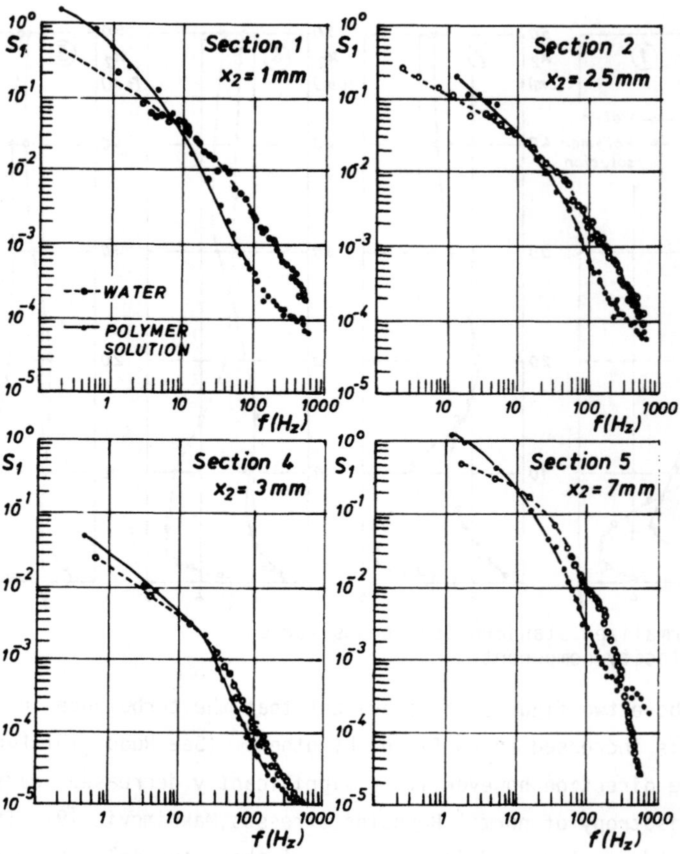

Fig. 8. Power spectrum density function for u_1

Results obtained in this study are in accordance with the ones obtained by Berner and Scrivner [18], by Shieh and Karaki [4] and Latto and El Riedy [3].

Conclusions

The experimental results enable the following conclusions to be expressed:

- Extremely high local values of drag reduction can be obtained at short distances with very low concentration of polymers if they are injected near the wall.
- The diffusion of polymers causes decrease in near wall concentrations and consequently lower drag reduction.
- Turbulence intensities of streamwise velocity component are increased whereas in spanwise direction are decreased, causing the anisotropy to increase.

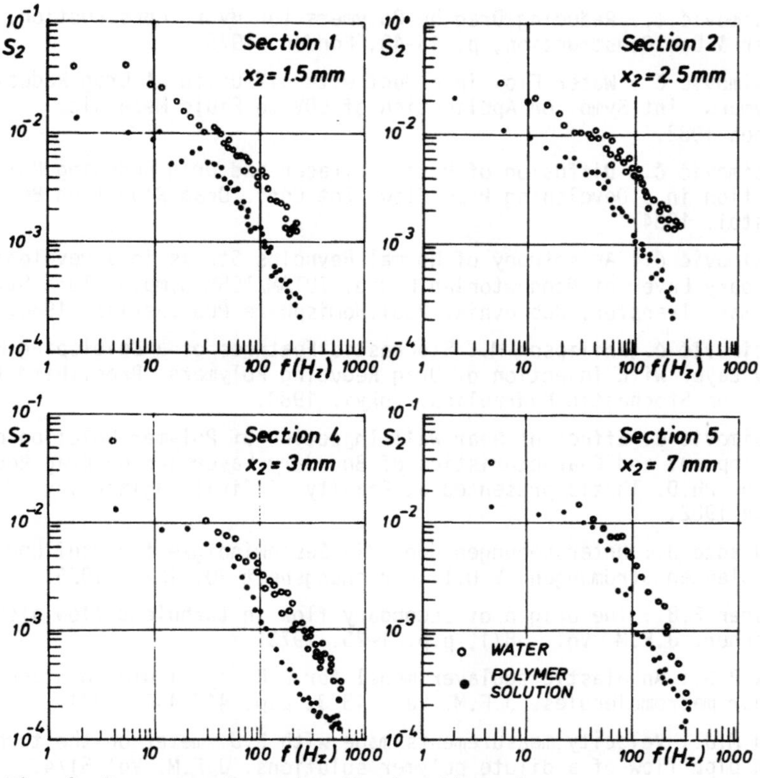

Fig. 9. Power spectrum density functions for u_2

- The power spectrum confirm that the vorticity structure in polymer solution flows is altered, the length of vortices in the flow direction is increased.
- The possible applications of this type of drag reduction could be justified if a relatively short duct is subject to peak discharges of short duration

References

1. Virk P.S.: Drag Reduction Fundamentals. AIChE Vol 21/4, pp 625-655 1975.
2. Hoyt,J.W.: Polymer Drag Reduction,A literature review 1975-6 Int.Symp.on D.R.BHRA-1977, A1-1, A1-19.
3. Latto B.,Chaban J.: On the Performance of Turbomachinery in the Presence of Aqueous Polymer Solution, Proc.I Int. Conf.on D.R.BHRA, Sept. 1974.
4. Shieh S.C.,Karaki S.: Turbulence in Developing Boundary Layer with Polymer Additives. Colorado State University,Fort Collins. 1974.
5. Tullis P.J.,Ramu K.L.V.: Viscous D.R. in Developing Pipe Flow. Hydro Machinery Lab. CSU, Fort Collins,Rep.No 34. 1973.

6. Maksimović Č.: Reducing Drag by Polymers for Hydro Construction. Water Power & Dam Construction, p. 45-48, London. 1978.

7. Maksimović Č.: Water Flow in a Duct with Diffusion of Drag Reducing Polymers. Int.Symp. on Application of LDV to Fluid Mechanics. Lisboa 1982.

8. Maksimović Č.: Diffusion of Passive Tracer and Drag Reducing Polymer Solution in a Developing Pipe Flow. Int.Conf. Drag Reduction'84, Bristol. 1984.

9. Maksimović Č.: Anisotropy of Normal Reynolds Stress in a Developing Boundary Layer of Nonnewtonian Fluid. IUTAM/ICHM Symp.on Turb.Struct. and Mass Transfer. Dubrovnik. Publ.Hemisphere Publ.Series. 1980.

10. Maksimović Č.,Philipson M.: Stochastic Features of a Developing Boundary Layer with Injection of Drag Reducing Polymers. Proc.Third Int. Symp. on Stochastic Hydraulics. Tokyi. 1980.

11. Maksimović Č. Effect of Near Wall Injection of Polymer Solution on Development and Characteristics of Boundary Layer and on Drag Reduction. Ph.D. Thesis presented at Faculty of Civil Engineering, Belgrade 1982.

12. Nikuradze J.: Untersuehungen über die Geschwindigkeitsverteilung in Turbulenten Strömungen. V.D.I. Forschungsheft 70, 1229. 1926.

13. Gessner F.B.: The origin of secondary flow in turbulent flow along a corner. J.F.M. Vol. 58/1, p.p. 1-25. 1973.

14. Virk P.S.: An elastic sublayer model for D.R. by dilute solution of linear macromclecules. J.F.M. Val. 43/3, p.p. 417-440. 1971.

15. Rudd M.J.: Velocity measurements made with L.D. meter on the turbulent pipe flow of a dilute polymer solutions. J.F.M. Vol.51/4. p.p. 673-685. 1972.

16. Khabakhpasheva, Pereplitsa: Velocity and Turbulent Fluctuation Fields in Water Containing Law Concentrations of Gigh Molecular Substances. Journal of Engineering Physics (In Russian), 14/4, p.p.319-320. 1968.

17. Singleton R. An Algorithm for Computing the Mixed Radix Fast Fourier Transform for Estimation of Power Spectra. IEEE Transactions. June 1969.

18. Berner C. and O.Scrivener: Drag Reduction and Structure of Turbulence in Dilute Polymer Solution. Proc. Astronautics and Aeronautics, 72, Viscous flow drag reduction (AIAA) p.p. 290-299. 1980.

Session 11

Today and Future Possibilities of Industrial Applications of Drag Reduction

J. POLLERT

Technical University of Prague
Faculty of Civil Engineering,
Department of Hydraulics and Hydrology,
Prague, Czechoslovakia

Summary

The paper sets out details of specific drag reduction applications and are divided from flow point of view into internal and external flow. These include: hydrotransport, transport of sludges in pipelines and sedimentation, heating circuits, irrigation, hydro-power stations, sewers, marine and hydrofoils and jet-cutting. Wherever possible field test or industrial utilization results are included and discussed.

Introduction

The drastic reduction in wall shear stress caused by the addition of small amounts of linear macromolecules to a flowing solvent, was first reported by Toms, some 36 years ago, and later supported by numerous experiments. A similar effect had been observed earlier, during the second world war, in study of gasoline flow following addition of aluminum soaps. The report of those findings, however, was not published at that time and only a patent appeared much later [1]. This case was probably the first industrial application of drag reductions.

The importance of "drag reduction" as it is called today - another name for it being "Toms´effect" - seems to have been overlooked for quite a number of years, and it was not until the late fifties that scientists started their interest in it.

Drag reduction really came alive, however, in the early sixties when the U.S. Navy became interested in its potential for military applications and supported the relevant research

The Influence of Polymer Additives
on Velocity and Temperature Fields
IUTAM Symposium Essen 1984
Editor: B. Gampert

in various academic and industrial laboratories, as well as in its numerous research centers. Hoyt [2] made the first significant contributions, having since become a key figure in drag reduction research, followed by Fabula [3], who discovered the most effective polymer reducer yet, polyetylenoxide, leading to drag reductions of 80 % in some cases. Similarly, activity in the various industrial research centers was dominated by Crawford [4] and Savins [5] equally responsible for the recent popularity of drag reduction as a field of study and experiments.

The years that followed these early efforts produced an enormous number of papers and reports on an international scale confirming and extending the details of the drag reduction effect have appeared since the last major review by Sellin et al. [6] in 1982. These papers and reviews have filled in many gaps in our knowledge of polymer flow properties, but much more remains to be learned. In particular, progress in the theoretical understanding of the mechanism of drag reduction has been very slow. The investigation efforts have been particularly encouraged during the last decade.

Evaluation of economic benefits

Justification for using drag reducing polymers in industrial flow situations will normally depend upon demonstrantable cost savings. In some cases the rapidity with which a polymer injection system may be installed to cope with an emergency following a sewer collapse, for example, could be the overriding factor. But in the majority of cases it will be a direkt comparison of the estimated total costs of alternative ways of solving the problem which will dictate the final choice.

In order to asses the financial advantages of using polymer additives in any particular flow system it is necessary to reconcile reccurent costs, polymer and power consumption and maintenance costs, with initial or phased capital expenditure. Different industries will have different ways of relating these two types of expenditure but in all cases the total costs will be expressable as an equivalent annual expenditure rate although this may vary through the expected life of the project.

Internal flow applications

Published results show that hydraulic transport of material has been found to be a technically feasible, enviromentally attractive and very economic method of transporting solid particles in different shape and size for short or long distances. The hydrotransport of solids in pipes is a particularly interesting possibility for the industrial applications of drag reduction effect. Despite intensive investigations of drag reduction its practical use for hydrotransport has only been treated in a few technical publications.

Liquid-solid systems vary considerably in the nature of their solid particle size distribution, concentration, and flocculation. For this reason polymer additives are normally employed as sedimentation aids. It is therefore necessary:

- to select the appropriate available polymer for the solid-liquid system under consideration
- to determine the optimum concentration of polymer for sedimentation as well as for drag reduction
- to select the most effective polymer addition process
- to carry out an economic evaluation of the process

In all cases a wrong choice of polymer can cause operation difficulties in the whole transport system, i.e. blockage of pipeline, considerable increase of investment for dewatering solid particles, etc.

The world over, the liquidation of solid wastes, especially fly-ash from thermal power plants, is a cause of growing problems because of low caloric value of coal leading to a high production of ash. The availability of convenient ash disposal lagoons close to power stations is decreasing and new more distant localities must be found involving increased capital and operating costs for the requisite pipelines.

With this in mind a programme of drag reduction experiments was initiated in Czechoslovakia in 1976 which has continued until now [7, 8, 9]. The main purpose of these experiments was to investigate polymer degradation, explore new ways of polymer dosing, influence of drag-reducing polymers on settling velocity solid particles, and establish that drag reduction was feasible with waterborne fly ash. The main field trials were carried out in a 250 mm diameter pipeline 4.1 km long. The additive was Separan AP 273 which was tested in six different concentrations, Fig. 1, Tab. 1.

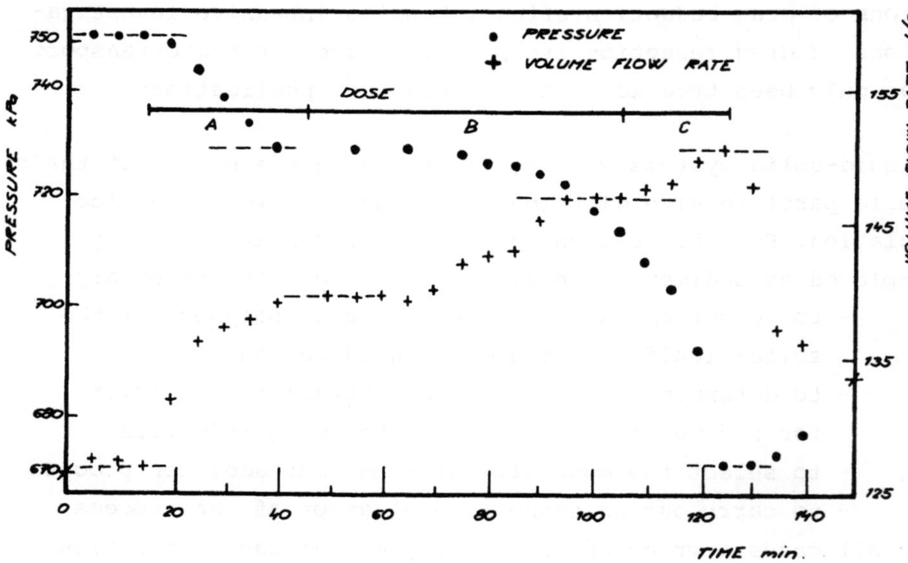

Fig. 1. Experiment No. 1 - Volume flow rate and pressure gauge record. Fly-ash concentration c_m = 4 %, pipe diameter D = 0.25 m, pipe length l = 4121 m, Separan AP 273 concentration: A - 9 ppm, B - 40 ppm, C - 83 ppm

On the basis of our present knowledge it is possible to utilize drag reducing polymers in the following main cases of hydraulic transport of fly-ash:

- a quick filling of existing ash disposal lagoons

without any reconstruction of pipeline and the pumping system and prolongation of pipelines along the lagoons

- increase of concentration and mean velocity of water fly-ash suspension in the pipe.

Modern methods of waste-water treatment depend on macromolecular additives used as coagulants to intensify sedimentation processes, to increase concentration of solid particles, and to improve separation processes as well. The hydraulic transport of sludge forms a part of the waste-water treatment technology. Sludge is pumped not only between individual treatment processes but also out of the water-treatment plant to the sedimentation fields, which are often distant. Even in this case the application of drag reduction is advisable from the point of view of saving electric energy.

Table 1. Full-scale tests for hydrotransport of fly-ash with polymer additives

Pipeline: Antonín - Bohemia D = 250 mm; l = 4121 m				
	Experiment No.			
	1	2	3	4
Volume flow rate Q_m (l/s)	127,2	127,9	125,7	128,1
Concentration of fly-ash C_v (%)	4,0	3,89	4,35	4,16
Density of mixture ρ_m (kg/m^3)	1013,2	1014,1	1020,1	1017,1
Average temperature of mixture in pipeline (°C)	27	27	28	26
Concentration of Separan AP 273 by weight (%)	0,000936 0,00395 0,00828	0,00737	0,00549	0,00423
Max. drag reduction $\Delta\lambda_{max.}$ (%)	15,7 24,3 32,0	24,7	26,4	27,6
Concentration of polymer solution (%)	0,875	0,625	0,625	

Laboratory and full scale experiments have been made with activated and digested sludge from the Central Waste-Water Treatment Plant in Prague. Laboratory experiments in a steel pipe of 32 mm diameter and full scale experiments were performed in a cast iron pipe of diameter 250 mm used for sludge transport from the Prague Central Plant to the sedimentation fields, Fig. 2, when maximum drag reduction was 15 - 20 % [10, 11].

Because of high polymer prices an effort is made to its multipurpose use - both in waste water treatment processes, and in hydraulic transport. Several additives were found to be more efficient in separation processes - cationactive, others in drag reducing processes - usually anionactive. Besides their different price, the anionactive ones being cheaper.

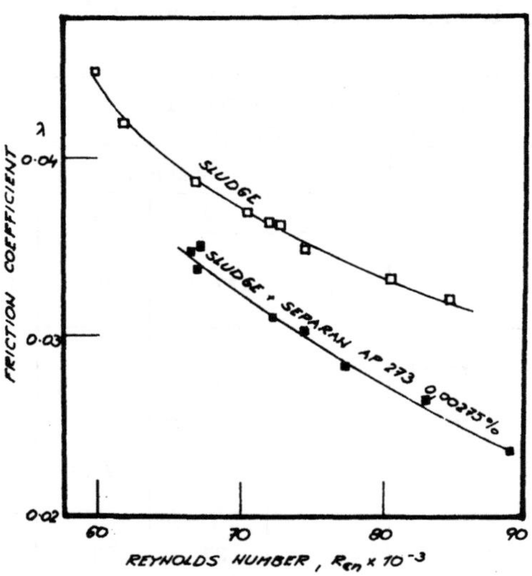

Fig. 2. Relationship between generalized Reynolds number Re_n and friction coefficient λ, cast iron pipeline 250 mm, l = 6615 m, weight concentration of digested sludge 5.5 %

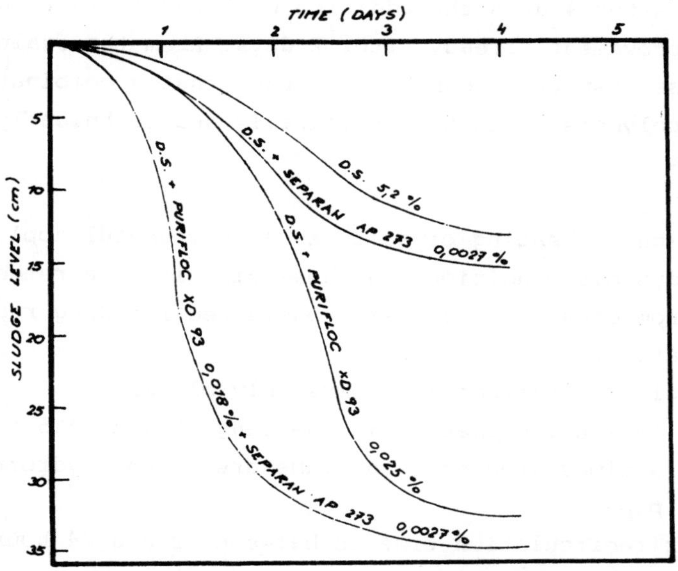

Fig. 3. Typical thickening curve for digested sludge = 5.2 % solids with different additives. All experiments were carried out after hydrotransport sludge - D = 250 mm, l = 6615 m

The main questions appearing with the multipurpose utilization of additives are as follows:

- what combination of various additives and what doses lead to the best final effect, both in separation and in drag reduction
- how the efficiency of additives is decreased by degradation.

In view of these questions field trials had been carried out using a combination of anionic Separan AP 273 and cationic Purifloc XD 8493. The function of Separan was to reduce drag in the hydrotransport of the sludge and the Purifloc to assist dewatering in drying beds. All polymers were added before hydrotransport. The four experimental drying beds used in these trials had surface areas of 1.5 m^2 each. In all 4 cases the sludge levels were recorded during the following 4 days. The results are shown in Fig.3. Although the tests in which Purifloc additives had been used showed similar results after 4 days the addition of Separan led to a striking improvement already after 2 days. From the evaluation of these results it can be concluded that a combination of the two polymers might be economically and technically advantageous.

Drag reduction in heating circuits is a very useful application. Hydrodynamic conditions in these circuits are rather different from other flow systems considered for drag reduction utilization:

- circulation systems are completely closed
- pipe internal diameters in the range 1.0 to 1.5 m is 3 to 5 times greater than diameters of the hydrotransport pipes
- water recirculation time is between 12 and 24 hours at Re = $0{,}5.10^5 = 5.10^5$
- water temperature range is between 40 - 140° C

These extraordinary conditions necessitate great care in selecting drag-reducing additives.

The enlarged thickness of the buffer layer which accompanies drag-reducing processes results in an increased thermal resistance between the wall and the bulk fluid so that the rate of heat transfer is reduced. This effect is important when considering industrial applications of drag-reducing additives which also include heat transfer processes. Most of the heat transfer studies of drag-reducing fluids have been made for internal pipe flows. For example, Debrule and Sabersky [12] have performed a series of experiments for the purpose of determining the heat transfer, and friction coefficients in smooth and rough tubes for tubes for dilute solutions of Polyox in water. Tests were conducted with solutions containung 10 ppm and 50 ppm of the polymer respectively. Typical results for the smooth tube showed that the friction coefficient was reduced by a factor of 3 and the heat transfer coefficient by a factor of 5. The results with the rough tube were of particular interest. Both the friction coefficient as well as the that heat transfer coefficient were reduced drastically, that is by factors of 6 and 10 resp., Fig. 4.

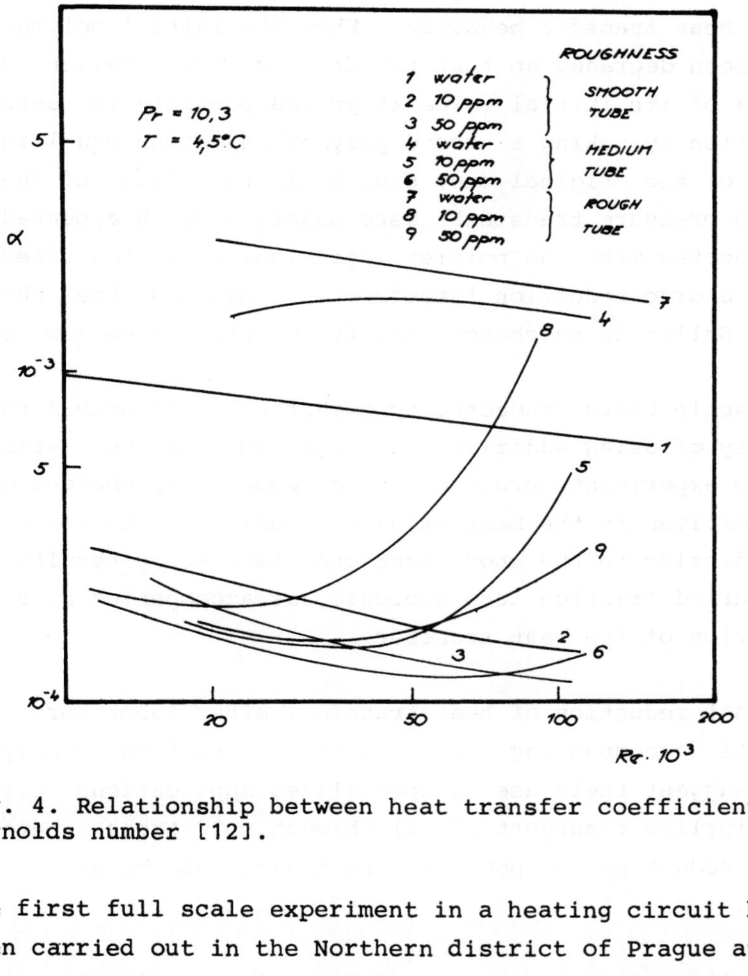

Fig. 4. Relationship between heat transfer coefficient α and Reynolds number [12].

The first full scale experiment in a heating circuit have been carried out in the Northern district of Prague at a local thermal station. A maximum drag reduction of 12.8 % was achieved. The low drag reduction for the polymer concentration 225 ppm Separan AP 273 was caused by mechanical degradation of the polymer in a great number of bends and fittings in the system.

For this reason a second test was carried out in Southern Bohemia in 1981, and again in spring, 1983. The first experiment was more succesful, some results of the second proved the first one. In order to preserve comparability with previous tests, the same polymer was used. Concentrations were 44 and then 22 ppm in a 0.3 m diameter pipe 5066 m long. The results showed a maximum drag reduction of 27 % with no change in heat transfer behaviour. When the initial polymer dose had been degraded so that the drag reduction fallen to about 50 % of its initial value it proved possible to restore the situation by making a second polymer injection equal to only half of the original one, Fig. 5. In the course of these tests some pressure transients were observed which appeared to be connected with the polymer injection. A "piston effect" caused by a drag reduction interface in a pipe had been observed by Sellin in surcharged and free surface flow sewers.

Two full scale tests conducted by author [13, 14] proved the feasibility of using additives in large pipe systems. During full scale experiments practically no temperature changes due to the additives in the heat exchanger were observed which is in contradiction to the above mentioned laboratory results where a marked friction loss decrease was accompanied by a deterioration of the heat transfer.

An important reduction of heat transfer, after laboratory experiments, accompanying the reduction of friction by polymers may suggest their use in specialized applications, e.g. for the pipeline transport of oil through cold areas, where good drag reduction and poor heat exchange would be an advantage.

Another full scale study was carried out in the district heating pipe system Ägesta and Farsta [15]. The test system comprised two pipes of diameter 0.6 m and total length of 7000 m. The investigations covered one calibration phase without additive, one phase when additive was dosed to the system, and finally one phase when the longterm effects caused by the polymer additive had been studied under 14 days. The mean velocities of flow were during the tests

2 - 3 m/s and the system temperature was 80° C.

The results of the present investigation show that the diminuation of friction losses is more pronounced reaching 80 %, by polymer Separan AP 273 addition. This is the mean value during the longterm test. This is a surprisingly great drag-reducing effect achieved with a concentration of 120 ppm. The study shows that the additive has more influence on friction losses than on singular head losses.

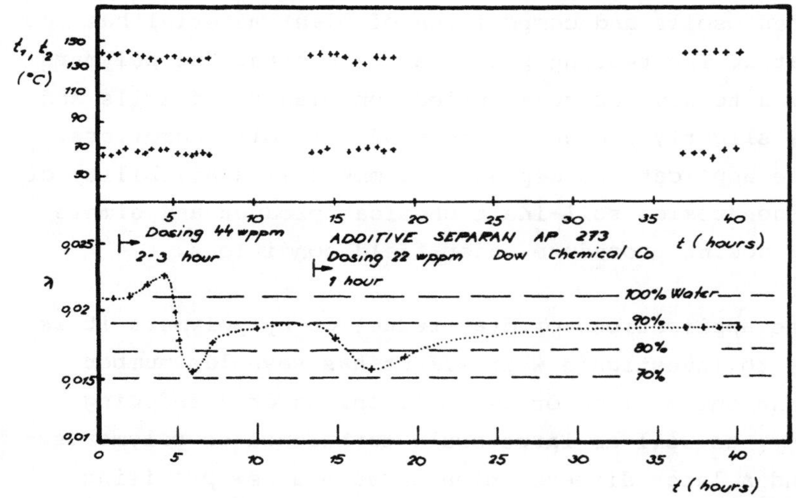

Fig. 5. Results of a polymer injection test on a district heating system's primary pipe circuit.

Another field of drag reduction applications studied in Czechoslovakia is the agricultural irrigation [16, 17]. In long distances with the initial pressure head of 0.8 to 0.9 MPa and required pressure head of 0.4 to 0.5 MPa at the sprinkler outlets. The drop of the hydraulic gradient is considerable. About 50 % of the initial pressure is used to overcome the pipe friction losses. Large scale experiments under field condition were carried out near Brno (southeast of Bohemia) after preliminary laboratory experiments with four different polymers. For field experiments the new SIGMA mobile sprinkler device had been used - plastic pipe D = 55 mm, l = 250 m, sprinkler nozzle with an internal diameter of 18 mm. Drag-reducing polymer Percol 140 had been used in concentration of 100 ppm. The flow at the nozzles was maintained in the range from 2.8 l/s to 6.9 l/s, Fig. 6 and drag reduction was ranging 20 - 30 %.

Besides of the hydrodynamic effect on drag reduction the polymer additives affect also plants and soils. For this reason the pot vegetation trials have been carried out and for several growing seasons since 1979 influence of Percol on agrochemical and microbiological properties of two entirely different soils and composition of plant material has been investigated. The testing plant has been corn. The polymers were proved to have no toxic effect on plants and soils and stimulate slightly the development of the soil microflora. The future applications depend on commercial availibility of a cheap, non-toxic, soil-inert chemical product and simple practical dosing procedure under field conditions.

For future applications of drag reduction by polymers it is necessary to investigate what effect has Reynolds number higher than one million or ten millions on drag reducing efficiency, as well as large scale turbulence on polymer solution and a large diameter pipe because a new promising area of possible application has recently appeared - i.e. water power stations.

By using only drag-reducing fluids in a pump turbine installation, a considerable power gain can be achived [18]. The gain with the pump-turbine model is more than 4 efficiency points with 120 ppm solution, Fig. 7. Another important finding is that working with fully mixed solutions, the best efficiency point is shifted to higher volume flow rate values and the efficiency curves are flatter at higher volume flow rates. This may be important for two reasons:

- peak demands can be met easier and demand variations do not affect the machine performance
- using drag-reducing additives becomes more economical

at peak demands since the power is scarce at that time.

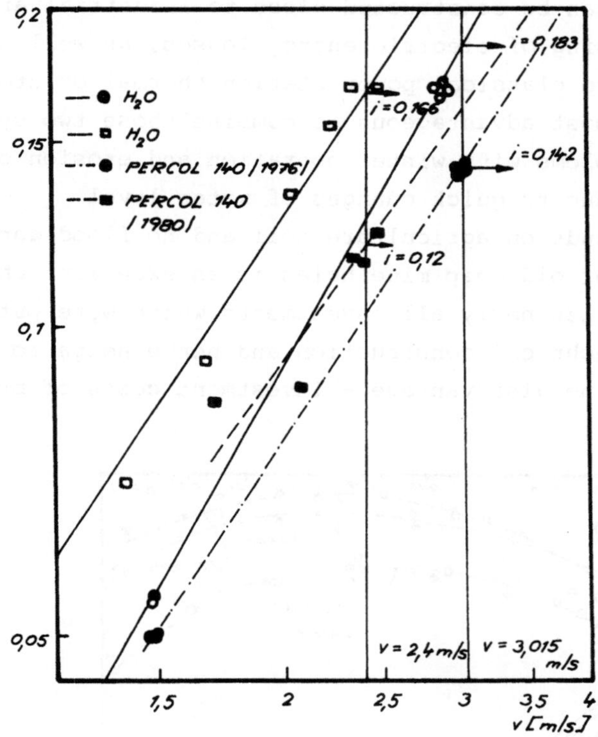

Fig. 6. The relationship between the hydraulic grade line slope and the mean flow velocity. Field experiments, D = 55 mm; 250 m.

It appears that the use of drag-reducing additives in hydraulic machines and hydraulic power stations is quite feasible. The additional investment is negligible. The power gain may be several hundred kW per kg of polymer at very low injection rates, hence, it may be economical at peak demands.

Higher stage of utilization of drag reducing polymers in water power stations is their use in pumped storage plants, resp. in deep storage plants, e.g. in old deep mine holes [19] for increasing of performance. The deep plants have many advantages in comparison with classical pumped storage plants:

1. Practically no limits in head and independence on terrain
2. Plant can be constructed close to industrial area - decreasing of electric energy losses, as well as close to classical power station thermal or atomic. It is most advantageous to combine those two systems.
3. No problems with winter operation and erosion of banks due to quick changes of water level
4. No demands on agriculture soil and no flood danger
5. Using of old deep mine holes is an excellent chance to utilize nerly all investments which were put into this technical construction and hence helps to exclude one disadvantage - investment costs of breaking works.

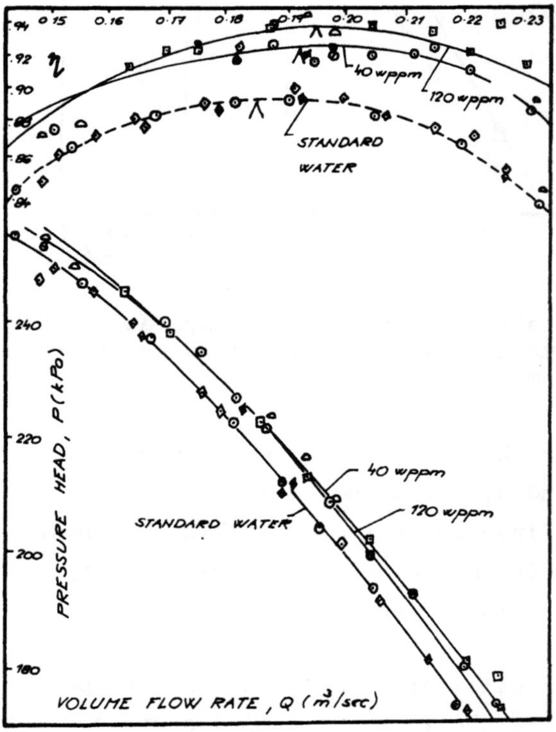

Fig. 7. Experimental results with a Francis pump-turbine working as a pump.

The equation for a pumped power station give only three possibilities how to change a power plant output:

$$P = \frac{1}{102} \cdot \rho \cdot Q \cdot H \cdot \eta_T \cdot \eta_Q \cdot \eta_H \cdot \eta_G \cdot \eta_{TR}$$

P = performance during turbine operation (kW)

ρ = density of water or other flowing media (kg/m^3)

H = head (m)

η_T= turbine efficiency, η_Q = discharge efficiency, η_H = head efficiency, η_G = generator efficiency, η_{TR} = transmission efficiency

1. by increasing of transition time, i.e. decreasing of idle discharge
2. by increasing of media density
3. by drag reducing polymers decreasing of all friction losses, resp. increasing of head in the system, Fig. 8.

Still another aspect of polymer flow is the discovery that drag-reducing polymers inhibit the inception of cavitation in a given flow situation. Cavitation is the sudden formation of vapour bubbles in the fluid, analogous to boiling.

Cavitation studies in polymer solutions have centered on jet cavitation, cavitation inception on rotating disks, propellers, pumps and turbines. Jet cavitation studies [20] show that the cavitation index can be reduced to a half of the previous value when small quantities of Polyox are present in the flow. The cavitation inception index σ_i is defined by

$$\sigma_i = \frac{P_i - P_v}{1/2 \ \sigma \ v^2}$$

where P_i is the static pressure at cavitation inception, P_v the vapur pressure of water, ρ the water density and v the flow velocity.

	Tunnel velocity (m/s)	Inception Cavitation Index
water	12.55	0.43
20 ppm PEO	13.40	0.50
50 ppm PEO	14.18	0.39
100 ppm PEO	13.70	0.41

Cavitation in polymer solutions is an attractive field study, since research may shed new light on old problems of Newtonian fluid cavitation, and also since the potential applications of cavitation reduction are many, e.g. combining with previous deep pumped storage plants.

Maksimovitch [21] has come with a new idea to use polymers in short diversion tunnels and drainage culverts commonly used during construction. Such tunnels may be from 2 to 10 m in diameter and from 10 to 100 diameters in length. The tunnel flow characteristic curves are in Fig. 9. Maksimovitch develops these ideas in relation to temporary diversion tunnels suggesting that a smaller and cheaper tunnel may be constructed because extra 20 or 30 % of discharge capacity can be obtained by using polymer injection if and when the need arises. The choice of discharge values will depend in any case on a proper assessment of the probalility of river flow occurences during the design life of the diversion tunnel.

Although most experimental work up to now has been carried out using water-based solutions, the succesful experiments in the Trans-Alaska Pipeline System ultimately led to the first commercial application of drag reduction in a crude oil pipeline [22, 23]. The injection of a polymer additive has accounted for about 1300 m^3/h of this increase, an amount representing over one percent of the United States´ total crude consumption. Conoco´s success with patented CDR Flow Improver encouraged crude oil producers for using this polymer for increasing capacity of undersea pipelines

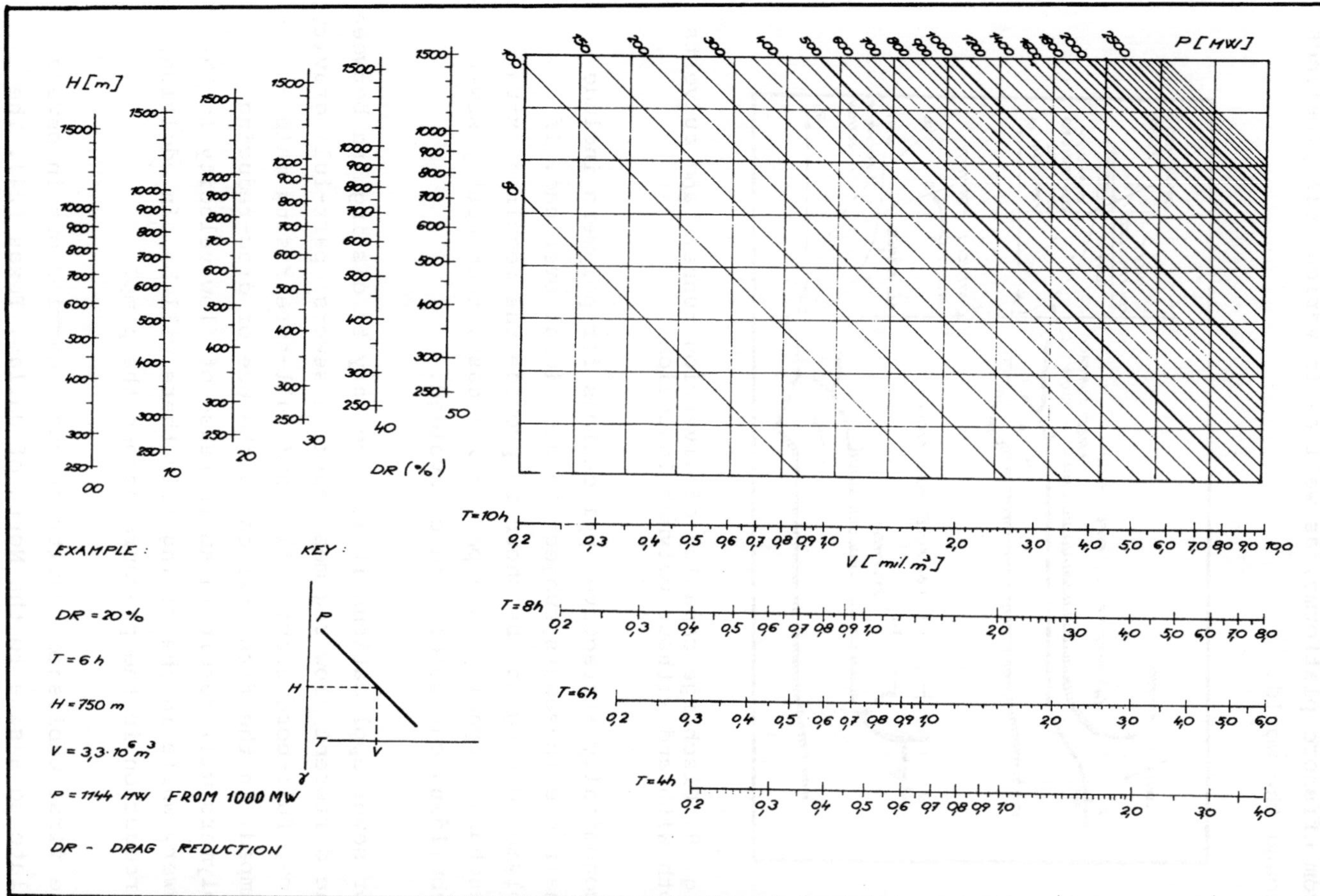

Fig. 8. Performance increasing of pumped storage plant

from offshore platforms, as well as in various oil pipelines around the world.

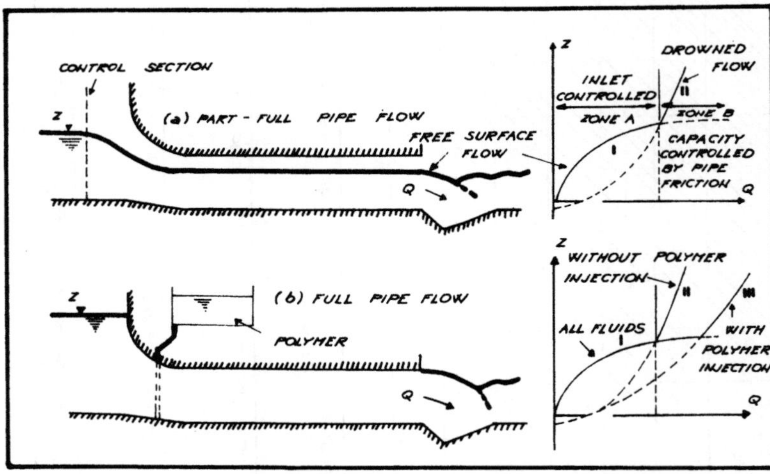

Fig. 9. Discharge capacity of diversion tunnels and culverts both with and without polymer injection

Economically attractive applications for polymers include their use in sewers subject to occasional overload. If polymer dosing can be shown to provide the required extra discharge capacity in a particular case, then costly sewer rebuilding can be delayed or avoided.

For sewer applications it is necessary to distinguish between the different flow regimes found in sewers: part-full gravity flow, full-bore gravity flow and full-bore rising main - pumped. In the first two cases the use of drag-reducing polymers will result in an increase of flow velocity in the sewer, while in the rising main there will be, in addition, a reduction in the pressure across the pump.

The first published results of tests carried out in sewers relate to a site to the North of Dallas, Texas [24]. The next report [25] appeared in 1971. Following the 1969 tests in Dallas a permanent and fully automatic polymer dosing station was installed [26] on the Bachman Creek sewer near the city. Before the publication of the Bachman Creek report the decision had been taken to construct a permanent polymer dosing system at Knowle, to the East of Bristol, in the U.K. This scheme involved a 305 mm diameter gravity flow sewer design to carry domestic sewage [27]. Another series of tests have been carried out in Bath [28] U.K. in which the drag-reducing polymer had been supplied in liquid emulsion form. This was premixed with a small quantity of water immediately before being injected into the discharge side of a pump supplying sewage to a 760 mm pumped main 8000 m long.

The main points that have to be assesed when considering polymer dosing are the duration of dosing period, the increase in sewer capacity achieved by polymer dosing and the length of sewer that is causing the bottleneck. Sellin [29] compared costs of a 305 mm sewer laid at a 1 : 1000 gradient that is surcharging and which would require 20 hours of polymer dosing per year at 20 ppm Polyox to give 20 % increase in sewer capacity in order to eliminate all the surcharging with a newly constructed 381 mm diameter sewer. The costs are given in Table 2.

Tab. 2.

Polymer dosing 1979 prices	
Capital costs of building and services	6000
Capital costs of equipment	3500
Total	9500
Operating costs polymer, maintenance, etc.	500/year
New sewer construction 1979 prices	
Capital cost 381 mm dia., 1000 m long	152000

<u>External flows</u>

A number of towing tank tests have confirmed that the drag of a ship model can be substantially reduced by adding polymers to the water, Dove [30], Canham [31]. The economics of drag reduction on a full size ship seems to place this method reducing ship power requirements until extremely cheap polymers can be obtained. In contrast to pipeline situations where the power saved by using polymers increases with pipe length assuming no degradation, the one-time ejection from a ship is not economical yet. The use of the additives will be restricted to very specialized ship applications with short-duration injections. Warships could carry the injection mechanism and a supply of additive to use only when bursts of speed are essential, e.g. during an action, or in pursuit of a submarine, or trying to get out of the line of approaching torpedos, etc.

Hydrofoil boats could become foil-borne with lower expenditures of power. There are indications that stall of hydrofoils can be delayed by injecting additives into foil boundary layer. Therefore at the same speed and for the same foil area, lift can be increased by increase of angle of attack beyond the angle of stall in water. This is in addition to any reduction in drag of the foils. For a better understanding of this phenomenon, behaviour of a thin flat jet of an aqueous solution polyethylenoxide 2000 ppm after leaving the injection slit into a rectangular duct with turbulent flowing water has been studied by Fruman and Galivel [32] and Kolář [33], Fig. 10.

With the polymer injection the pressure distribution along the duct is radically changed. Measured pressure differences between the flow with and without injection of the polymer solution are in agreement with the jet swell, its maximum being 5 times the slit thickness 8.5 mm from the orifice in the 10 mm high duct, 80 mm wide at v_i = 0.686 m/s, v_e = 1.91 m/s as measured on photographs of the visualized flow. On the picture is possible to see schematical velocity distribution, pressure lines with water, polymer injection and without it and observed jet swell of injected polymer solution like results of experimental works.

Striking changes occur in the visual appearance of a jet of water discharging into air when small amounts of a high polymer such as PEO are added to the flow. This occurs in laminar, transitional, and turbulent jet flow. Although the polymer jet appears more coherent and less dispersive than a pure-water jet to the eye, the overall carrying power of the jet is not greatly enhanced. However, the coherence of the polymer jet does add significantly to the penetration and cutting ability of small jets used in jet-cutting applications.

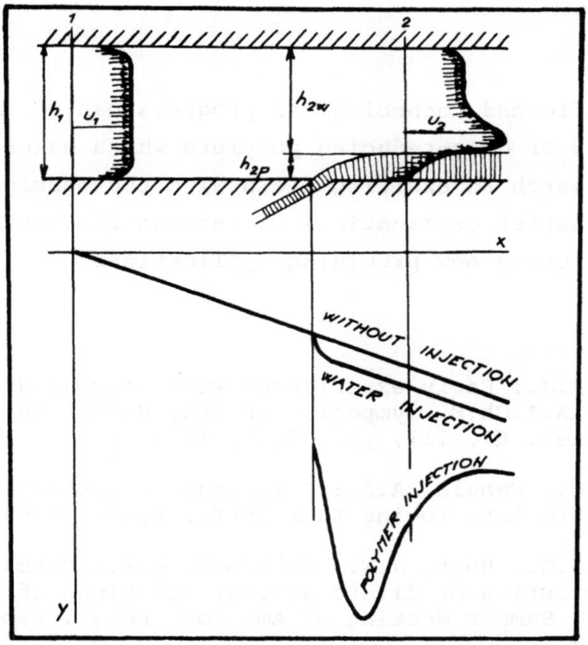

Fig. 10. Probable velocity distribution, pressure lines with water of polymer injection and without it. Observed jet swell of injected polymer solution.

Slanec [34] succesfully introduced drag reducing polymer-water solution in rock disintegration by jet cutting for forming of trench for additional sealing diaphragm at Nýrsko earth dam southeast of Bohemia. The concentration of Praestol 2830 was 100 ppm. The main results are summarized in the next points:

1. increased jet coherence and higher dynamic froces on disintegrated rock
2. diminuation of nozzle wear in comparison with pure water due to erosion, resp. cavitation effect
3. increased reliability of used piston pumps - diminuation of number of breakdown during the operation
4. decreasing friction losses from 0.7 MPa to 0.2 MPa, and thus increasing jet outlet velocity.

Investigations have been going on in the field of better nozzle shaping and precise routing of jet.

Conclusions

The scientific and technological progress can be expected from studies of drag-reducing polymers which makes this area of research an exciting field with the possibilities of finding better explanations of various flow mechanisms and of developing new promising applications.

References

1. Mysels, K.J.: Early experiences with viscous drag reduction. A.I.Ch.E. Symposium on Drag Reduction, Symposium Series, 67, 111, pp. 45-49, 1971

2. Hoyt, J.W., Fabula, A.G.: Frictional resistance in towing tanks. 10th Int. Towing Tank Conf., London 1963

3. Fabula, A.G., Hoyt, J.W., Crawford, H.R.: Turbulent-flow characteristics of dilute aqueous solutions of high polymers. Summer Meeting of Am. Soc. Phys., Paper K5, June 1963

4. Crawford, H.R., Pruitt, G.T.: Drag reduction of dilute polymer solutions. 56th Ann. Meeting A.I.Ch.E., December 1963

5. Savins, J.G.: Drag reduction characteristics of solutions of macromolecules in turbulent pipe flow. Petrol. Engrs. 4, 3, pp. 203-213, September 1964

6. Sellin, R.H.J. et al.: The effect of drag reducing additives on fluid flows and their industrial applications. Part 1 and Part. 2 J. of Hydraulic Research 20, No. 3, 1982

7. Pollert, J.: Hydrotransport of fly ash-water mixture and drag reduction. 2nd Int. Conf. Drag Reduction, Cambridge, Paper B3, 1977

8. Pollert, J., Kolář, V., Havlik, V.: Drag reduction in hydraulic transport in pipes. IV Seminar on Transport and Sedimentation of Solid Particles, Wroclaw, Paper A21, 13 pp. 1980

9. Kolář, V., Stastna, J.: Modelling of solid-liquid-polymer flow. 17th Congress of the IAHR, Paper A-55, Baden-Baden, 1977

10. Pollert, J.: The technical economic studies for drag reduction utilization in Czechoslovakia. CNRS Round Table, Strasbourg, 1978

11. Pollert, J., Kolář, V.: Sludges and their transport by pipelines. Symposium of Council of Mutual Economical Help, Prague, Paper 16, 1977 - In Russian

12. Debrule, P.M., Sabersky, B.H.: Heat transfer and friction coefficients in smooth and rough tubes with dilute polymer solutions. Int. J. Heat and Mass Transfer, Vol. 17, 5, 1974

13. Pollert, J., Kolář, V., Havlik, V.: Drag reduction and its engineering applications. Acta Polytechnica Series of Technical University in Prague, Section Civ. Eng., Int. Conference pp. 133-142, 1982

14. Pollert, J., Urbânek, M.: Possibilities of decreasing friction losses in primary heating plant distribution networks by means of polymer additives. Vodnî hospodârstvi Series B, 32, 10, pp. 275-278, 1982 - In Czech

15. Ahrnbom, L., Hagstrand, U.: Toms'effect in district heating system. Full scale demonstration test. Varmeteknisk Forsknung, projekt P-56, 1980

16. Eliáš, V., Vocel, J.: Polymer additives for sprinkler irrigation. Communications of the Inst. of Hydrodynamics, No. 8, Prague, October 1978

17. Eliáš, V., Vocel, J.: Long term field experiments with polymer additives for sprinkler irrigation. Communications of the Inst. of Hydrodynamics, No. 10, Prague, October 1980

18. Bilgen, E., Vasseur, P.: Feasibility of utilization of drag reducing additives in hydraulic machines. 2nd Int. Conf. Drag Reduction, Paper B4, Cambridge 1977

19. Pollert, J., Holata, M.: Mean of operation pumped water stations with circulated liquid media of special quality. Prague Czech Patent Office, April 1984 - In Czech

20. Hoyt, J.W.: Recent progress in polymer drag reduction. Coloques Int. du CNRS, No. 233 - Polymêres et Lubrification Brest 1974

21. Maksimovič, Č.: Reducing drag by polymers for hydro-construction. Water Power and Dam Construction, pp. 53-58, June 1978

22. Burger, E.D., Chorn, L.G. Perkins, T.K.: Studies of drag reduction conducted over a broad range of pipeline conditions when flowing Prudhoe Bay crude oil. J. Rheology, 24, 5, pp. 603-626, October 1980

23. Burger, E.D., Munk, W.R., Wahl, H.A.: Flow increase in the Trans Alaska Pipeline using a polymeric drag reducing additive. S.P.E. Paper 9419, 1980

24. Western Company: Polymers for sewer flow control. U.S. Fed. Water Pollution Contr. Admin. Rep. WP-20-22, 1969

25. Hercules Inc.: A feasibility study of the use of polymers to relieve sewer overloading in Milwaukee, Wisconsin. Rep. of the Environmental Serv. Div., Hercules Inc., 1971

26. Chandler, R.W., Lewis, W.R.: Control of sewer overflows by polymer injection. Dallas Water Utilities-Bachman Creek Project U.S. Environmental Protection Agency, Report: EPA 600/2-77-189, September 1977

27. Sellin, R.H.J.: Drag reduction in sewers: first results from a permanent installation. J. Hydraulic Res., 16, 4, 1978

28. Sellin, R.H.J., Ollis, M.: Polymer drag reduction in large pipes and sewers: Results of recent field trials. J. Rheology, 24, 5, pp. 667-684, 1980

29. Sellin, R.H.J.: Use of polymer additives to relieve overloaded sewers. Proc. Inst. Civ. Eng., 74, November 1983

30. Dove, H.L.: The effect on resistance of polymer additives injected into a boundary layer of a frigate model. Proc. 11th Int. Towing Tank Conf., Tokyo, 1966

31. Canham, H.J.S., Catchpole, J.P., Long, R.F. Boundary lyer additives to reduce ship resistance. The Naval Architect, J. Rina, No. 2, p. 187, July 1971

32. Fruman, D.H., Galivel, P.: Anomalous effects associated with drag-reducing polymer injection into pure-water

turbulent boundary layers. Symposium on Viscous Drag Reduction, Dallas Texas 1979, Published AIAA, pp. 332-350, 1980

33. Kolář, V.: Effects of polymer injection from a thin slit on flow along a wall. Vodohospodársky cas., 28, 4, pp. 382-395, 1980 - In Czech

34. Slanec, K.: Disintegration of rock by water-polymer jet. Research Report Fac. of Mechan. Eng., Technical University Prague 1983

Session 12
Pressure Fields, Cavitation

Effect of Polymer Solutions on Tip Vortex Cavitation

D.H. FRUMAN, D. BISMUTH and S. AFLALO

Ecole Nationale Supérieure de Techniques Avancées
Centre de l'Yvette - Chemin de la Hunière
91120 PALAISEAU (France)

Summary

The effect of polymer solutions, ejected at very small flow rates at the tip of a finite span hydrofoil, on the inception and extent of tip vortex cavitation is investigated. The results show that the dessinent cavitation numbers are significantly reduced by polymer additives as compared with those in pure water. Moreover, for operating conditions well below critical, the presence of polymer in the vortex core results in the nearly complete elimination of the vapour cavity. These effects can be interpreted on the basis of a modification of the velocity field due to the viscoelastic properties of the polymer solutions. L.D.A. measurements indicate a large decrease of the tangential velocity component in the transition region between the inner solid body rotation core and the outer potential vortex flow. Also, a change of the slope of the velocity within the core is observed. This phenomenon is, besides its potential practical applications in naval hydrodynamics, a very good exemple of the influence of polymer additives on the velocity field of a high Reynolds number flow.

Introduction

The effects of the viscoelastic properties of polymer solutions in rotating flows have been recently summarized by Bird and Curtiss [1]. These effects are not limited to highly concentrated solutions but also occur in extremely diluted solutions as shown by the velocity modifications observed during "vortex inhibition" investigations [2]. Rotating or vortex flows exist in many practical situations such as cyclone separators, tip vortex on wings and propellers, etc... In this paper, we describe the effects of polymer solutions on tip vortices and, in particular, on the cavitation resulting from the low pressures generated within the core. This subject has received some attention in recent times and Inge [3,4] has made an important experimental and theoretical contribution.

Experimental

Tests were conducted in a cavitation tunnel fitted with a rectangular test section 150 mm high and 80 mm wide. Tip vortices were generated using a 60

The Influence of Polymer Additives
on Velocity and Temperature Fields
IUTAM Symposium Essen 1984
Editor: B. Gampert

mm half span, s, and a 40 mm maximum chord, t, elliptical wing with a NACA 16020 symmetrical cross section. Ejected fluids, 1000 ppm POLYOX WSR 301 solutions, are brought to the wing tip, through small diameter tubes imbeded into the foil, by using a pressurized reservoir. Ejection flow rates are determined using a precalibrated rotameter.

Different techniques were used to determine the effect of the polymers on the cavitation and the velocity changes. High speed cinematography and backscattered light diffusion allowed a qualitative understanding of the cavity modifications, as well as the quantitative determination of the dessinent cavitation numbers. L.D.A. measurements were performed in the absence of cavitation using a DISA backscatter optic and its associated electronics.

RESULTS

General

Figure 1 shows high speed movie camera images (4000 images/s) of the tip vortex for three different ejection conditions. For a pure water flow (no ejection), Figure 1(a), the cavitating vortex core is seen as a continuous trace which is initiated at the wing tip. For an ejection flow rate of $7.4\ 10^{-7} m^3/s$, Figure 1(b), the cavity extends only a limited distance downstream of the wing tip and it occurs intermitently. Further increase of the ejection flow rate, $2.6\ 10^{-6} m^3/s$, leads to the complete disapearence of the cavity, Figure 1(c), even in the immediate vicinity of the tip.

In Figure 2 we present the backscattered light signature of a point on the cavity situated at 0.2 m downstream of the tip and illuminated by a laser beam. The spikes in Figure 1(a) are reasonably believed to be the response of individual bubbles of different diameters separated by time intervals of the order of about 2 ms. For an ejection flow rate of $9\ 10^{-7}\ m^3/s$ most of the spikes have disapeared and the time interval between bubbles has increased to about 8 ms. Finally, for the largest ejection flow rate, $1.48\ 10^{-6}\ m^3/s$, there is only one sharp spike during the total acquisition period of 10 ms. Most of the randomness observed in Figure 2(a) is not present in Figures 2(b) and (c).

Therefore, our previous [5] observations based on visual inspection and still photographs of the flow are fully confirmed by the present high speed movie camera images and light scattered signatures. Moreover, these results

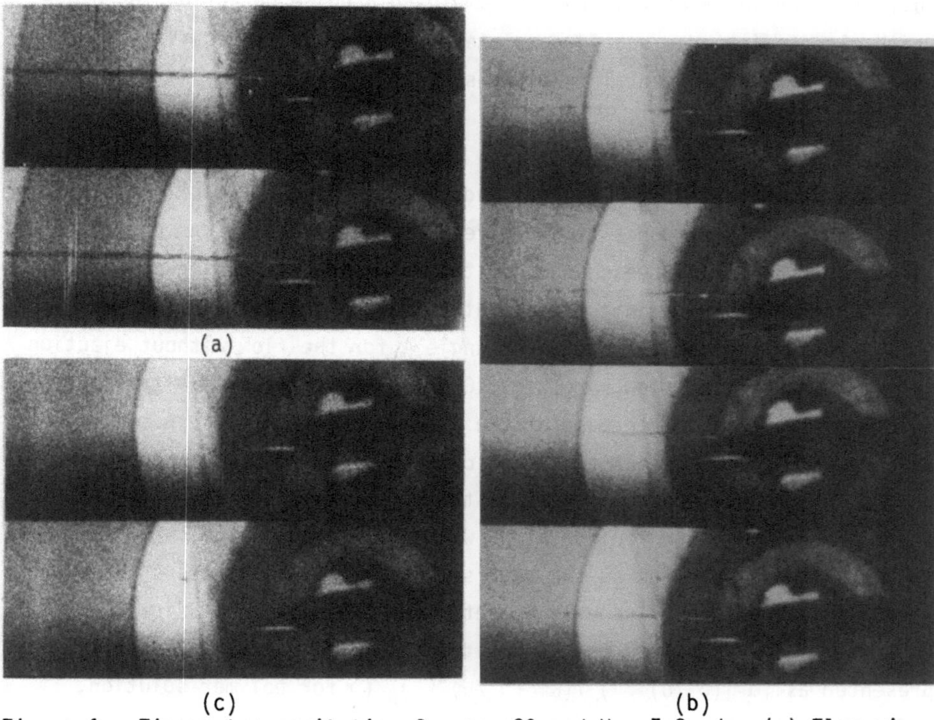

Figure 1 - Tip vortex cavitation for $\beta = 8°$ and U = 7.8 m/s. (a) Flow without ejection ; (b) Flow with polymer ejection, 1000 ppm, $q_i = 7.4\ 10^{-7}\ m^3/s$ (c) Flow with polymer ejection, 1000 ppm, $q_i = 26\ 10^{-7}\ m^3/s$.

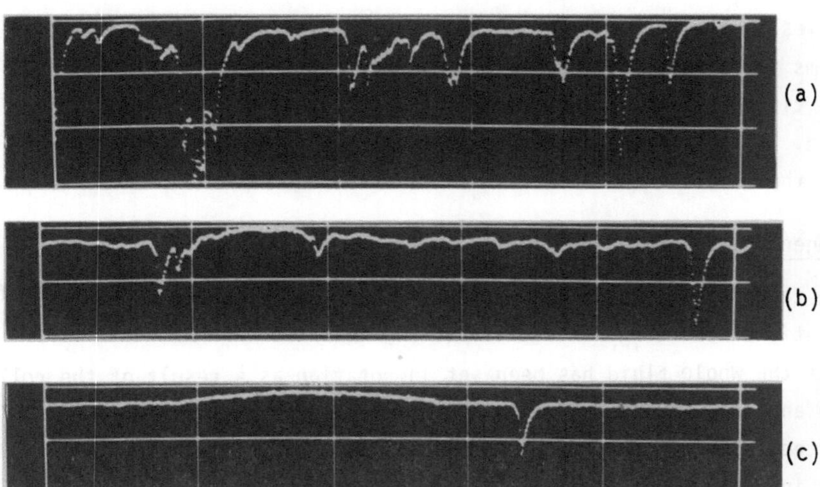

Figure 2 - Backscattered light signature from tip vortex during for $\beta = 9°$ and U = 4.46 m/s. (a) Cavitation without ejection ; (b) With polymer ejection $q_i = 9\ 10^{-7}\ m^3/s$ and (c) with polymer ejection $q_i = 1.48\ 10^{-6}\ m^3/s$.

fully confirm those of Inge [3] who employed a noise detection technique which also indicated a nearly complete elimination of the cavity under rather different experimental conditions.

Critical Cavitation Number

The results presented in Figures 1 and 2 remain of a qualitative nature. Some quantitative evidence of the effect of polymer solutions can be obtained by considering the changes of the dessinent cavitation number, σ_d, as a function of the ejection flow rate. In Figure 3 we show the values of $\sqrt{\sigma_d}/U$ as a function of the incidence angle α for the flow without ejection and with an ejection flow rate of 9 10^{-7} m^3/s. The ejection of the polymer solution results in a decrease of the slope of the linear regression lines passing through the rather scattered data points. The scatter can be explained by the strong sensitivity of the tip vortex cavitation to the presence of nuclei in the recirculating water. As noted by Bovis [6] the amount of nuclei depends on the test section velocity and pressure, which have been varied simultaneously to determine the data points in Figure 3. In spite of this scatter, the composite plot, Figure 4, of all our results presented as $(\alpha^{-1}(\sigma_d/U)^{1/2})_p/(\alpha^{-1}(\sigma_d/U)^{1/2})_w$ (p for polymer solution, w for water) versus the non dimensional polymer mass flow rate $\overline{q}_i = q_i c/Ust$, where U is the free stream velocity, shows the significant effects associated with polymer solution ejections. For relatively low values of $\overline{q}_i$ the efficiency of the ejection is greater for large incidence angles, while, for a large $\overline{q}_i$ the reverse occurs. A critical value of $\overline{q}_i$ seems to be associated to each incidence and therefore to the strength of the tip vortex which can be, in turn, directly related to the lift of the wing. It is interesting now to consider the flow modifications responsible for the reduction of the critical cavitation number.

Tangential Velocity Profiles

The tangential velocities were measured at a station .20 m downstream from the tip of the wing. This distance was selected because it was expected that the whole fluid has been set in rotation as a result of the roll-up at the tip and that vortex diffusion was still limited [7]. Measurements were conducted, for non cavitating conditions, at a free stream velocity of 5 m/s, two incidence angles, 5 and 10°, and five ejection conditions.

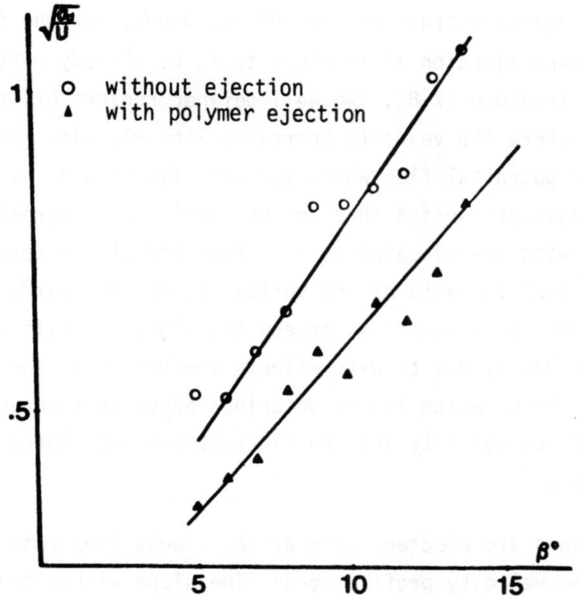

Figure 3 - Effect of polymer solution ejection on the dessinent cavitation number. Polymer concentration 1000 ppm, $q_i = 9\ 10^{-7} m^3/s$

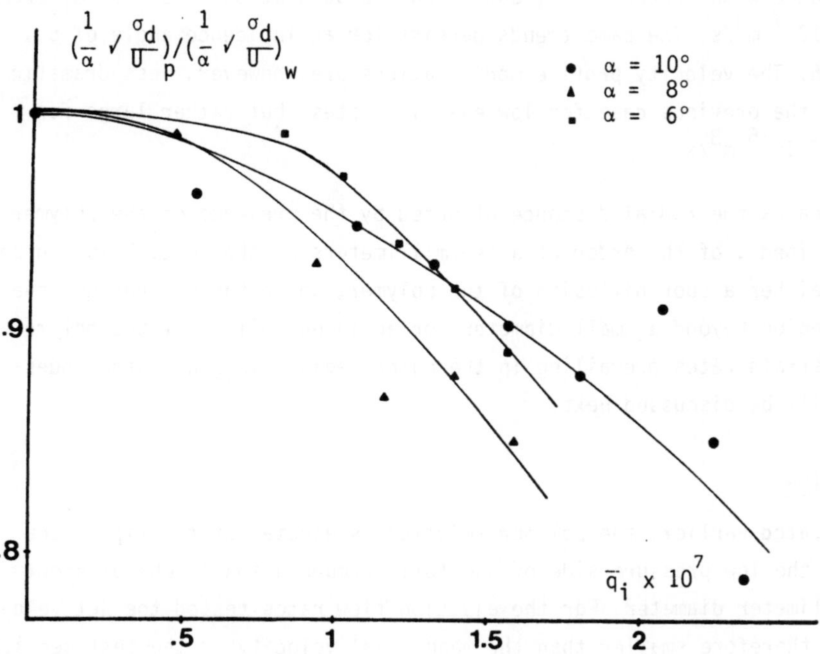

Figure 4 - Effect of ejected polymer rate on the dessinent cavitation number for different incidence angles.

Figure 5 shows the velocity profiles for 10° incidence. For the flow without polymer solution ejection it is clear that, as already discussed extensively in the literature [7,8], two well defined regions exist ; the inner viscous core where the velocity increases linearly with radial distance and the outer potential flow where the velocity behaves as the inverse of the radial distance. Notice that in our particular experimental set-up, the tip of the wing was situated at 2 cm from one of the test section vertical walls and that the axis of the vortex was moved inwards by a distance of about 11 mm. The dissimetry between the right and left side of the velocity profile is partly due to wall effects and partly to the wing wake. Moreover, the transition region is not described accurately enough because of the rapid changes on velocity and the fluctuations associated with a slight core wandering.

When polymer solutions are ejected, even at the lowest flow rates, several major changes in the velocity profile occur. The slope within the vortex core is slightly but noticeably changed, being reduced for increasing ejection flow rates. The velocity peak in the transition region is significantly reduced with increasing q_i but reaches a saturation level for q_i exceeding $7\ 10^{-7}\ m^3/s$. The same trends persist for an incidence angle of 5°, Figure 6. The velocity profile modifications are, however, less dramatic than in the previous case for low ejection rates, but rather large for $q_i = 2.2\ 10^{-6}\ m^3/s$.

In all cases the radial distance affected by the presence of the polymer is very limited ; of the order of a few millimeters at the most. This can be due to either a poor diffusion of the polymer, which has not reached the outer region beyond a small diameter, or an insensibility of the polymer to the strain rates prevailing in the outer region. This and other questions will be discussed next.

DISCUSSION

As indicated earlier, the polymer solution is ejected at the tip of the wing on the low pressure side of the foil through a small tube of about one millimeter diameter. For the ejection flow rates tested the jet velocity is therefore smaller than the mean axial velocity in the test section. It can be thus safely considered that the polymer solution is first confined within the vortex as a result of the vortex roll-up at the tip and next

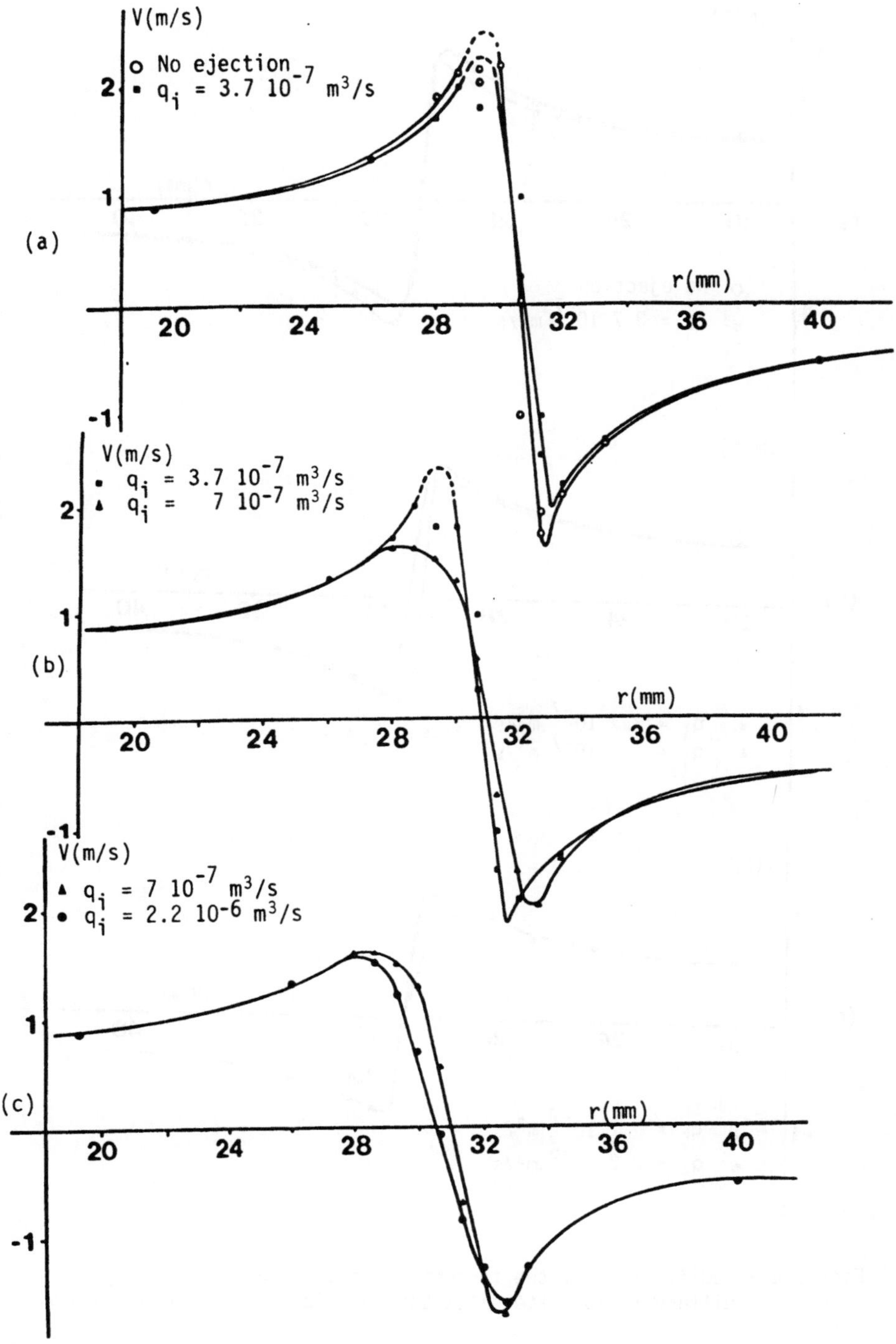

Figure 5 - Modification of the tangential velocity profile during different flow rate ejections (q_i) for $\beta = 10°$ and $U = 5$ m/s.

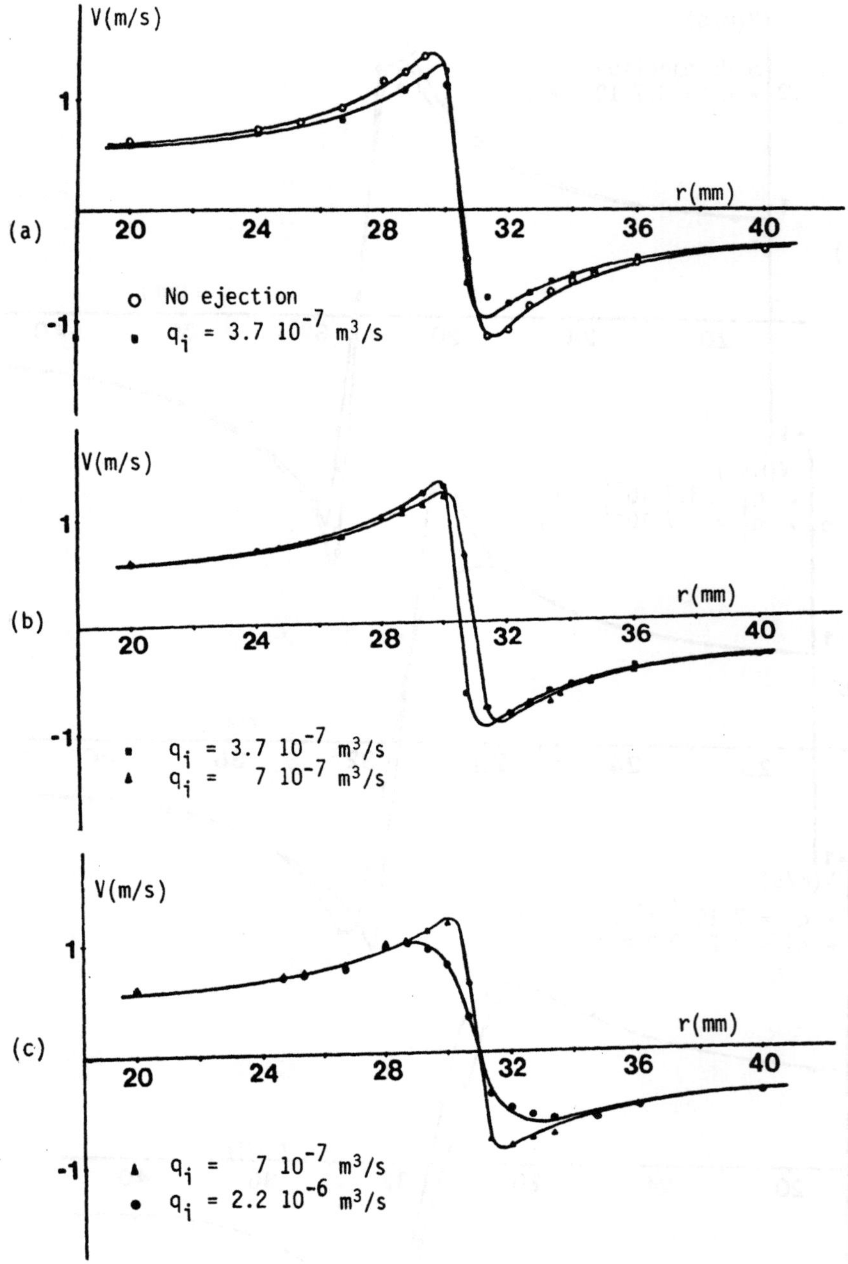

Figure 6 - Modification of the tangential velocity profile during different flow rate ejections (q_i) for β = 5° and U = 5 m/s.

convected downstream by the axial flow. The radial spreading of the polymer solution will result from the molecular and turbulent diffusion and from the radial convection due to the lack of symmetry of the flow centered on the axis of the vortex.

A considerable body of experimental evidence is available showing that the spreading of polymer solutions, with concentrations similar to the one used in this work, ejected in the center of a pipe with a cocurrent flow of water, is dramatically reduced as compared to water ejections. Indeed, Vleggaar and Tels [9] and Bewersdorff [10] have shown that the polymer solutions display a thread-like behaviour and remain coherent for very long distance-to-diameter ratios in spite of some large scale spatial fluctuations. Moreover, the present results are qualitatively similar to those obtained when the polymer solutions were ejected about 2 cm upstream of the wing tip [5] through a small (0.5 mm) capillary tube even if the diffusion might be probably enhanced in the latter case. Diffusion, laminar or turbulent, does not seem to be a predominant factor on the spreading of the polymer solution.

The net effect of the two lateral walls of the test section on the non centered vortex is to induce a maximum radial velocity of 0.15 m/s directed upwards. Because of the tangential velocity the polymer will fill up an increasingly larger circular area when convected downstream. The maximum radius occupied by the polymer at the measuring station will be 0.62 mm, comparable to the dimensions where velocity modifications occur.

Based on the previous arguments it does not seem plausible to believe that the polymer has significantly diffused. This is consistant with our observations which indicate that once the cavitation has been inhibited in the neighbourhood of the tip it does not develop much farther downstream. Also, Figure 1(b) demonstrates that for relatively low ejection flow rates a cavity may exist near the tip but it is completely inhibited downstream once the polymer has reached radial distances where it becomes "active".

Let us now turn our attention to the changes of the velocity profiles and its consequences on the pressures. The equation of motion in the radial direction is

$$\frac{\partial p}{\partial r} = \frac{V^2}{r} + \frac{\partial \sigma_{rr}}{\partial r} + \frac{\sigma_{rr} - \sigma_{\theta\theta}}{r}$$

where p is the pressure, V the tangential velocity and σ_{ii} (i = r, θ) the extra stresses. Since the values of σ_{ii} are unknown we will integrate only the first term of the second member from the wall to the center of the vortex. The computed values of $(p_w - p_o)_p/(p_w - p_o)_w$, are shown in Figure 7. These results indicate that, for constant U and α, the pressure at the center of the vortex can increase up to 60 % of the pressure difference (wall-center) of the Newtonian case. This is very much consistant with the data presented in Figure 4 since, for constant velocity and incidence, we also obtain a maximum increase of the critical pressure leading to the onset of cavitation of about 40 %. The contribution of normal stresses, if any, is probably small as compared with the contribution due to tangential velocity modifications.

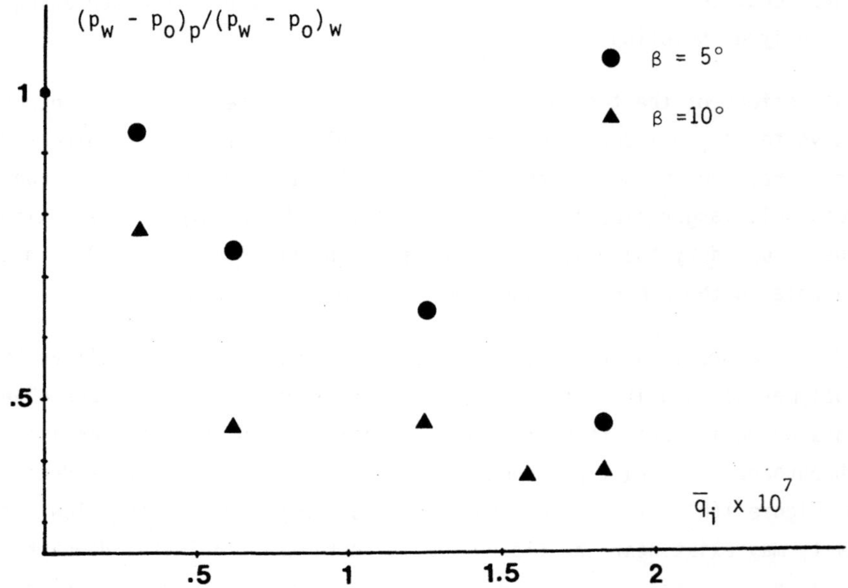

Figure 7 - Effect of polymer ejection on the pressure difference (wall-center).

INTERPRETATION

The process of tip vortex roll-up and decay is a rather complicated phenomenon even in the simplest case of flow without polymer ejection. Here we will limit ourselves to considerations of order of magnitude based on our experimental findings and to a qualitative understanding of the velocity profile modifications.

Within the core region, the polymer molecules will be subjected to an intense streching as a consequence of the strain rates prevailing in the inner region. The order of magnitude of these is given by $\Gamma/2\pi a^2$ where Γ is the circulation of the vortex and a the core radius. In the absence of ejection, our results show that they will be of 6 10^4 and 9 10^4 s^{-1} for 5 and 10° incidence angle respectively, thus nearly independent of it. Their order of magnitude is larger than the inverse of the relaxation time, $\tau = 0(10^{-3}$ s), of dilute, about 10 ppm, polymer solutions. Under such circumstances, the Deborah number, De = $\Gamma_{\tau}/2\pi a^2$, at the measuring station is 0(10). Since the core radius is an increasing function of the distance from the tip, it can be assumed that in its vicinity De will be probably larger, even if the polymer solution spreading is much reduced.

It is by now well substantiated that dilute polymer solutions have the property of modifying the flow to accommodate for stretching rates exceeding the inverse of their relaxation time. This is particularly noticeable in elongational flows such as those occurrying near a stagnation point [11] or at the entrance of an orifice [12]. Since vorticity exists in the viscous core region molecules will be streched and rotated ; while in the potential outer region vorticity is zero and molecules will not rotate. It is thus speculated that elastic effects will be initially predominant in the innermost region of the potential flow and then, because of the necessary matching of the velocity, in the viscous core region. Work is in progress to experimentally and theoretically validate this hypothesis.

CONCLUSION

Experimental evidence on tip vortex cavitation inhibition by polymer solution ejection at the tip of a finite span hydrofoil has been presented. The phenomenon is shown to be directly related to the modification of the tangential velocity profiles of a non cavitating vortex during polymer solution ejection as compared to pure-water flow. Indeed, integration of the inertia term of the equation of motion shows that the pressure in the center of the vortex is increased by up to 60 %, all other conditions being equal.

ACKNOWLEDGMENTS

The authors wish to express their deep appreciation to Mr Philippe Genoux for making possibly this work through a grant of the Direction des Recherches, Etudes et Techniques, Ministère de la Défense, France.

REFERENCES

1. Bird, R.B. and Curtiss, C.F. : Fascinating polymeric liquids, Physics Today. (1984) 36-43.

2. Chiou, C.S. and Gordon, R.J. : Vortex inhibition : Velocity profile measurements. Aiche Journal. 22,5 (1976) 947-950.

3. Inge, C. and Bark, G . : Tip vortex cavitation in water and in dilute polymer solutions. Technical Report TRITA-MEK-83-12, Royal Inst. of Technology, Stockholm, Sweden. (1983).

4. Inge, C. : Effect of polymer additives on tip vortex cavitation. Technical Report TRITA-MEK-83-05, Royal Inst. of Technology, Stockholm, Sweden. (1983).

5. Fruman, D.H. : Tip vortex cavitation inhibition by polymer additives. Proceedings of the A.S.M.E. Cavitation and Polyphase Flow Forum, New-Orleans. (1984).

6. Bovis, A. : Etude asymptotique du phénomène de cavitation. Cavités non sphériques. E.N.S.T.A. Research Report 130, Paris. (1980).

7. Baker, G.R. ; Barker, S.J. ; Bofah, K.K. and Saffman, P.G. : Laser anemometer measurements of trailing vortices in water. J. Fluid Mech.65,2 (1974) 325-336.

8. Escudier, M.P. ; Bornstein, J. and Zehnder, N. : Observation and LDA measurements of confined turbulent vortex flow. J. Fluid Mech. 98,1 (1980) 49-63.

9. Vleggaar, J. and Tels, M. : Drag reduction by polymer threads. Chem. Engin. Sc. 28 (1973) 965-968.

10. Bewersdorff, H.W. : Effect of a centrally injected polymer thread on drag in pipe flow. Rheologica Acta, 21, 4/5 (1982) 587-589.

11. James, D.F. and Acosta, A.J. : The laminar flow of dilute polymer solutions around circular cylinders. J. Fluid Mech. 42, 2 (1970)269,288.

12. Ouibrahim, A. : Effect of capillary tube diameter in laminar flows of polymer solutions. 2nd Inter. Conf. on Drag. Reduction, Cambridge (1974) F4-43-56.

The Influence of Non-Newtonian Liquids on the Cavitation Thresholds and on the Pressure Pulse Velocity in Two-Phase Mixtures

H.J. RATH and P. Desch

Dept. of Mechanics and Fluid Mechanics, Production Engineering
University of Bremen, Bad Gasteiner Straße, IW 1
D-2800 Bremen, West Germany

Summary

This paper shows the influence of non-Newtonian liquids on the acoustic cavitation thresholds. Furthermore a theoretical model is developed describing the influence of the non-Newtonian liquid on the pressure pulse velocity and on the damping effect in Two-Phase mixtures, where the carrier fluid is assumed to be a non-Newtonian liquid in which are distributed small gas bubbles of variable radius.

Acoustic Cavitation Thresholds

Cavitation is a rather general term used to describe the rupture of a liquid when subjected to sufficiently large negative pressures. Furthermore cavitation is a group of phenomena which are associated with the occurence of cavities or bubbles in liquids. From the practical point of view we know two main research areas of cavitation, namely the hydraulic cavitation and the acoustic cavitation. The hydraulic cavitation is an unwanted phenomenon and occurs in pumps, turbines, ship propellers and hydrofoils for example.

The acoustic cavitation appears in sound fields at sufficiently high sound pressure amplitudes and has found some practical applications in ultrasonic diagnostics and in industrial and chemical engineering. It is important to define criteria for the onset of cavitation. The first criterion is the appearance of visible cavities or bubbles in the liquid. It should be noted that for our investigations we will use this criterion. Other criteria are the onset of nonlinearity in the liquid coupling between a sending and receiving transducer, and the onset of noise in the liquid. For the acoustic cavitation gases must

The Influence of Polymer Additives
on Velocity and Temperature Fields
IUTAM Symposium Essen 1984
Editor: B. Gampert

be present in the liquid in the form of nuclei. The second condition is that a nuclei grow to a visible size under the action of sound waves only if the sound pressure amplitude exeeds a threshold value. The fundamental nature of vibratory cavitation is the transient growth and collapse of individual gas bubbles in liquids[1],[2].

We consider an isolated gas bubble in an unbounded incompressible liquid initially at rest. We neglecting diffusions effects and changes in temperature. The following differential equation describes the nonlinear oscillations of the spherical gas bubble boundary[3],[4]

$$\frac{d^2R}{dt^2} + 3/2\frac{1}{R}\left(\frac{dR}{dt}\right)^2 - \frac{\frac{2\sigma}{R_o}+p_o}{\rho_f}R_o^3\frac{1}{R^4} + \frac{2\sigma}{\rho_f R^2} + \frac{p}{\rho_f R} + g(R,\dot{R}) = p_A/(\rho_f R)\sin\omega t \quad ,(1)$$

in which R, R_o, ρ_f,σ,p,p_A, ω the momentary radius of the gas bubble, the equilibrium radius of the gas bubble, the density of the liquid, the surface tension, pressure of the liquid, the sound pressure amplitude and the frequency of the sound field respectively. The fuction $g(R,dR/dt, \mu)$ is the damping function due to the viscosity of the liquid. For the calculation on the cavitation thresholds we use for a first approach a very simple model, the power law model

$$g(n,m,R,\dot{R}) = \frac{4m(2\sqrt{3})^{n-1}}{n\ \rho_f}\left|\frac{\dot{R}}{R}\right|^{n-1}\frac{\dot{R}}{R}\ \frac{1}{R} \qquad (2)$$

where n is the flow behaviour index and m is the consistency index of the fluid. This model describes pseudoplastic($n<1$), dilatant ($n>1$) and Newtonian ($n=1$) flow behaviour.For the optical acoustic cavitation threshold we define a maximal visible radius of 50 μm. Esche [5] has used this value of the radius for his experimental investigations.The estimation of the optical cavitation thresholds are carried out by a numerical integration of equation(1), using a Runge-Kutta method.For example for a constant frequency of the sound field we increasing the sound pressure amplitude until a stationary maximal amplitude of 50μm is reached, so we are interested in the steady-state solution of equation(1). Figure 1 shows the cavitation thresholds-that is the critical sound pressure amplitude p_{a_k} - as a function of the frequency of the sound field for an equilibrium radius of the bubbles of R_o=5µm. The experimental boundary lines are given by Esche, obtained for expriments in water, so the experimental lines correspond to a Newtonian fluid. It is well known

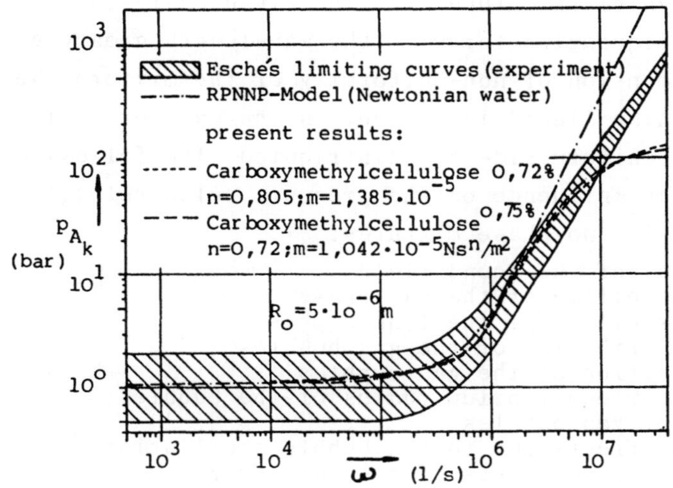

Fig.1. Cavitation thresholds for an equilibrium radius of R_o=5µm and a visible cavitation criterion of R_{max}= 50µm.

that for higher frequencies the cavitationthresholds depend on the frequency of the ultrasonic sound field. In comparison to the Newtonian case the non-Newtonian behaviour changes the surface tension and the viscosity. We see that for larger frequencies there is a big difference between the cavitation thresholds of a Newtonian and a non-Newtonian liquid.

Pressure Pulse velocity of Pressure waves of high frequency in non-Newtonian liquids containing small gas bubbles

In connection with studies on cavitation and pipe flow systems in chemical engineering it is desirable to have mathematical expressions for the pressure pulse speed and the damping of propagating pressure waves in Two-Phase flow mixtures. The propagation of pressure waves is one of the characteristic phenomena in Two-Phase flow. It is well known that pressure waves propagating in a gas- liquid mixture shows extremely different behaviour from that in a Single-Phase flow [6].Our study deals with pressure waves of finite amplitude propagating in a gas- liquid Two- Phase mixture in an elastic tube. There are a lot of parameters which have an influence on the pressure pulse velocity of propagating pressure waves of high frequency in Two-Phase mixtures through elastic tubes [7].The given analysis is restricted to sound frequencies for wich the wavelength greaty exceeds the average dimension of non-uniformity of the mixture. We assume a non-Newtonian liquid in a tube. In the carrier fluid small gas bubbles of uniform size are distributed. The following parameters have an influence on the pressure pulse velocity of pressure waves in two-phase mixtures:

* frequency of the sound field
* compressibility of the gas-phase
* compressibility of the liquid-phase
* size distribution of the gas bubbles
* void fraction of the gas bubbles
* initial or equilibrium radius of the bubbles
* number of gas bubbles
* inertia effects due to the bubble oscillations
* rheological constants of the carrier fluid
* surface tension
* shape of the gas bubble boundary
* elasticity of the pipe wall material
* relative velocity between the two phases
* diffusions effects
* temperature of the mixture

It is very difficult to consider all these parameters. So it is assumed that the gas bubbles have a spherical shape and we neglegting diffusions effects and changes of the temperature of the mixture. The mathematical model is based on a one-dimensional approach to the flow of a liquid-gas mixture, so we average the flow properties across the cross-section of the tube. Furthermore for the gas-phase and liquid-phase velocity we assume an equal mixture or pseudofluid velocity.

For the translational motion we use the following continuity equations and the momentum equation

$$\frac{\partial}{\partial t}(\rho_g \alpha A) + \frac{\partial}{\partial x}(\rho_g \alpha A u) = 0 \quad (3)$$

$$\frac{\partial}{\partial t}[\rho_f (1 - \alpha)A] + \frac{\partial}{\partial x}[\rho_f (1 - \alpha)A u] = 0 \quad (4)$$

$$[\rho_g \alpha + (1-\alpha)\rho_f][\frac{\partial u}{\partial t} + u\frac{\partial u}{\partial x}] + \frac{\partial p}{\partial x} = 0 \qquad (5)$$

where $\alpha, u, p, \rho_g, \rho_f$, A, x,t denote respectively void fraction of the gas, mixture velocity, pressure in the liquid, density of the gas, density of the liquid, momentary cross-section of the tube, coordinate along the axis of the tube and the time. We neglecting the inertia of the pipe wall and consider only small changes of the cross-section of the tube using a linear relationship between stress and strain.

$$\frac{1}{A}\partial A = \frac{D\ \mu^*}{E\ s}\ \partial p \qquad (6)$$

in which E is Young's modulus of elasticity of the pipe material, s is the wall thickness, μ^* is the pipe constraint factor and D is the mean diameter of the tube.

For the radial oscillations of the bubbles we use equation (1), where we should try to get from eq.(1) a linear differential equation with coefficients which depend on the amplitude of the oscillations. In many flows of practical interest the volumetric concentration of the gas bubbles is small. Therefore we should assume that $\alpha \ll 1$ and furthermore $\rho_g \ll \rho_f$. Then we get the following set of differential equations.

$$\frac{\partial}{\partial t}\{\rho_f(1-\alpha)A\} + \frac{\partial}{\partial x}\{\rho_f(1-\alpha)Au\} = 0 \qquad (7)$$

$$\rho_f(1-\alpha)\ \frac{\partial u}{\partial t} + u\frac{\partial u}{\partial x} + \frac{\partial p}{\partial x} = 0 \qquad (8)$$

$$\ddot{R} + f(R,\dot{R}) = 0, \qquad (9)$$

with the quantities

$$f(R,\dot{R}) = \frac{3}{2}(\dot{R}^2/R) + B_1/R^2 + B_2/R + B_3 \mathrm{sign}(\dot{R}/R)\left|\frac{\dot{R}}{R}\right|^{n}\frac{1}{R} - B_4/R^4 \qquad (10)$$

$$B_1 = 2\ \sigma/\rho_f \quad ; B_2 = p/\rho_f ; \ B_3 = 4m(2\sqrt{3})^{n-1}/(n\rho_f) \qquad (11)$$

$$B_4 = p_{go}R_o^3/\rho_f ; \ B_5 = \frac{D\ \mu^*}{E\ s} ; \ \rho = \rho_f(1-\alpha) \ ; p_{go} = p_o + 2\sigma/R_o . \qquad (12)$$

We consider now the nonlinearity of eq.(9) and eq.(10) in an approximate way . Using the method of harmonic balance [8]we expect a harmonic solution of eq.(9) and so it is suitable to to put up for the solution the following expression

$$R = \bar{R}_o(1 + \gamma\cos\omega t) \quad \text{and also} \quad \dot{R} = -\bar{R}_o\gamma\omega\sin\omega t \tag{13}$$

in which $\bar{R}_o$ depends from the amplitude γ. Now we expanding the function $f(R,\dot{R})$ in a Fourier series

$$f(R,\dot{R}) = f(\bar{R}_o(1+\gamma\cos\omega t), -\bar{R}_o\gamma\omega\sin t) \approx \frac{a_o}{2} + a_1\cos\omega t + b_1\sin\omega t \tag{14}$$

$$f(R,\dot{R}) = \frac{a_o}{2} - \frac{a_1}{\gamma} + \frac{a_1}{\bar{R}_o\gamma} R - \frac{b_1}{\bar{R}_o\gamma\omega}\dot{R} \tag{15}$$

in which the coefficients are defined by

$$a_o = \frac{1}{\pi}\int_0^{2\pi} f(R,\dot{R})\,d\omega t \;;\; a_1 = \frac{1}{\pi}\int_0^{2\pi} f(R,\dot{R})\cos\omega t\,d\omega t \tag{16}$$

$$b_1 = \frac{1}{\pi}\int_0^{2\pi} f(R,\dot{R})\sin\omega t\,d\omega t \tag{17}$$

The coefficients a_o and a_1 we calculate analytically while for the calculation of b_1 we must use a numerical integration. For the Fourier coefficients we get the following quantities.

$$a_o = 3\omega^2\bar{R}_o(1-\sqrt{(1-\gamma^2)}) + \frac{2\,B_1}{\bar{R}_o^2(1-\gamma^2)^{3/2}} + \frac{2B_2}{\bar{R}_o\sqrt{(1-\gamma^2)}} - \frac{(2+3\gamma^2)B_4}{\bar{R}_o^4(1-\gamma^2)^{7/2}} \tag{18}$$

$$a_1 = 3\omega^2\bar{R}_o\left[\frac{\sqrt{1-\gamma^2}-1}{\gamma} + \gamma/2\right] - \frac{2B_1\,\gamma}{\bar{R}_o^2(1-\gamma^2)^{3/2}} + \frac{2\,B_2}{\bar{R}_o\,\gamma}\left(1 - \frac{1}{\sqrt{1-\gamma^2}}\right) + \frac{B_4\gamma(4+\gamma^2)}{\bar{R}_o^4(1-\gamma^2)^{7/2}} \tag{19}$$

$$b_1 = -\frac{2\,B_3\,\gamma^n\omega^n}{\pi\,\bar{R}_o}\int_0^{\pi}\left|\frac{\sin\omega t}{1+\cos\omega t}\right|^{n+1} d\omega t \tag{20}$$

Using these coefficients we get the differential equation

$$\ddot{R} - \frac{b_1(\gamma)}{\bar{R}_o\gamma\omega}\dot{R} + \frac{a_1(\gamma)}{\bar{R}_o\,\gamma}R + \frac{a_o(\gamma)}{2} - \frac{a_1(\gamma)}{\gamma} = 0 \tag{21}$$

in which $\bar{R}_o$ depends on the quantity γ. The natural frequency of the undamped system is given by

$$\omega_0^2(\gamma) = \frac{a_1(\gamma)}{\bar{R}_o\ \gamma} \quad . \tag{22}$$

For =0 we get the linear frequency of free oscillations of a gas bubble:

$$\omega_{0M}^2 = \frac{1}{\rho_f R_o^2}[3(p_o + \frac{2\sigma}{R_o}) - \frac{2\sigma}{R_o}] \tag{23}$$

For the natural frequency of the damped system we obtain

$$\omega_D = \omega_0(\gamma)\sqrt{1 - D^2(\gamma)} \tag{24}$$

in which the damping coefficient is given by

$$D(\gamma) = \frac{1}{2}(-\frac{b_1(\gamma)}{\bar{R}_0\gamma\omega_0})\frac{1}{\omega_0(\gamma)} \quad . \tag{25}$$

It may be noted that for a Newtonian liquid we get the equation for the natural frequeny in a closed form.
Now we try to formulate the wave equation to detemine the phase velocity and the damping of the system. For the pressure wave propagation we assume linear waves $p_g/p_{g_o} = 1 + \varepsilon\xi$, in which ε is a small number. For the wave equation we need the expressions

$$\frac{\rho_g\alpha}{1-\alpha} = \frac{\rho\ \alpha_0}{1-\alpha_0}\ ;\quad p_g/\rho_g = p_{g_o}/\rho_g\ ;\ R = R_o(p_{g_o}/p_g)^{1/3} \tag{26}$$

With help of equations (7),(8),(9)and (21) we obtain for the first order in ε the wave equation

$$\begin{aligned}&\frac{\partial^2\xi}{\partial t^2} - \frac{B_5B_6}{3\ \alpha_0}(R_0\frac{\partial^4\xi}{\partial t^4} - R_o\frac{b_1(\gamma)}{\omega_0\bar{R}_o\gamma}\frac{\partial^3\xi}{\partial t^3} + 3B_7\frac{\partial^2\xi}{\partial t^2}) + \\ &\frac{B_6}{3\alpha_0\rho_f\ (1-\alpha_0)}(R_o\frac{\partial^4\xi}{\partial x^2\partial t^2} - R_o\frac{b_1(\gamma)}{\omega_0\bar{R}_o\gamma}\frac{\partial^3\xi}{\partial x^2\ \partial t} + 3B_7\frac{\partial^2\xi}{\partial x^2}) = 0\end{aligned} \tag{27}$$

This is a partial differential equation of fourth order with the quantities

$$H_1 = (\sqrt{1-\gamma^2} - 1)/\gamma^2\ ;\ H_2 = 2H_1(1 - R_o/\bar{R}_o) - 1 \tag{28}$$

$$H_3 = \frac{3\omega_0^2\bar{R}_O}{2}(1-\sqrt{1-\gamma^2}) + \frac{2\sigma}{\rho_f\bar{R}_o^2(1-\gamma^2)^{3/2}} + \frac{(2+3\gamma^2)P_{go}R_o^3}{2\rho_f\bar{R}_o^4(1-\gamma^2)^{7/2}} \quad (29)$$

$$H_4 = 3\omega_0^2\bar{R}_o\left(\frac{\sqrt{1-\gamma^2}-1}{\gamma^2}+1/2\right) - \frac{2\sigma}{\rho_f\bar{R}_o^2(1-\gamma^2)^{3/2}} + \frac{R_o^3P_{go}(4+\gamma^2)}{\rho_f\bar{R}_o^4(1-\gamma^2)^{7/2}} \quad (30)$$

$$B_6 = \rho_f\bar{R}_o\sqrt{1-\gamma^2}/H_2 \;;\; B_7 = \frac{H_4R_o}{3\bar{R}_o} + \frac{2H_1R_o}{3\bar{R}_o\,H_2}\left(\frac{H_4R_o}{\bar{R}o} + H_3 - H_4\right) \quad (31)$$

Using soltutions of the wave equation of the form

$$\xi = \exp i(k\,x - \omega t) \quad (32)$$

we obtain from (27) the dispersion equation

$$\frac{k^2}{\omega^2} = \frac{-[1+\frac{B_5B_6}{3\,\alpha_0}(R_o\omega^2 - i\frac{R_o\omega b_1(\gamma)}{\omega_0\,\ddot{R}_o\,\gamma} - 3B_7)][3\alpha_0(1-\alpha_0)\rho_f]}{B_6B_7} \,. \quad (33)$$

The phase speed of the system corresponds to the real part of the dispersion equation (33) while the imaginary part of (33) is the damping of the system. It may be noted that for a Newtonian liquid neglecting the surface tension and assuming $\gamma=0$ we get from the disperion equation the relationship given by Wijngaarden[7].

Figure 2 shows the phase velocity as a function of the frequency of the sound field for three different values of void fraction α_0 and for an equilibrium radius of 20^{-6}m and $\gamma=0$. The pipe factor is zero, which corresponds to a mixture flow without pipe walls. We see that for the case of resonance we have a very strong dependence of the phase velocity from the void fraction. The influence of the pipe factor B_5 on the pressure pulse velocity is plotted in Figure 3. The phase velocity is a function of the frequency of the sound field. We see that with increasing values of the pipe factor the phase velocity decreases. Figure 4 shows the influence of the void fraction and the frequency on the damping effect of the system. It may be noted that the imaginary part of the dispersion relation correspons to the damping of the pressure waves. Figure 4 shows the strong dependence of the damping parameter from the void fraction (see also [9]).

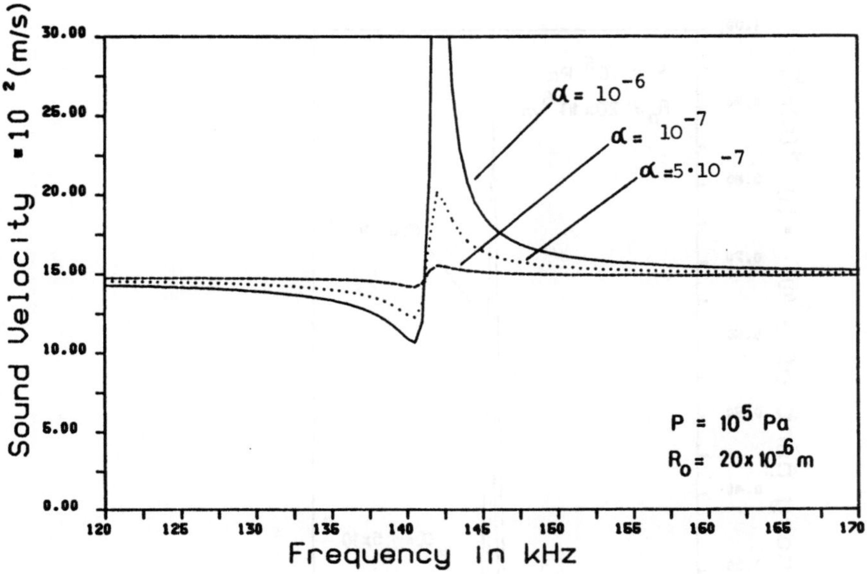

Fig.2. Phase velocity versus frequency of the sound field for different values of void fraction

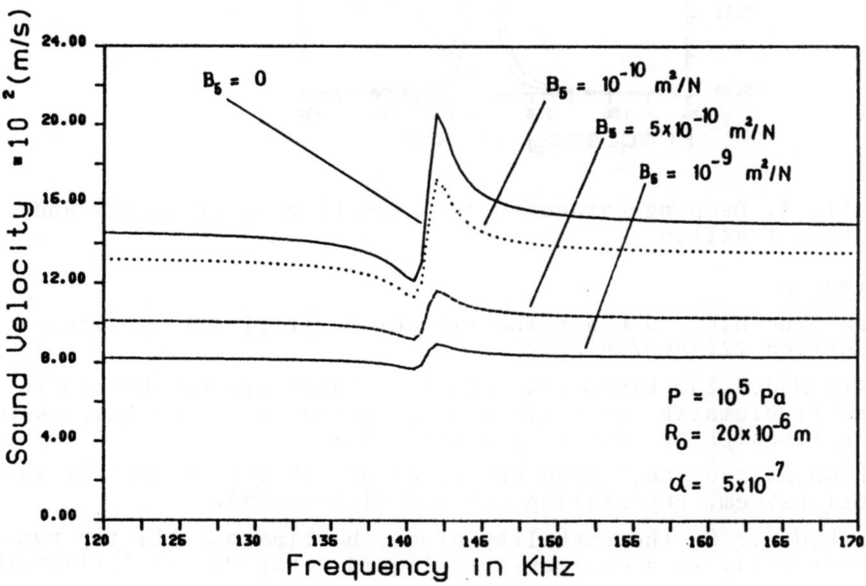

Fig.3. Phase velocity as a function of the frequency of the sound field for different values of the pipe factor

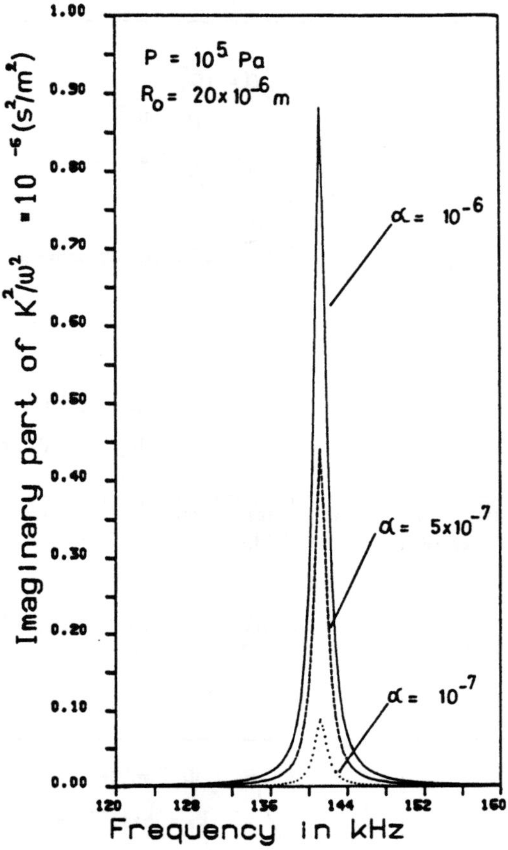

Fig.4. Damping parameter as a function of frequency and void fraction

References

1. Lauterborn,W.:Zu einer Theorie der Kavitationsschwellen. Acustica 22(1969/70)813.
2. Rath,H.J.: Zur Kinematik sphärisch kavitierender Gasblasen und Problematik der Kavitationsschwellen in einer kompressiblen Flüssigkeit. Acustica 43(1979)242.
3. Shima,A.;Tsujino,T.:The behaviour of bubbles in polymer solutions.Cem. Engineering Science 31(1976)863.
4. Rath,H.J.: On the stability of gas bubbles oscillating nonspherically in a compressible liquid.J.Eng.Mathem.15(1981)161.
5. Esche,R.:Untersuchungen zur Schwingungskavitation in Flüssigkeiten.Acustica 2(1952)208.
6. Rath,H.J.:Unsteady pressure waves and shock waves in elastic tubes containing bubbly air-water mixtures.Acta Mechanica 38 (1981)1.

7. Wijngaarden,van,L.: One-dimensional flow of liquids containing small gas bubbles. Annual Rev. Fluid Mechanics 4(1972) 369.

8. Magnus,K.: Schwingungen.Teubner Verlag,Stuttgart(1969).

9. Rath,H.J.; Desch,P.: Propagation of pressure waves in Non-Newtonian fluids containing small gas bubbles. (to be published in Acta Mechanica (1975))

Session 13
Heat and Mass Transfer

Nucleate Boiling in Drag-Reducing Polymer Solutions

D. D. PAUL

Batelle's Columbus Laboratories
Columbus, Ohio

S. I. ABDEL-KHALIK

Department of Nuclear Engineering
University of Wisconsin - Madison

Abstract

This paper summarizes the important findings of all our previous work concerning nucleate boiling in drag-reducing polymer solutions. Nucleate boiling curves are presented for aqueous solutions of three different molecular weight grades of hydroxyethyl cellulose and three different types of polyacrylamide. The boiling curves have been measured using an electrically heated platinum wire submerged in a saturated pool of liquid at atmospheric pressure. When compared to the boiling curve of the pure solvent, in this case water, all of the polymer solution boiling curves showed higher temperature differences as the polymer solution viscosity increased. These results differ from those reported by earlier investigators for pool boiling of dilute polymer solutions on heated flat plates.

In addition, for one of the hydroxyethyl cellulose solutions and for one of the polyacrylamide solutions, the boiling bubble dynamics along with the entire heated length of the wire have been photographed using a high speed movie camera. When compared with the pure solvent boiling at the same heat flux, both polymer solutions showed a considerable decrease in the number density of active nucleation sites, slightly smaller average bubble departure diameters, and slightly higher average bubble departure frequencies. The decrease in the active nucleation site density for the polymer solutions is primarily responsible for the higher temperature differences needed to maintain the same boiling heat flux in our experiments.

Introduction

Kotchaphakdee and Williams [1] reported in 1970 that the nucleate boiling performance of water was significantly enhanced by the addition of small amounts of the drag-reducing polymers, polyacrylamide and hydroxyethyl cellulose. In 1978, Miaw [2] repeated these tests with the same two polymers and reported similar results. Gannet and William [3] in 1971 showed that the boiling performance of organic systems could also be improved

The Influence of Polymer Additives
on Velocity and Temperature Fields
IUTAM Symposium Essen 1984
Editor: B. Gampert

by the addition of polymers. All of these studies were performed using heated flat plates.

The objective of this investigation has been to examine the nucleate boiling performance of aqueous solutions of three different molecular weight grades of hydroxyethyl cellulose (Natrosol 250 MR, 250 HR, and 250 HHR manufactured by Hercules, Inc.) and three different types of polyacrylamide (Separan AP-30, NP-10P, and MGL manufactured by Dow Chemical Co.). A horizontal, electrically heated platinum wire, 0.30 mm in diameter and 30.0 mm long has been used. The wire is thin enough that bubbles completely envelope the wire as they grow. Nucleate boiling curves for each polymer solution have been measured at polymer concentrations corresponding to relative viscosities of 1.08 and 1.32. In addition, slow motion films of the entire heated length of the platinum wire have been analyzed in order to compare the bubble dynamics of the pure solvent with those obtained in the Natrosol 250 HHR and Separan AP-30 solutions.

Boiling Curves for Drag-Reducing Polymer Solutions

The experimental apparatus and procedure used in this investigation are identical to those reported earlier in our studies of boiling along a heated platinum wire (refs. [4] and [5]). The results presented in this section are a summary of our previously published work on the nucleate boiling curves of drag-reducing polymer solutions [5].

All of the polymer solutions used in this investigation were freshly prepared and allowed to age 24 hours before testing. This procedure insures that the solutions are homogeneous. An alternative method, employed by Miaw [2], is to inject a highly concentrated solution into hot distilled water just prior to testing. Studies by Hinch and Elata [6] have shown that this leads to highly inhomogeneous solutions. In addition, the dissolution of dissolved gases in the injected solution could lead to spurious results for boiling studies.

For the six polymers examined, the solutions' viscosities have been measured as a function of concentration [7]. These data have been used to determine intrinsic viscosity, Huggins constant and, where appropriate, the viscosity-average molecular weight for each polymer as shown in Table I. The rheological properties of these polymers [8-10] have been reported elsewhere. The concentrations of polymers used in this investigation correspond to relative viscosities less than or equal to 1.32; this is low enough to insure that shear thinning effects are negligible.

Table 1. Intrinsic Viscosities, Huggins Constants, and Viscosity-Average Molecular Weights for the Polymers Examined in this Investigation

Polymer	Intrinsic Viscosity $(wppm)^{-1}$	Huggins Constant	Molecular Weight
Natrosol 250 MR	.000805	0.98	4.6×10^5
Natrosol 250 HR	.00120	0.72	4.3×10^5
Natrosol 250 HHR	.00135	0.87	8.4×10^5
Separan AP-30	.00820	0.37	unknown
Separan NP-108	.00330	0.00	unknown
Separan MGL	.00104	0.37	4.9×10^6

In addition to solution viscosity, interfacial surface tensions have also been measured as a function of polymer concentration [7]. In general, the polyacrylamides did not significantly change the surface tension of water over the concentration range of interest. On the other hand, the hydroxyethyl celluloses did lower the interfacial surface tension of water by about 5 percent over the concentration range of interest. Other than solution viscosity and surface tension, all other physical properties of the solvent have been assumed to remain unchanged at the very low polymer concentrations used in these experiments.

To provide a common basis for comparisons, all of the boiling curve data for the polymer solutions have been run at constant relative viscosities. Figures 1(a) and 1(b) show the nucleate boiling curves for water and the six aqueous polymer solutions all at relative viscosities of 1.08 and 1.32, respectively. The polymer concentrations required to give these relative viscosities may be calculated from the intrinsic viscosities and Huggins constants given in Table 1. For all of the polymers tested, the nucleate boiling curves were shifted to higher temperature differences when compared to the boiling curve of pure water. The shifts all showed the same trend, i.e., as the relative viscosity increased the relative temperature shift increased. These results differ from those reported by earlier investigators [1-3] who observed temperature shifts in the opposite direction to those reported here. Our experiments have been conducted using electrically heated platinum wires whereas previous investigators used heated flat plates. While the hydrodynamics associated with bubble growth and departure on

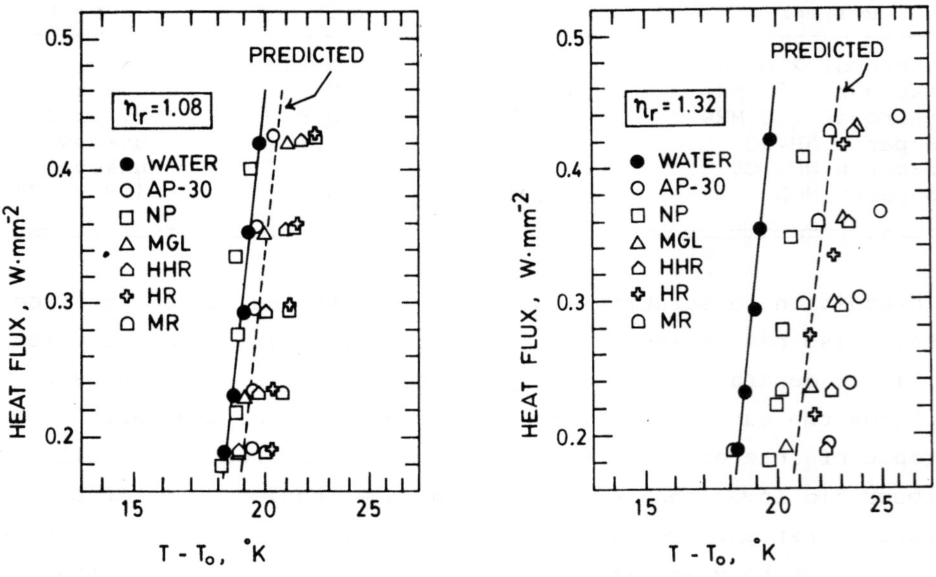

(a) (b)

Fig.1. The nucleate boiling curve data of water and the six polymers of interest at concentrations corresponding to relative viscosities of (a) 1.08 and (b) 1.32.

wires are different from those on flat surfaces, it is not clear why the results should be opposite.

In order to examine how changes in the physical properties of the solution are expected to affect nucleate boiling heat transfer, the Rohsenow [11] pool boiling correlation for platinum surfaces has been used to predict the relative temperature shift. For a given heat flux, the proportionality relationship between the temperature difference and changes in solution properties gives:

$$(\Delta T)_r = \sigma_r^{1/6} \eta_r^{2/3} \quad . \tag{1}$$

Since the polymeric additives examined change the surface tension slightly, and the relative surface tension in Equation 1 is raised to the one-sixth power, this effect is ignored inasmuch as it is considerably less than the viscosity effect.

The curve shifts predicted by the Rohsenow correlation due to changes only in solution viscosity are shown in Figures 1(a) and 1(b). Although there is considerable variation between polymer type, both the magnitudes and directions of the predicted shifts agree reasonably well with the experimental results for polymer additives. Thus, most of the observed changes in the boiling curves for our experiments can be explained based only on how the polymer additives change the solution viscosity.

Boiling Bubble Dynamics in Drag-Reducing Polymer Solutions

In addition to the boiling curve measurements, the bubble dynamics have been photographed along the entire heated surface of the platinum wire at a saturated boiling heat flux of 0.358 $\pm$ 0.006 $W.mm^{-2}$ for the Natrosol 250 HHR and Separan AP-30 solutions. Both polymer solutions have been tested only at a relative viscosity of 1.08. Slow motion films of the bubble dynamics have been analyzed to determine the average number density of active nucleation sites, and the frequency distribution of bubble departure diameters.

A nucleation site is considered active as soon as a bubble appears on the heated surface and is considered inactive the moment the bubble departs from the surface. At the same boiling heat flux, the time-averaged active nucleation site density values for the polyacrylamide and hydroxyethyl cellulose solutions are 270 and 330 sites per meter, respectively; these are considerably lower than the water value of 405 sites per meter. These differences are best explained by realizing that the addition of polymers to distilled water results in a unique fluid-surface interaction for each type of polymer. When compared to water, the decrease in the active nucleation site density for the polymer solutions could be the primary reason for the increased temperature difference needed to give the same boiling heat flux in our experiments.

The departure diameter for each bubble leaving the heated surface has also been measured. If two bubbles combine before leaving the surface, their departure diameters are measured separately at the moment just before they combine. By taking these measurements over a known period of time, typically 1000 frames, a distribution of bubble departure diameters is obtained.

Figure 2(a) shows the frequency distribution of bubble departure diameters for pure water boiling at 0.352 $W.mm^{-2}$. The histogram portion of the graph represents the actual experimental data. The curve drawn on the plot is a fitted asymptotic expansion of the normal frequency function F(D) (see [4]), defined such that F(D)dD is the number of bubbles leaving the surface per nucleation site per unit time with diameters between D and D + dD. Values of the average bubble departure frequency per nucleation site <F>, the average bubble departure diameter <D>, the standard deviation of bubble departure diameters σ_D, the skewness coefficient γ_1, and the kurtosis coefficient γ_2 are given on each plot.

Figures 2(b) and 2(c) show the frequency distributions of bubble departure diameters for the polyacrylamide and the hydroxyethyl cellulose solutions, respectively. In both cases, the polymer solution's average bubble departure diameters (<D> = 1.19 mm for the polyacrylamide and <D> = 1.13 mm for the

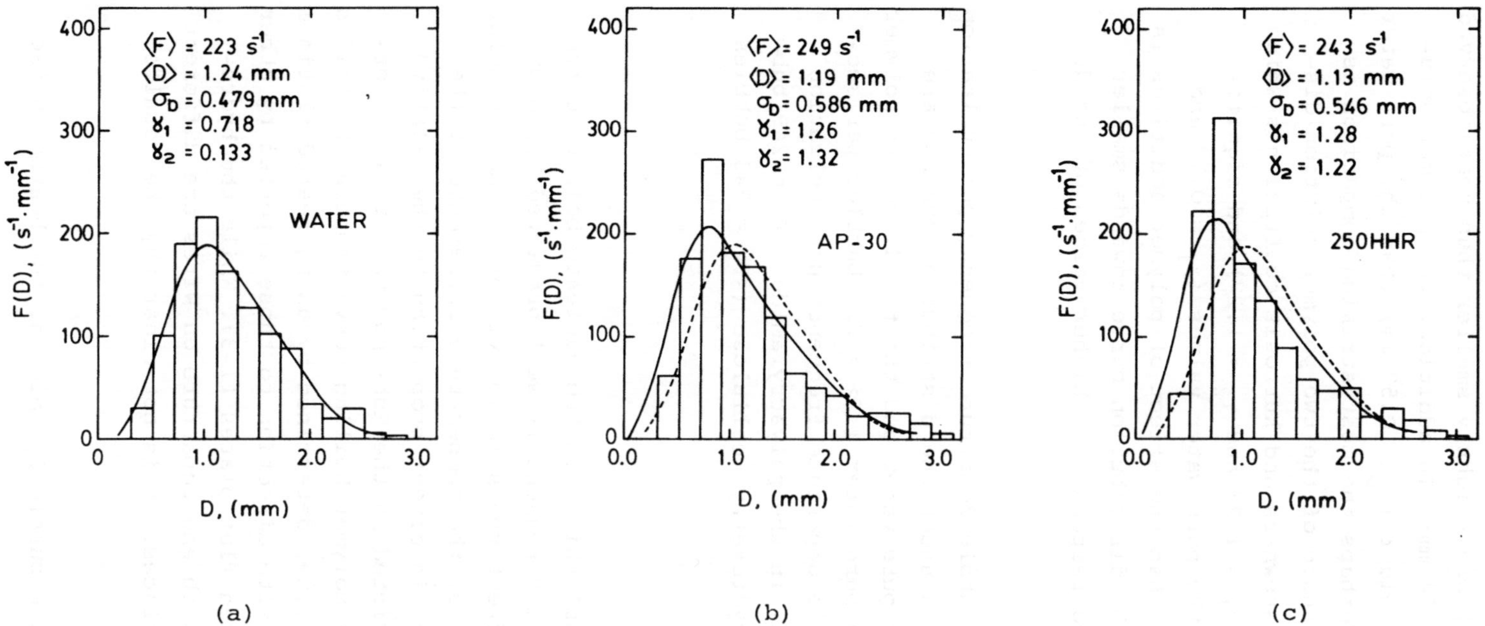

Fig.2. The frequency distributions of bubble departure diameters for: (a) pure water boiling at a heat flux of 0.352×10^6 $W.m^{-2}$, (b) a polyacrylamide solution ($\eta_r = 1.08$) boiling at a heat flux of 0.364×10^6 $W.m^{-2}$, and (c) a hydroxyethyl cellulose solution ($\eta_r = 1.08$) boiling at a heat flux of 0.357×10^6 $W.m^{-2}$. The dotted lines drawn in Figs. 2(b) and 2(c) corresponds to the distribution for pure water.

hydroxyethyl cellulose) are slightly smaller than that observed in pure water (<D> = 1.24 mm). In addition, the polymer solution distributions are wider ($\sigma_D \approx 0.57$ mm) than the pure water case ($\sigma_D = 0.48$ mm). Perhaps the most striking comparison is the similarity in the shape of the two polymer solution distributions; both have skewness and kurtosis coefficients which are nearly identical ($\gamma_1 \approx 1.27$ and $\gamma_2 \approx 1.27$) and significantly different from the pure water values ($\gamma_1 = 0.72$ and $\gamma_2 = 0.13$). This shows that the effect of polymer additives is to shift the peak in the distribution curve towards smaller bubble diameters and correspondingly higher values of the departure frequency.

Values of the average bubble departure frequency per nucleation site <F>, i.e., the area under the distribution curves, are given on each plot for pure water and the two different polymer solutions. Compared to pure water (<F> = 223 bubbles per second), the average bubble departure frequency per nucleation site increased slightly in the polyacrylamide (<F> = 249 bubbles per second) and hydroxethyl cellulose (<F> = 243 bubbles per second) solutions.

Conclusions

The observed temperature shifts in the nucleate boiling curves for polymer solutions are reasonably well predicted based only on how the polymers affect the solution viscosity. As the solution viscosity increases, the temperature difference of the boiling curve increases in direct proportion to the relative temperature shift predicted by the Rohsenow pool boiling correlation. However, the polymer boiling curve temperature shifts from the pure solvent line, determined by using heated platinum wires, are in the opposite direction to those reported by other investigators for heated flat plates [1-3]. While the hydrodynamics of bubble growth and departure on wires are different from those on flat surfaces, it is not clear why the results should be opposite.

From measurements of the bubble dynamics at the heated surface hydroxyethyl cellulose) are slightly smaller than that observed in pure water (<D> = 1.24 mm). In addition, the polymer solution distributions are wider ($\sigma_D \simeq 0.57$ mm) than the pure water case ($\sigma_D = 0.48$ mm). Perhaps the most striking comparison is the similarity in the shape of the two polymer solution distributions; both have skewness and kurtosis coefficients which are nearly identical ($\gamma_1 \simeq 1.27$ and $\gamma_2 \simeq 1.27$) and significantly different from the pure water values ($\gamma_1 = 0.72$ and $\gamma_2 = 0.13$). This shows that the effect of polymer additives is to shift the peak in the distribution curve towards smaller bubble diameters and correspondingly higher values of the departure frequency.

Values of the average bubble departure frequency per nucleation site <F>, i.e., the area under the distribution curves, are given on each plot for pure water and the two different polymer solutions. Compared to pure water (<F> = 223 bubbles per second), the average bubble departure frequency per nucleation site increased slightly in the polyacrylamide (<F> = 249 bubbles per second) and hydroxyethyl cellulose (<F> = 243 bubbles per second) solutions.

Conclusions

The observed temperature shifts in the nucleate boiling curves for polymer solutions are reasonably well predicted based only on how the polymers affect the solution viscosity. As the solution viscosity increases, the temperature difference of the boiling curve increases in direct proportion to the relative temperature shift predicted by the Rohsenow pool boiling correlation. However, the polymer boiling curve temperature shifts from the pure solvent line, determined by using heated platinum wires, are in the opposite direction to those reported by other investigators for heated flat plates [1-3]. While the hydrodynamics of bubble growth and departure on wires are different from those on flat surfaces, it is not clear why the results should be opposite.

From measurements of the bubble dynamics at the heated surface of the wire, the time-averaged number density of active nucleation sites has been shown to be considerably less in the polymer solutions than in the pure solvent. This could be responsible for the increased temperature differences required in the polymer solutions in order to give the same boiling heat flux as the pure solvent.

Nomenclature

D	bubble departure diameter;
F	frequency distribution function (see [4]);
$(\Delta T)_r$	relative temperature difference, i.e., the ratio of the boiling curve temperature difference in the polymer solution to that in the pure solvent;
T	wire temperature;
T_o	saturation temperature of liquid.

Greek

γ_1	skewness coefficient of frequency distribution function ([4]);
γ_2	kurtosis coefficient of frequency distribution function ([4]);
η_r	relative viscosity, i.e., the ratio of the polymer solution viscosity to that of the pure solvent;
σ_D	standard deviation of bubble departure diameters;
σ_r	relative surface tension, i.e., the ratio of the polymer solution surface tension to that of the pure solvent.

Others

$\langle\ \rangle$	represents average of quantity enclosed.

References

1. Kotaphakdee, P. and Williams, M.C.: Enhancement of nucleate pool boiling with polymer additives. Int. J. Heat Mass Transfer 13 (1970) 835-848.

2. Miaw, C.-B.: A study of heat transfer to dilute polymer solutions in nucleate pool boiling. Ph.D. Thesis, University of Michigan 1978.

3. Gannett, H.J. and Williams, M.C.: Pool boiling in dilute non-aqueous polymer solutions. Int. J. Heat Mass Transfer 14 (1971) 1001-1005.

4. Paul, D.D. and Abdel-Khalik, S.I.: A statistical analysis of saturated nucleate boiling along a heated wire. Int. J. Heat Mass Transfer 26 (1983) 509-519.

5. Paul, D.D. and Abdel-Khalik, S.I.: Nucleate boiling in drag-reducing polymer solutions. Journal of Rheology 27(1) (1983) 59-76.

6. Hinch, E.J. and Elata, C.: Heterogeneity of dilute polymer solutions. J. Non-Newtonian Fluid Mech. 5 (1979) 411-425.

7. Paul, D.D.: Nucleate boiling in drag-reducing polymer solutions. Ph.d. Thesis, University of Wisconsin-Madison 1982.

8. Bruce, C. and Schwarz, W.H.: Rheological properties of ionic and nonionic polyacrylamide solutions. J. Polymer Sci. A-2, 7 (1969) 909-927.

9. Tomita, Y. and Mochimaru, Y.: Normal stress measurement of dilute polymer solutions. J. Non-Newtonian Fluid Mech. 7, (1980) 235-255.

10. Naik, S.C.; Pittman, J.F.T.; Richardson, J.F.: The rheology of hydroxyethyl cellulose solutions. Trans. Soc. Rheology 20(4) (1976) 639-649.

11. Rohsenow, W.M. and Hartnett, J.P. (eds.) Handbook of Heat Transfer. New York: McGraw-Hill 1973.

Mass Transfer in Elongational Flows of Dilute Polymer Solutions

A. AMBARI, C. DESLOUIS, B. TRIBOLLET

L.P. N° 15 du CNRS "Physique des Liquides et Electrochimie", associé à l'Université Pierre et Marie Curie,
4, place Jussieu, 75230 Paris Cedex 05, France.

Summary

The analysis of mass transfer from a small cylindrical wire or from a rotating disk in a dilute polymer solution by using an electrochemical technique is reported here. Steady state measurements showed a large decrease of the mass transfer rate with respect to the newtonian behaviour, starting from critical velocity values. This phenomenon was ascribed to the fast stretching of the polymer coils which occurs when the elongational gradient equals the reciprocal of the polymer relaxation time.
Fluctuations arise at the transition and result from an hydrodynamic instability induced by the viscosity increase during elongation.

Introduction

The drag reduction phenomenon by linear high polymers is now widely considered as determined by the presence of local elongational strains in turbulent flows and by their action on these compounds.

Therefore, an increasing interest in this field arose concerning the possibility of analyzing the behaviour of the polymers in the presence of more simple and well characterized elongational flows. Pure elongational flows are not so easy to devise, and for those involving a stagnation point at a wall, the use of classical experimental techniques appears more difficult.

Electrochemical techniques, based on mass transfer measurements with a redox reaction, enable an easy investigation of the near wall region and were thus selected to this end.

Two flow geometries with prevailing elongational strains were studied within the scope of this work : The rotating disc in a free volume and the uniform flow around a small cylinder.

In previous works, it has been experimentally established that mass transfer is very disturbed starting from a critical flow velocity U^* for the cylinder [1], or from a critical angular velocity Ω^* for the disc [2] ; those critical values were shown to depend on the physicochemical parameters of the polymer solution.

The Influence of Polymer Additives
on Velocity and Temperature Fields
IUTAM Symposium Essen 1984
Editor: B. Gampert

The first aim of this work was to precise the relationships between the molecular properties of the polymer and the flow-characteristics in dilute solutions, and the second to propose an explanation of the instabilities generated beyond the critical velocities.

Analysis

According to theoretical predictions by de Gennes [3] and Hinch [4] namely, for an elongational gradient S high enough (exceeding a critical value S^*) a linear flexible chain polymer, originally under a coil shape, could be steeply stretched into a fully extended shape. This coil $\rightleftarrows$ stretch transition is expected when the criterion $S.\tau_Z > 1$ is fulfilled.

$$\tau_Z = \eta_S R_F^3 / k_B T \tag{1}$$

is the Zimm relaxation time of the polymer coil in an infinitely dilute solution [5], η_S is the solvent viscosity and R_F is the Flory gyration radius.

For both flow geometries (disc or cylinder) the elongational gradients can be readily evaluated since the flow fields are well known.

For the disc system, an earlier analysis of V. Karman [6], allowed us to derive the rate of strain tensor, and thus to calculate an elongational gradient equal to 2 F.Ω [2], where F is the dimensionless radial velocity component which is only dependent on the reduced axial coordinate $\eta = y\sqrt{\frac{\Omega}{\nu}}$. This elongational gradient reaches a maximum value of 0.36 Ω for an approximate distance of $\sqrt{\frac{\nu}{\Omega}}$ roughly equal to the boundary layer thickness.

For the cylinder upstream in the plane of symmetry (P) (Fig. 1) the macromolecules "see" a pure elongational flow (vorticity is zero in this plane by virtue of symmetry). By considering the velocity field established by Lamb[7] the following maximum elongational gradient is found for $Re < 5$:

$$S_{max} = \frac{2}{9} \frac{\sqrt{3} \quad U_\infty}{a(\ln(8/Re) - 0.077)}$$

at a distance($\sqrt{3}$.a)from the cylinder axis,independent of U_∞. a is the cylinder radius.

Experimental

1. Electrochemical method

The principle consists in measuring the electrical current I flowing through

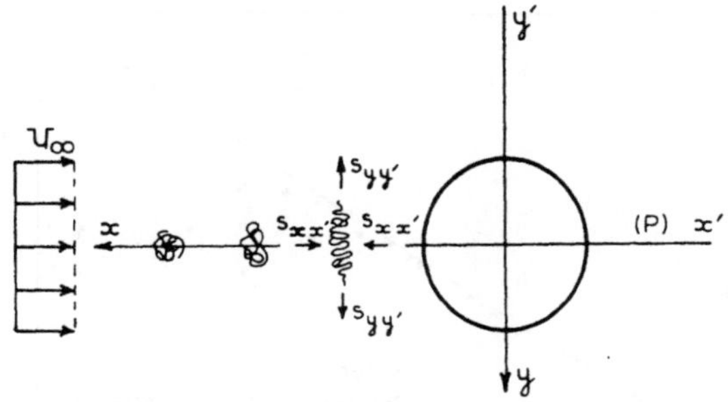

Figure 1 : Sketch of deformation of a macromolecule convected toward the cylinder.

a platinum electrode, which results from a very fast redox reaction : $Fe(CN)_6)^{3-} + e^- \rightleftarrows Fe(CN_6)^{4-}$ where $[Fe(CN_6)^{3-}] = [Fe(CN_6)^{4-}] = 10^{-2}$ M with a large excess of supporting electrolyte KCl(M) so as to eliminate the migration current due to the electric field. The mass flux J is linked to the current I by the relationship J = I/n F, where n is the number of electrons involved in the reaction (here n = 1), and F the Faraday number.

The disc system consisted of a truncated cone with a small base angle, the cross section of which generated the flow under investigation. It was made of an insulating material (Plexiglass) and equipped, in the disc plane, with a small circular electrode (ϕ = 80 μm) located at r = 2.7 cm from the rotation axis. The external disc radius was equal to 3 cm.

For the flow around a cylinder, a platinum wire (ϕ = 50 μm) constituted the cylinder and acted as a microelectrode. The wire was placed along a diameter in a tube and was covered by a varnish except for a small portion around the tube axis.

2. Time average mass transfer results

(*i*) *Cylinder*

In Figure 2 the data have been plotted in terms of directly accessible parameters i.e. the limiting diffusion current I versus the flow velocity U. For the solvent, we find the relationship : $I \propto U^{0.28}$. In Figure 2 we have also plotted the data obtained with dilute PEO solutions ($M_w \simeq 5 \times 10^6$).

It must be noticed that mass transfer is practically unchanged for U velocities below a critical value U^* slightly depending on the polymer concentration (50, 100 and 200 ppm). Beyond U^*, differences are observed with respect

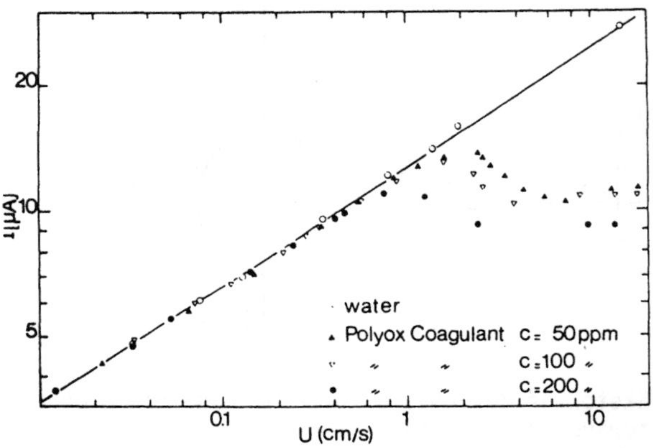

Figure 2 : Limiting diffusion current I vs the flow velocity U in water and in PEO coagulant solutions (small cylinder).

to the corresponding newtonian fluid behaviour : the mass transfer rate reaching a maximum value followed by a decrease and finally a plateau expressing a saturation regime. The higher the concentration, the lower the velocities U^*, which mark the onset of the non newtonian behaviour. Moreover, it must be emphasized that as soon as the deviation as regards to the newtonian domain is observed, fluctuations of the limiting diffusion current I take place.

The experimental values of S^* which mark the deviation from the Newtonian behaviour, have been precisely determined in the 50, 100 and 200 ppm solutions , by considering the U^* values where a large decrease of the power of the mass transfer fluctuations is observed. The obtained values for S^{*-1} (2.10^{-2}s, 2.10^{-2}s, 6.10^{-2}s) are of the same order of magnitude as the Zimm relaxation time, $\tau \sim 10^{-2}$s, according to Eq.(1). Similar behaviours for heat transfer in the same geometrical conditions were obtained by James and Acosta [8] but they did not mention the existence of fluctuations at the transition.

(*ii*) *Disc*

The values of the limiting diffusion current I have been plotted versus the angular velocity Ω of the disc in Figure 3. In the solvent alone, the equation $I \propto \Omega^{1/2}$ is fulfilled up to $\Omega \sim 1800$ rpm at least for which the first Stewart vortices arise. In the presence of PEO "coagulant" (c = 50 and 200 ppm) the same features as those reported with the cylinder were qualitatively displayed : A deviation with respect to the reference solvent curve is noticed starting from small Ω values (about 300 rpm). In addition, when the velocity is increased, instabilities superimposed to the average value are observed. The critical Ω^* -and therefore S^* - values were determined in such

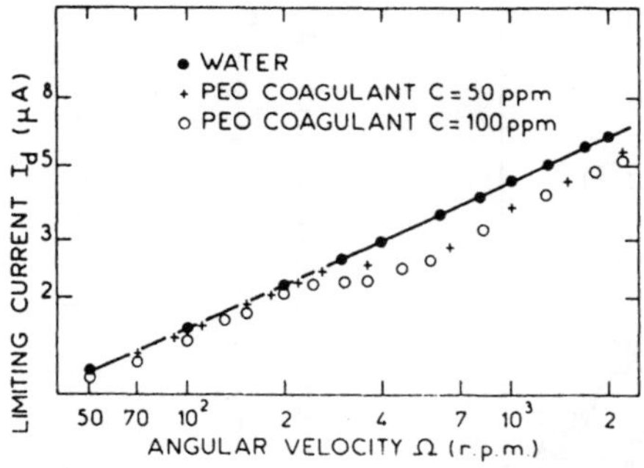

Figure 3 : Limiting diffusion current I vs the angular velocity Ω in water and in PEO coagulant solutions (disc system).

a way that a large increase of the power of the fluctuations is measured (see the following sections). The obtained S^{*-1} values, i.e. $(0.36\,\Omega^{*})^{-1}$, lie between 3 and 4.10^{-2}s and are therefore consistent with the results found for the cylinder.

3. Statistical analysis of the instabilities

In view of a quantitative treatment, these fluctuations have been recorded and whence analyzed in terms of power spectrum densities (psd). It is known that, in the framework of a linear theory and assuming the system as ergodic, the mass flux psd $W_{\tilde{i}}$ and the velocities psd $W_{\tilde{u}}$ can be linked through the following relationship [9].

$$W_{\tilde{i}} = ||H(jf)||^2 W_{\tilde{u}} \tag{2}$$

H(jf), which is the transfer function between sine wave variations of low amplitude of these quantities, has been calculated for the rotating disk system [10].

From the structure of the convective diffusion equation it comes out that an asymptotic high frequency solution can be obtained in the form $||H(jf)||^2 \propto f^{-3}$, whatever the geometry, since the convective terms become then negligible as regards to the diffusive terms.

(*i*) *Cylinder*

The different experimental psd have been calculated and are displayed as an example in Figure 4 for a 200 ppm PEO concentration. These spectra are characterized by an horizontal plateau at low frequencies, followed at increasing f values by a power law domain, $W_i \propto f^{-3}$. The cut-off frequency f^{*}

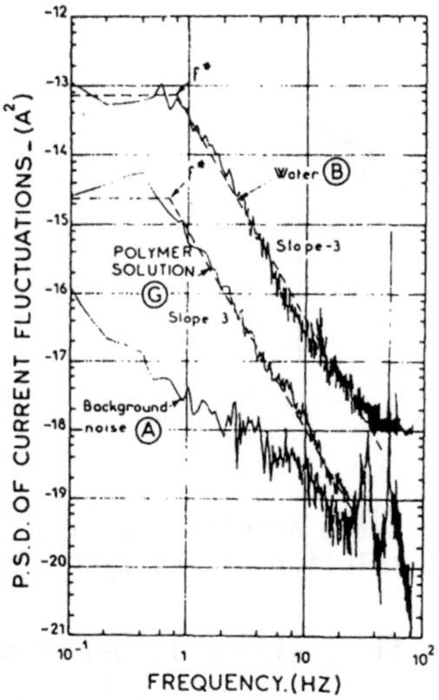

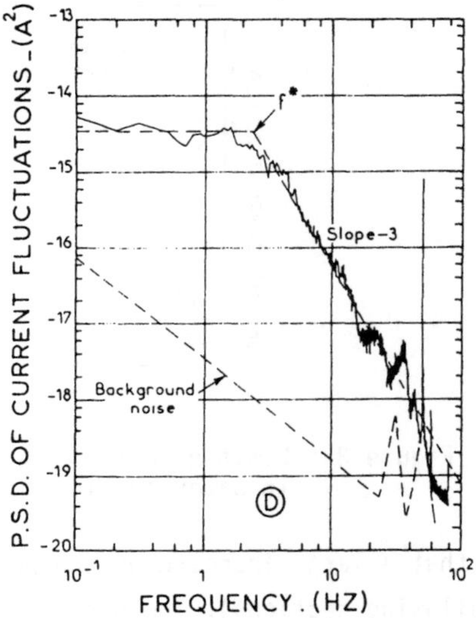

Figure 4 : psd of the mass transfer rate instabilities with the small cylinder. Circled letters correspond to the same conditions as in Fig. 5.

(defined by the intercept between the low frequency plateau and the straight line characterizing the f^{-3} regime, both in dotted lines) corresponds to the characteristic frequency associated to the response of the overall diffusion layer.

In Figure 5, the power of the mass transfer fluctuations (energy spectrum), defined as the integral of the psd in the frequency domain investigated, has been plotted as function of Re. In pure water, and for Re < 5, the experimental data (point A) lie within the thermal noise relative to the impedance of the system. Above this Re value, a sharp increase of the power of the fluctuations is observed in connection with the onset of secondary flow.

In polymer solution, the experimental value of the fluctuations power is higher by about three decades than that in pure water at the onset of the non newtonian behaviour.

(ii) Disc

The experimental psd relative to the solvent and to PEO solutions (50 and 100 ppm) are shown in Fig. 6 as an example for Ω = 1000 rpm. The solvent curve (dashed line) provides the background noise which is only determined by parasitic secondary flows due to the finite volume of the cell and also by the electronic regulation.

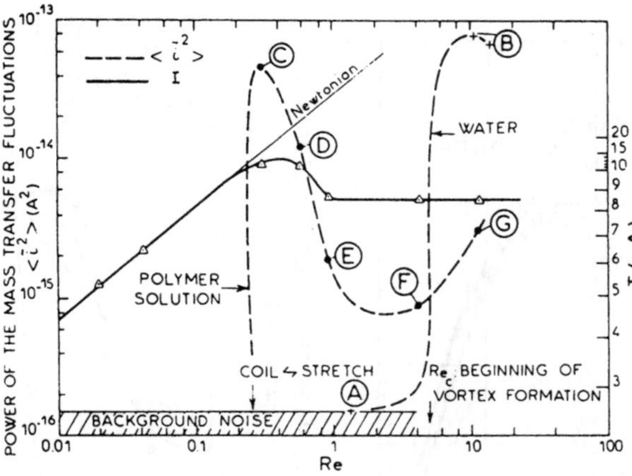

Figure 5 : Power of the mass transfer fluctuations in (A^2) for water (+) and the PEO coagulant solution. c = 200 ppm (⁜) (small cylinder). The steady curve I_d vs Re(Δ) (full lines) was plotted in Fig. 2

For the PEO solutions a low frequency horizontal plateau is observed as for the cylinder system, and is followed at higher frequencies by two power law domains, the difference of which exponents is equal to 3. The cut-off frequency for the mass transfer attenuation (H(jf) function) in the solvent would be thus of 335 Hz. This value seems to better correspond to the power law domain at higher frequencies, so the first one, at intermediate frequencies directly marks the onset of an hydrodynamic attenuation.

In Figure 7 are reported the power of the mass transfer fluctuations for the solvent (◘) and for the PEO solutions. Between 800 and 1500 rpm a roughly linear dependence of this power versus the PEO concentration is verified.

The accompanying time dependent fluctuations, for both systems, are obviously induced by hydrodynamic instabilities. These instabilities may be considered either as a direct or an indirect consequence of the abrupt conformation change of the polymer chains.

As à direct consequence, the non linear nature of this transition generates local jumps in the fluid velocity, since a certain volume of solvent follows the local motions of the chain.

As an indirect consequence, the polymer stretching gives rise to an increase

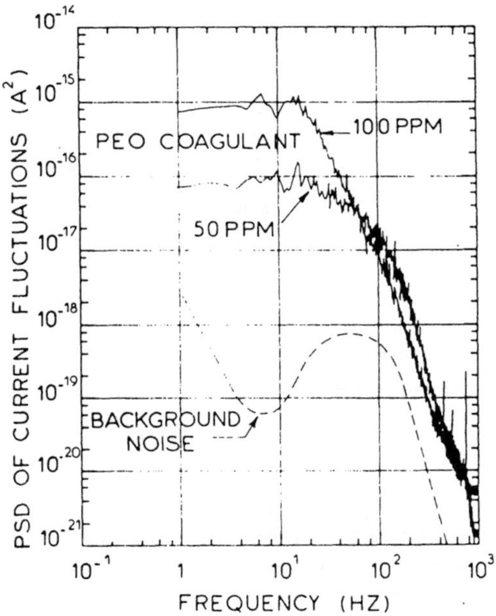

Figure 6

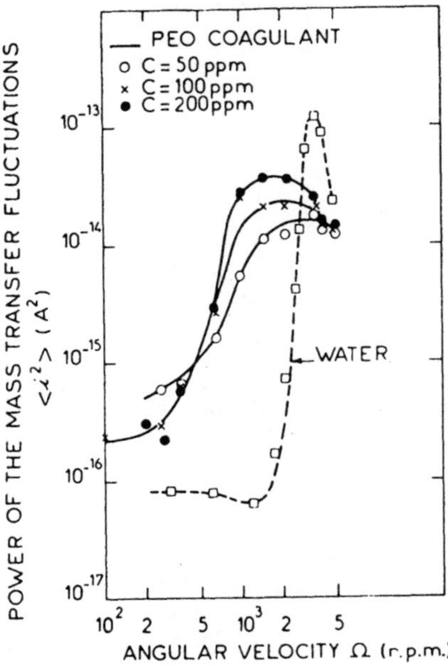

Figure 7

viscosity in the vicinity of the wall : The particular configuration due to the stagnation point (i.e. the more viscous layer in the near wall region with a viscosity gradient at the liquid-liquid interface) leads to an hydrodynamic unstable situation such as that described by Yih [11] for instance.

The analysis of the psd supports this interpretation :

Concerning the cylinder, only one power-law domain of the type $W_{\tilde{i}} \propto f^{-3}$ is detected and therefore the cut-off frequencies f^* graphically determined in Fig. 4 can be related only to the value δ of the diffusion layer thickness by the relationship $f^* \sim D/\delta^2$. Now, it is known that the local δ value monotonously increases starting from the stagnation point, to the opposite point on the cylinder downstream. For this reason, the half upstream portion of the cylinder always determines for the major part, the mean current value. The case is different for the fluctuations according to the place where they burst. The variations of f^* can therefore be explained in terms of a change in the location of the instabilities around the cylinder.

Indeed, from the first set of experiments carried out with water, it can be concluded that the high level of the fluctuations power in condition B (Figure 4), is due to the Karman vortices which are known to appear for $Re > Re_c$ (with $Re_c \sim 5$). Since they arise in the wake, i.e. in a region where the δ value is high, a low f^* value must therefore be found from the previous consideration : the graphical evaluation in Figure 4 B yields 0.8 Hz approximately for f^*.

On the contrary, in polymer solutions, the fluctuations start at much smaller Re values than in water (about 30 times below between B and C Figure 5). However, the f^* values are equal (C $\rightarrow f^* \sim 0.6$ Hz) or higher (D $\rightarrow$ F , $f^* \sim 2$ Hz) than that measured in water. A reasonable assumption which implies a smaller δ value can then be put forward. Since the Karman vortices cannot be established in this creeping flow (Re ~ 0.3), one must assume that the instabilities are not generated in the wake, but actually in the upstream region where the local δ value is small. This hypothesis is consistent with the above interpretation based on the onset of a conformation change upstream.

It is worth noticing that f^* remains nearly constant in a wide range of the Re number ($0.6 < Re < 4.3$) in the same way as the average current. In this range, it must then be inferred that the diffusion layer upstream and therefore the upstream hydrodynamic boundary layer keep constant thicknesses. This layer could be identified with the region of increased viscosity previously mentioned. An evaluation of both the hydrodynamical and diffusion layer thicknesses -δ_o and δ - can be tried in the following way : From the mean value of 2 Hz found for f^* and with $D \sim 7.10^{-6}$ $cm^2.s^{-1}$, one obtains $\delta \sim 6$ µm. Besides, δ and δ_o can be dimensionally related through : $\delta_o \sim \delta .\ Sc^{1/3}$. In our conditions, this leads to a value of about 0.1 mm for δ_o. This last value, which is of the order of magnitude of the wire diameter, substantiate the thickening of the upstream velocity profile.

The explanation can be more simply developped with the disc system since
- The boundary layer thickness is uniform.
- The function H(jf) has been calculated and, for the microelectrode used, the cut-off frequency was determined as $f^* = 216\ \Omega\ D^{1/3} / \nu^{1/3}$.

It appears then from the psd analysis of Fig. 6 that the mass transfer cut-off frequency is lowered in the presence of PEO with respect to the solvent, which accounts for a viscosity increase (D being unchanged). This result is consistent with the decrease of the steady state current (Fig. 3). The beginning of the psd attenuation is therefore only assigned to flow perturbations. The instabilities generated, exhibit a flat spectrum up to some tens of hertz, and then a regular damping where no characteristic frequency eventually ascribed to a molecular transition is detected. They can thus be considered as hydrodynamic instabilities.

Conclusion

From the study presented here it can be concluded that the mass transfer in elongational flows is strongly affected by the presence of a small quantity of flexible polymers in solution, even at very low Reynolds numbers.

The results led us to postulate that this behaviour is due to the presence of a stagnation point. In fact, the flow at the upstream of this stagnation point is strongly elongational and therefore able to give rise to a strong deformation of macromolecules corresponding to a coil ⇄ stretch transition.

The electrochemical method used, was proved to be very sensitive and it has been shown, for the first time, that fluctuations are associated with the transition. The onset of these fluctuations has been explained in terms of an hydrodynamic instability induced by a viscosity stratification which is generated by the coil ⇄ stretch transition.

References

1. Ambari, A.; Deslouis, C.; Tribollet, B.: Coil-stretch transition of macromolecules in laminar flow around a small cylinder. Chem. Engng. Comm. (to be published).
2. Abel, H.; Ambari, A.; Deslouis, C.; Tribollet, B. : Elongation des macromolécules et instabilités induites en écoulement axisymétrique. Table ronde sur les écoulements élongationnels, La Bresse (1983).
3. de Gennes, P.G. : Coil-stretch transition of dilute flexible polymers under ultrahigh velocity gradients. J. Chem. Phys. 60 (1974) 5030.
4. Hinch, E.J. : Mechanical models of dilute polymer solutions for strong flows with large polymer deformations. Colloque Intern. CNRS : Polymères et Lubrification, Brest (1974) 241-47.
5. de Gennes, P.G. : Scaling concepts in polymer physics, Part A. Cornell Univ. Press, Ithaca and London (1979).
6. Von Karman, Th. : Über laminare und turbulente Reibung. Zeit Angew. Math. und Mech. 1 (1921) 233-251.
7. Lamb, B.M. : Hydrodynamics, 6th Ed. New York (1945).
8. James, D.F.; Acosta, A.J. : The laminar flow of dilute polymer solutions around circular cylinders. J. Fluid Mech. 42 (1970) 269.
9. Bendat, J.S.; Piersol, A.G. : Random data : analysis and measurement procedure. Wiley-Interscience, John Wiley and Sons, p. 136 (1971).
10. Deslouis, C.; Gabrielli, C.; Fanchine Sainte-Rose, Ph.; Tribollet, B. : Electrohydrodynamical impedance on a rotating disk electrode. J. Electrochem. Soc. 129 (1982) 107.
11. Yih, C.S. : Instability due to viscosity stratification. J. Fluid Mech. 27 (1967) 337.

MIX
Papier aus verantwortungsvollen Quellen
Paper from responsible sources
FSC® C105338
FSC
www.fsc.org

If you have any concerns about our products,
you can contact us on
ProductSafety@springernature.com

In case Publisher is established outside the EU,
the EU authorized representative is:
Springer Nature Customer Service Center GmbH
Europaplatz 3, 69115 Heidelberg, Germany

Printed by Libri Plureos GmbH
in Hamburg, Germany